Remote sensing: an operational technology for the mining and petroleum industries

Remote sensing: an operational technology for the mining and petroleum industries

Papers presented at the conference 'Remote sensing: an operational technology for the mining and petroleum industries', organized by the Institution of Mining and Metallurgy and held in London from 29 to 31 October, 1990

 The Institution of
Mining and Metallurgy

Published at the office of

The Institution of Mining and Metallurgy
44 Portland Place London W1 England
© The Institution of Mining and Metallurgy 1990

ISBN 978-94-010-9746-8   ISBN 978-94-010-9744-4 (eBook)
DOI 10.1007/978-94-010-9744-4

Cover photographs show (green image with pink mines) the St.
Austell china clay mining district of Cornwall—a composite image
produced by SPOT PAN imagery with Landsat Thematic Mapper
and Chuquicamata mine, Chile, imaged by the French SPOT
satellite in 1987

Foreword

The Institution of Mining and Metallurgy has always played an important role in the promotion of new technologies within its areas of interest, and the Council decided in 1988 to support a conference on the applications of remote sensing in the mining industry. Following initial meetings of an Organizing Committee drawn from a range of disciplines and countries, it was agreed to widen the scope to include the petroleum industry, and to emphasize operational rather than experimental applications of remote sensing. It was decided to hold the conference in London at the end of October, 1990, and organization proceeded with the cooperation of the Remote Sensing Society, the Petroleum Exploration Society of Great Britain, the Geological Remote Sensing Group of the Geological Society of London, and the Environmental Research Institute of Michigan, which arranges regular conferences on geological remote sensing in North America.

Remote sensing is now a well-established technique in many branches of the oil and mineral industries, particularly in exploration. The use of digital imagery from satellites and airborne scanners was a logical supplement to air-photo interpretation, which has been used in geology for more than fifty years. Over the seventeen years since the launch of the first Landsat satellite remote sensing has progressed from an experimental curiosity to an operational technique. There are currently at least seven operational remote sensing satellites in orbit, and this is an appropriate time to review the current state of the technology as applied to the mineral and oil industries, to assess the cost-effectiveness of remote sensing and to examine the trends and possible consequences of current research.

This volume includes papers from around the world, including China and the Soviet Union, and beside the papers on exploration techniques and case histories there are sections on the use of remote sensing in logistical planning, hydrogeology and in studies of the environmental aspects of mining operations. Gold exploration receives special attention, and there is considerable emphasis on the integration of remotely sensed imagery with other data sets in geographic information systems. Advances in high spectral resolution remote sensing, particularly in China and the United States, are presented in a series of papers.

Keynote addresses on the importance of remote sensing to the oil and mineral industries will be delivered by leading managers from major metal and hydrocarbon production companies, and the current state of the art of remote sensing in these two branches of the industry will be reviewed by acknowledged leaders in their fields.

Thanks are due to members of the Organizing Committee, and to the Secretary and Conference Office staff of the IMM, for the hard work that has led to the preparation of this volume, and to authors and referees of the papers presented here. It is hoped that those who attend the London conference, and the many others who will read these proceedings in the future, will find this collection of papers a milestone in remote sensing, indicating the scope and diversity of applied remote sensing in the mineral and oil industries at the beginning of the 1990s.

Christopher Legg
Chairman, Organizing Committee

August, 1990

Organizing Committee

Executive Panel
C.A. Legg (Chairman) (National Remote Sensing Centre, United Kingdom)
K.E. Bowden (RTZ Mining & Exploration, Ltd., United Kingdom)
Dr. G.M. Lawrence (BP Exploration Co., Ltd., United Kingdom)

Advisory Panel
J. Aarnisalo (Outokumpu Oy Exploration, Finland)
R.L. Bedell (BUMINCO, Burundi)
Dr. A.P. Crósta (Universidade de Campinas, Brazil)
Dr. S. Drury (The Open University, United Kingdom)
J. Falconer (Centre for Surveying, Mapping and Remote Sensing, Kenya)
Dr. D. Greenbaum (British Geological Survey, United Kingdom)
J. Komai (Earth Resources Satellite Data Analysis Centre, Japan)
J. McMahon Moore (Imperial College of Science, Technology and Medicine, United Kingdom)
T. Nishidai (JAPEX Geoscience Institute, Japan)
E. Ortega (Minas de Almadén y Arrayanes S.A., Spain)
Dr. K. O'Sullivan (CRA Exploration Pty, Ltd., Australia)
E.R. Peters (The Robertson Group plc, United Kingdom)
Dr. A. Phillips (Environmental Resources Analysis, Ltd., Republic of Ireland)
N.P. Press (Nigel Press Associates, Ltd., United Kingdom)
Dr. D.A. Rothery (The Open University, United Kingdom)
Dr. Y. Scanvic (Bureau de Recherches Géologiques et Minières, France)
M. Taylor (Brintex Exhibitions, Ltd., United Kingdom)
Professor Tong Qingxi (Institute of Remote Sensing Applications, People's Republic of China)
Dr. P. Volk (Gesellschaft für Angewandte Fernerkundung mbH, Federal Republic of Germany)

Acknowledgements

The Institution of Mining and Metallurgy wishes to acknowledge the cooperation of
   The Environmental Research Institute of Michigan (ERIM)
   The Remote Sensing Society
   The Petroleum Exploration Society of Great Britain
   Geological Remote Sensing Group
in the organization of the conference, and the generous financial support of
   British Aerospace (Space Systems) Limited
   BP Exploration Company Limited
   RTZ Mining and Exploration Limited

Contents

Processing and interpretation technology

Pennine exploration dataset re-examined

M. W. C. Barr B.A., M.A., Ph.D., F.G.S.
Hunting Technical Services, Ltd., Hemel Hempstead, Hertfordshire, England

SYNOPSIS

A method of combining spatially referenced exploration data was evaluated using lead mineralisation in the northern Pennines of England as the test topic. The method allows subdivision of the search area into zones where mineralisation is more and less likely based on the spatial correlation between known mineral deposits and each dataset. It was successful in highlighting the Alston Block as a promising area for lead deposits even where only 6 of the principal mines were used to guide the technique but was less effective at predicting the distribution of mineralisation within the Askrigg Block. Its advantages include the ability to compare the strength of correlations between different data and known mineralisation, to guide the user towards metallogenic concepts of use to exploration and to highlight areas where exploration data are deficient and should be collected or upgraded.

INTRODUCTION

Several exercises have been carried out which demonstrate the value of integrating spatially referenced exploration data on image processing and GIS equipment[1-3]. The ability to register and display these data overlaid, side-by-side or by using more sophisticated combinations easily and quickly has led to the recognition of spatial correlations and consequently to insights into the controls of mineralisation which might otherwise have been missed. Nevertheless, these techniques do not represent a major advance over what can be done by manual techniques, for example by overlay of a series of transparent maps. There is a need for a technique which goes beyond the establishment of a relationship by simple inspection and which

provides an objective measure of its strength.

One such method has been widely demonstrated by Bonham-Carter et al.[4] and shows considerable promise, not least because it is comprehensible to geologists without extensive statistical training. The purpose of the work described here is to investigate how useful this method might be to the practical guidance of exploration programmes.

TEST AREA

The topic chosen to test the method is lead mineralisation in the northern Pennines of England (Fig. 1) . This choice was made firstly because the area is well explored and the distribution of lead very well known allowing the robustness of the method to be tested, secondly because the controls on the mineralisation have been widely modelled and thirdly because exploration data is readily available.

The geology and lead mineralisation of the area is extensively described in the literature[5-7]. Lead sulphides associated with fluorite, barite and minor zinc occur in veins cutting the lower Carboniferous (Dinantian and Namurian) rocks of the area. The mineralisation is concentrated in the Alston and Askrigg Blocks, partly fault-bounded zones of shallow water sedimentation rich in limestones in the lower Carboniferous, surrounded by deeper basins in which shales predominate (Fig. 1). Mineralisation probably took place mostly in the lower Permian, associated with late Variscan fracturing.

Current models of the metallogenic process[7] stress the importance of source, transport mechanism and environment of deposition in the manner of a hydrocarbon play. Several well researched lines of evidence suggest that the sources of the Pb, Ba, F and Zn in the Pennine mineralisation were evolved formation brines within the surrounding Upper Dinantian-Namurian shale basins. These shales were dewatered during the Variscan orogeny, carrying metals to the margins of the basins. Mixture of these brines with sulphate-rich formation waters of the adjacent platform sediments overlying the blocks localised the mineralisation. High heat-production Caledonian granites embedded in the lower Palaeozoic substructure of the blocks played a two-fold role in this process, firstly by bouying the blocks up during extensive fracturing and secondly by acting as a source of heat encouraging circulation of mineralising formation waters.

Fig. 1 Northern Pennine test area. Solid lines, principal fault zones; stipple, Permo-Triassic cover. Corners annotated with British National Grid coordinates in kilometres.

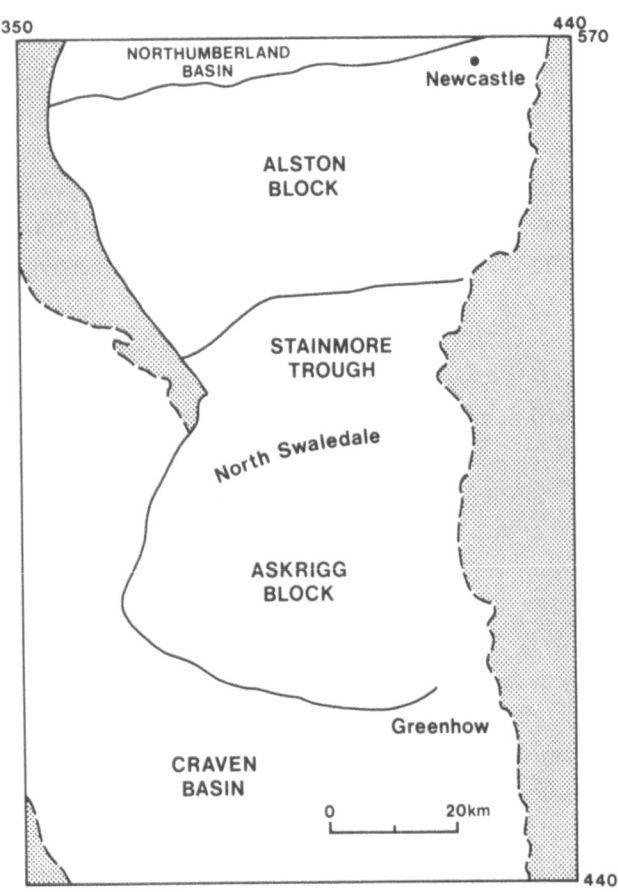

For the purpose of this exercise, we shall assume only limited knowledge about the nature of the mineralisation and its likely origin; that it occurs in veins, that it may have originated in lower Carboniferous basins and that there may be an association with buried granites. In the same way, the exploration data used are limited to those that might reasonably be expected at the start of an exploration undertaking; a geological map at 1:250,000 scale, Landsat MSS imagery and gravity data (see Appendix).

These data are compared with two subsets of the known distribution of lead mineralisation; firstly, a random selection of old mines and other occurrences represented by 10 per cent of 1km grid squares in which such occurrences are located (called hereafter the representative selection); secondly, a sample of variable size strongly biassed towards those mines from which historical production was derived (called hereafter the biassed selection). These samples are designed to simulate two possible treatments of previously known mineral occurrences at the stage of exploration where ground is being sought or exploration is about to start. Further details of the datasets used are presented in the Appendix.

METHOD

Fig.2a shows the representative selection of lead occurrences overlaid on the Bouguer gravity field. (In this and the following illustrations themes are coded as dark blue or purple for low values, through light blue, green and yellow to red for high values.) Inspection shows a good correlation of occurrences with lows in the gravity field suggesting that concentrations of further mineralisation might be expected in areas where the gravity field is low. However, it is not immediately apparent whether the search should be carried out only where gravity is very low (Fig. 2b) or over somewhat larger areas corresponding to a more relaxed thresholding of the gravity data (Fig. 2c). In the first case, the area to be searched is gratifyingly small; on the other hand, many known occurrences are excluded and there must be a considerable chance that further worthwhile mineralisation will be missed by concentrating the search only within the thresholded area. In the second case, rather few occurrences are excluded from the thresholded area and the chance of missing worthwhile mineralisation by ignoring the excluded areas is correspondingly reduced; on the other hand, the area that has to be searched is a large proportion of the whole.

The technique offers an objective method of balancing these competing interests. It is based on the concepts of prior and posterior probability and Bayes' theorem. From knowledge of the gravity field at mineralised localities (Fig. 2a), it is possible to compute the probability that a locality will have gravity below a given threshold given that it is mineralised. Bayes' theorem allows the inversion of this statistic into something of greater interest to the explorationist, the probability that a locality will be mineralised, given that it satisfies some condition of the data, for example that its gravity field falls below a particular threshold. The calculation requires knowledge of the likelihood of a locality being mineralised in the absence of exploration data (the prior probability) and the calculation of the conditional probability of the occurrence of mineralisation given that it satisfies a condition of the data, for example that its gravity falls below a particular threshold. Of equal importance is the computation of the probability of mineralisation, given that it does not satisfy the condition.

The calculation involved is set out in Bonham-Carter et al.[4]. Intermediate products, which following these authors are here called w+ and w- are the natural logarithms of the likelihood ratios which relate the prior and posterior probabilities. The posterior odds, which are closely related to the posterior probability, are given by the product of the prior odds and the likelihood ratio.

The prior odds are based on the known mineral occurrences only and are an underestimate since there is an expectation that more occurrences will be found. The posterior odds are also therefore an underestimate. In one sense therefore, the likelihood ratio is more interesting than the posterior probability of mineralisation. It is a measure of the way in which the chances of finding mineralisation have increased (or decreased) by considering the exploration data.

Table I lists these statistics for the three thresholds set in Fig. 2. Where the threshold is restrictive (Fig. 2b), includes a relatively small area, but in proportion a relatively large number of occurrences, then w+ and the corresponding

TABLE I   PROBABILITY WEIGHTINGS AND LIKELIHOOD RATIOS FOR THRESHOLDS SET IN FIGURE 2

				Likelihood Ratio	
	W+	W-	W diff	In pattern, exp(w+)	Outside pattern, exp (w-)
Low Threshold	1.34	-0.06	1.40	3.8	0.9
High Threshold	0.36	-0.32	0.68	1.4	0.7
Optimum Threshold	1.67	-0.36	2.03	5.3	0.7

likelihood ratio relating to the thresholded area are large and positive. The weight relating to the excluded area (w-) is slightly negative corresponding to a likelihood ratio slightly less than unity, reflecting the fact that many of the mineral occurrences have been excluded from the thresholded area. Where the threshold set is inclusive (Fig. 2c), then w+, though still positive, is small, reflecting the fact that rather fewer of the mineral occurrences in proportion have been included compared to Fig. 2b, but w- is more strongly negative, corresponding to the very few mineral occurrences that have been excluded compared to the area that has been excluded. Either of these thresholds could be set and the result would still be statistically valid; indeed, if the objective were to select ground for release then there might be merit in setting the threshold of Fig. 2c which identifies ground where there is very little chance of finding mineralisation based on this method. However, balancing the competing interests of high w+ and low w- is achieved where their difference (wdiff) reaches a maximum, corresponding to a threshold which maximises the efficiency of the method as a search tool (Fig. 2d).

So long as the patterns created by thresholding different data sets are conditionally independent, then the probability weights are additive. It is therefore possible to generate a series of maps like those of Fig. 2 where areas included in the pattern have value w+ and those excluded w-, and to add them together to show the likelihood distribution based on several kinds of exploration data. Areas where data are absent are assigned a value of zero. In these areas the posterior probability is equal to the prior probability and the likelihood ratio is unity. It is therefore possible to combine datasets which are incomplete or which do not cover exactly the same area.

RESULTS

The optimum threshold of the gravity data using a representative selection of lead occurrences is shown in Fig. 2d and the corresponding probability weights in Table I. Many of the lead occurrences in the Alston Block are included in the thresholded area but rather few of those in the Askrigg Block, especially those of the North Swaledale mineralised belt.

Both the gravity anomalies and the mineralisation are believed to be related to the presence of buried granites. There is merit therefore in testing the distribution of occurrences with the locations of the granites directly rather than through the intermediary of their effect on the gravity field. Fig. 3a shows proximity to the buried granites and the best threshold of this distance measure using the method. The pattern is similar to that based on the gravity data but the power of discrimination is considerably less (lower w+ and smaller wdiff). It appears that the work involved in modelling the outline of the buried granites resulted in little advantage for the task of finding lead mineralisation.

Inspection of Fig. 2a suggests that lead occurrences may be concentrated in regions where the gravity field is changing rapidly. This might be because of the localisation of hydrothermal circulation at the margins of the granites or their cupolas or along major faults, each of which could be represented in this way in the gravity field. Accordingly, a map representing the first horizontal derivative of the gravity field was prepared and the optimum threshold determined (Fig. 3b). The correlation was less good than expected, reflected in a rather small wdiff. Nevertheless, the technique has confirmed and indeed quantified the concentration of mineral occurrences in regions of high gravity gradient and at the same time achieved a useful subdivision of the area into more and less prospective zones.

Because the lead mineralisation occurs in veins, there may be a correlation between the density of fractures (including veins) visible on remotely sensed imagery and the distribution of occurrences. Accordingly, a lineament interpretation of Landsat MSS imagery was carried out and a lineament density map prepared (Fig. 3c). As with the gravity first horizontal derivative, mineralisation is slightly more likely in regions where

the theme has a high value (w+ positive but small). However, the optimum threshold includes a large proportion of the area, suggesting that the correlation being measured is one of a lack of mineral occurrences with zones of low lineament density rather than of concentrations of occurrences with zones containing many lineaments. This is undoubtedly due largely to the absence of occurrences in the Permian and younger rocks which are also poorly fractured. A similar result, although different in detail was achieved using fracture density based on the faults shown on BGS 1:50,000 scale maps.

The optimum discrimination with regard to stratigraphy was achieved by including both the Dinantian and Namurian in the pattern (Fig. 3d). This is because all the selected occurrences fall within the outcrop of rocks of this age range except two which plot within the Whin Sill, and w- has as a result a very large negative value.

A map showing proximity to shale basins (Fig. 3e) was generated on the basis of the hypothesis that these basins are the source of the mineralising solutions. While the basins themselves are poorly mineralised and have therefore a negative w+, a maximum occurs in wdiff where ground in their immediate vicinity is included in the pattern. This is caused by concentrations of occurrences adjacent to the Craven faults in the Greenhow area and in north Swaledale. The result is surprising because many lead occurrences are located over the buried granites which tend to be located far from the margins of the basins.

The work described in the preceding paragraphs was repeated using a number of different variations. Knowledge that the lead mineralisation is largely restricted to the Dinantian and Namurian was used to mask out data falling in the outcrop of other stratigraphic subdivisions and to restrict the search to the outcrop of the Dinantian and Namurian. Although the weights resulting were different, the optimum thresholds and therefore the way in which the remaining area of Dinantian and Namurian rocks was subdivided into parts within the thresholded patterns and parts outside them was almost identical for each dataset.

Selections of mineral occurrences biassed towards those areas with the greatest historical production were used in place of a representative selection (see Appendix). This series of tests was prompted by the view, often expressed by exploration geologists, that the geological factors which control major deposits of economic minerals are different from those which control mere shows. This implies that for the method under discussion, it is not proper to treat all occurrences in the same way; major deposits should be treated separately from shows. Under these circumstances, it is very likely that only a few major occurrences will be available to guide the investigator and therefore it is important that the method continues to work well with only a few training sites.

Table II shows the result of a series of tests using the Bouguer gravity as the thematic data. The optimum threshold, corresponding to the maximum in wdiff shows remarkable stability as the number of occurrences is reduced. Moreover, it is closely similar to the optimum threshold indicated by the unbiassed selection of mineral occurrences (Table I). The main effect of reducing the number of occurrences used is to increase the value of wdiff, and especially of w+ at the optimum threshold (Table II).

Broadly similar results were obtained by comparing the optimum thresholds obtained for the first horizontal derivative of the gravity data, and the lineament density data for the representative selection of mineral occurrences and the smallest sample (6) of occurrences biassed by historical production. For each of these datasets, the results suggest that the relationship between data and occurrences is not greatly affected by the size of the latter and that reducing the number of occurrences considered, at least down to 6, does not greatly change the outcome.

The results of the comparison using stratigraphic and basin proximity did show marked differences. Where only 6 occurrences were considered, the thresholding of the Namurian by itself led to a maximum value of wdiff. Because all of the 6 occurrences were in either the Namurian or the Dinantian, the inclusion of both these stratigraphic intervals in the pattern led to an infinitely negative w- and therefore an infinitely large wdiff. The result exposes a general weakness of the method where no mineral occurrences are excluded from the pattern. Similarly, where the threshold area is so small that it contains no occurrences, then w+ and wdiff are both infinitely negative. These limitations of the method become increasingly constraining as the number of occurrences considered is reduced.

When only 6 occurrences were considered, the threshold of the basin proximity data corresponding to the maximum value of wdiff included nearly all of the area outside the basins, a result sharply contrasted with that using the representative selection of mineral occurrences. Fig. 4 shows the way the probability weights change with the threshold applied to the basin proximity data. The saw-tooth nature of the curves is a consequence of the small number of occurrences

TABLE II   COMPARISON OF W+, W- AND W DIFF FOR THRESHOLDED GRAVITY USING PROGRESSIVELY FEWER LEAD OCCURRENCES BIASSED TOWARDS HISTORICAL MINERAL PRODUCTION

Number of Occurrences	W+				W-				W diff			
	35	25	15	6	35	25	15	6	35	25	15	6
Threshold												
68		2.36	1.98			-.18	-.12			2.54	2.10	
85	2.14	2.12	2.07	2.54	-.37	-.36	-.33	-.65	2.51	2.48	2.40	3.19
94				**2.64**				**-1.73**				**4.37**
95	**2.19**	**2.16**	2.20	2.59	**-.75**	**-.72**	-.76	-1.73	**2.94**	**2.89**	2.96	4.31
97			**2.20**				**-.91**				**3.10**	
102	1.94	1.92	1.96	2.18	-.96	-.94	-1.02	-1.70	2.89	2.86	2.98	3.87
112				1.33					-1.82			3.14
118		0.69	0.75			-.96	-1.19			1.65	1.94	
129			0.33				-1.31				1.65	

Probability weights at the maximum value of w diff are shown in BOLD

Fig. 4. Plot of probability weights against threshold for basin proximity data and 6 mines.

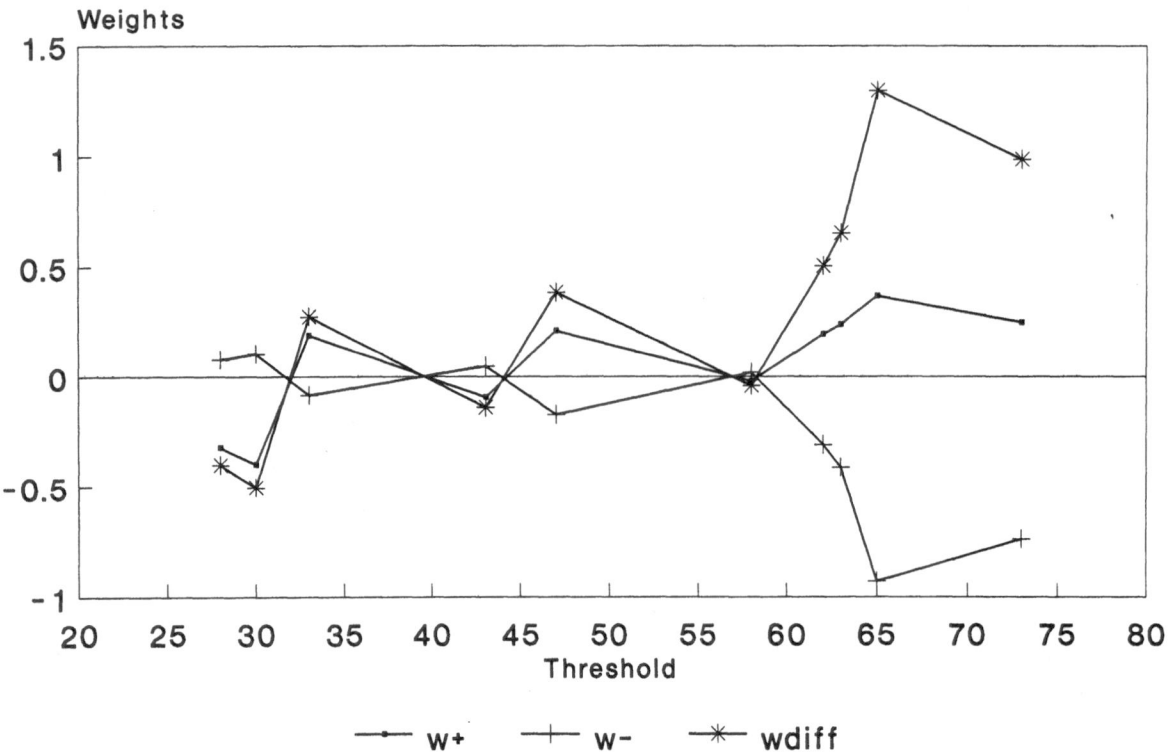

considered. As the threshold passes each of them, there is a sharp change in the rate and in places the direction of change of values of the weights. As a consequence, the threshold at which the maximum value of wdiff occurs is strongly dependent on details of the locations of the occurrences used when the correlation between them and the data is not strong. It is believed that this is the reason for the large change in the value of the optimum threshold between the representative set of occurrences and the biassed set of 6 mines.

Two examples of combining the results of thresholding the data are reproduced in Figs. 5a and 5b and compared with the known distribution of lead in the northern Pennines (Fig. 5c) and the 6 mines used to represent the biassed distribution (Fig. 5d). The annotated colour bar on the left of the likelihood maps keys the colours to the value of the likelihood ratios which are on a logarithmic scale. Where the likelihood ratio is 1, the chance of finding lead mineralisation has been unaffected by consideration of the exploration data. Where the likelihood ratio is greater than 1, the chances of finding mineralisation have been improved and conversely for those areas where it is less than 1. Strips of colour around the margins of the maps are caused by the fact that the exploration data used do not cover exactly the same area; for example the

generation of the lineament density map involved convolution with a 41x41 pixel filter, resulting in a density image which is 40 pixels (10km) smaller than the area covered by the Landsat data from which it was generated.

Both Figs. 5a and 5b have been successful in highlighting the rocks overlying the Weardale Granite as prospective for lead mineralisation. This is the consequence of the strong correlation between gravity lows and the distribution of lead occurrences in both selections. The main difference between them is that the maximum value of the likelihood ratio in the latter is some 10 times that of the former. Both maps show a reduced probability for lead mineralisation in zones outside the lower Carboniferous blocks largely as a result of the combination of weightings for gravity, gravity first horizontal derivative and stratigraphy. The main difference between them results from both the Dinantian and Namurian being assigned a positive weight in Fig. 5a while only the Namurian was in Fig. 5b. As a result, much of the Askrigg Block has been included with non-prospective marginal areas in Fig. 5b. Neither figure has been successful in predicting the distribution of lead mineralisation within the Askrigg Block. In both cases, the most likely ground indicated lies near the centre of the Wensleydale Granite where lead mineralisation is remarkable by its absence[6]. In the case of Fig. 5b, there is justification for this since the mineral occurrences used are concentrated in the Alston Block from which the great majority of historical production came. In the case of Fig. 5a, the failure to predict concentrations of occurrences in north Swaledale and in the Greenhow area must be considered a failure of the application of the method and is discussed below.

DISCUSSION

The method establishes and measures an objective spatial correlation between mineral occurrences and exploration data and says nothing about causality which must be provided by the geologist carrying out the work. Nevertheless, experience with the method suggests that it is particularly useful at encouraging the good use of data as the following examples show.

During the course of comparing the lead distribution with the gravity data, it was noticed that there appeared to be concentrations of occurrences at the margins of gravity lows. This led to speculation that hydrothermal circulation and lead deposition may have been enhanced at the margins of buried granites. Three new datasets (first horizontal derivative of gravity, proximity to buried granites and proximity to granite cupolas) were generated and the strength of the correlation tested using the method. In the event, the correlation was judged to be strong enough to be of value for exploration, whether or not the speculation about its cause is justified. The ability to compare the distribution of mineral occurrences with exploration data stimulates ideas about the environment of mineralisation and because the method is simple to use these ideas can be tested quickly and a judgement more informed than a mere hunch made about their value.

Knowledge that the lead mineralisation occurs in veins is not in itself spatial in nature and cannot be used directly. Nevertheless, it prompted the generation of spatial data (lineament density) which could be tested by the method. The poor correlation between these data and the mineral occurrences and the appreciation that lineament density is a poor representation of the density of fractures in the bedrock led to the generation of an alternative dataset, the density of faults shown on BGS 1:50,000 scale maps which was also tested,

with similar results. The subdivision of the area at the optimum threshold led to the realisation that what was being measured was a lack of occurrences in poorly fractured Permian and younger rocks rather than any marked concentration of occurrences in zones of strong fracturing. To this extent, the data are useful for helping to exclude ground where the chance of finding lead is low. The method can therefore encourage the generation or collection of spatial data of value to exploration based on insights into the nature or genesis of the mineralisation which are not in themselves spatial in nature.

One further general feature of the method which was judged particularly useful was the ability to compare the weights arising from the thresholding of different datasets and so to reach a more informed conclusion about which are likely to be useful for exploration.

Two aspects of the study relate more directly to the lead mineralisation of the northern Pennines. Firstly, the differences in the outcome of the trials based on a representative selection of lead occurrences and on one biassed towards historical production were not large. In all cases except the stratigraphic subdivision of the Carboniferous, the correlations were in the same direction; ie with low gravity, high rate of change of gravity, high lineament density and closeness to shale basins. Where the discriminatory power of the data was good (high wdiff), the optimum thresholds established were closely similar. One might conclude from this (but cannot prove) that the geological controls for both small and large lead deposits in this environment are similar.

Secondly, the optimum thresholds for the more discriminatory datasets show remarkable stability as the number of occurrences used is reduced. This suggests, but again does not prove, that large deposits in the Pennine environment are likely to occur where occurrences of any size are most densely distributed.

Several aspects which require particular care were also revealed by the study. The failure to predict the main locations of mineralisation on the Askrigg Block and especially in north Swaledale arises from the reduced influence of the Wensleydale Granite on the location of mineralisation compared to the Weardale Granite. This highlights the need to include in such studies only mineral occurrences which are believed to be closely similar in origin, if necessary by subdividing them into groups.

The study failed to make a satisfactory discrimination on stratigraphic grounds particularly where the biassed selection of mineral occurrences was used. Much of the historical production came from the Dinantian and, compared to its outcrop, a very large proportion from the Great or Main Limestone at the base of the Namurian, especially in the Alston Block. Re-examination of the locations used for the occurrences show that many of them plot on the outcrop of the Millstone Grit , even though the production which they represent was derived from the underlying Great Limestone and Dinantian strata. The lesson to be learned is to know the data being used and to apply the method with intelligence.

Some difficulties were also encountered where very few occurrences were included in the study. The methods can only be applied where at least one occurrence falls within the pattern and at least one outside it. Outside these limits, one or other of w+ and w- becomes infinitely large or small. Where only a few occurrences are included in the study, these limitations can become restrictive.

CONCLUSIONS

The main advantages of the method are in its potential for informing the processes of geological reasoning which lead to the selection of ground for further investigation on a regional scale, based on exploration data. It allows the strength of spatial correlations between known mineral occurrences and exploration data to be tested and compared. It highlights opportunities for the collection or generation of new data or the improvement in the quality of existing data which could be used to distinguish promising ground. By allowing the spatial correlation between occurrences and data to be investigated and quantified quickly and easily, it may prompt new insights into controls on the mineralisation, the utility of which can be tested using the same method.

The study was successful in highlighting the lead mineralisation in the Alston Block, less so, for the reasons discussed, in the Askrigg Block. The outcomes of the method using a representative selection of mineral occurrences and only 6 of the principal mines in the area were broadly the same. The method appears in this case to be robust with regard to the number of occurrences considered, which is of importance where there are grounds for believing that the factors which control the location of major deposits are different from those which control small occurrences.

ACKNOWLEDGEMENTS

This paper is published with the permission of Hunting Technical Services Limited. The digital geological map used was kindly made available for the study by the British Geological Survey.

APPENDIX

SOURCES OF DATA

Much of the data used form part of the Pennine Dataset[8] and were already registered to the British National Grid and in a raster format. Derived data were generated using I^2S System 600 and ERDAS software with new routines developed by Hunting Technical Services Limited.

Mineral Occurrences

The locations of mines, adits and other evidence of mineralisation shown on BGS geological maps of the area at scales of 1:50,000 and 1:63,360 were plotted on overlays. Evidence of lead mining was collated from these and other reference sources[8]. The data were aggregated into 1km grid squares each of which was labelled as lead "absent", "occurrence" or "production". These labels were assigned values of 0,1 and 2 respectively and inserted into a raster image of the area with 250m grid cell resolution. This forms the basis of the lead distribution shown in Fig. 5c.

A selection of 10 per cent of the production grid cells and 10 per cent of the occurrence grid cells was made using a random number generator. The selected data were represented by groups of 4 pixels, corresponding to squares 500m on a side located in the northeast corner of the original 1km grid cells. This selection was used as the representative sample of the distribution of lead in the area.

For the biassed selections of occurrences, the lead production in the northern Pennines was ranked by sub-area[5,6] and the number of mines by which each should be represented assigned on the basis of the proportion of total production which each sub-area represents (Table III). The principal mines and vein complexes in each sub-area were selected from the same references and also ranked by volume of historical production. The largest mines in each sub-area were included in the smallest selection. Progressively smaller mines were included in the larger selections.

Geological Map

The map was provided in digital form by the British Geological Survey. It was prepared by raster scanning the separations used to make the published geological map at 1:250,000 scale using a 125m pixel and joining the sheets together. For convenience in thresholding the data during the study, the class values representing each geological unit were reassigned chronostratigraphically, progressing from low values for old rocks to high ones for young rocks.

Landsat Imagery

The Landsat Multispectral Scanner data used were acquired from the Space Department RAE as a mosaic of 1970s images registered to the National Grid using a 50m pixel.

Lineament Density

An image was prepared from the weighted sum of Bands 5 and 7 of the Landsat data described above, the weights used being chosen interactively to subdue the agricultural field pattern and contrasts between upland and lowland areas. A linear contrast stretch was applied to the product, followed by Wallis filtering to brighten areas of blanket bog, and a high pass filter. The processing was designed to subdue agricultural field patterns and emphasise fine spatial detail caused by topography. The data were written to film, enlarged to 1:150,000 scale and interpreted for lineaments. The interpretation was digitised and converted to raster format using a 50m pixel. The data were smoothed with a 5x5 pixel convolution kernel, and subsampled every 5th sample and line. The subsampled data were convolved with a 41x41 pixel smoothing filter to produce the data illustrated in Fig. 3c.

Fault Density

The faults and veins shown on BGS 1:50,000 scale maps were digitised and rasterised using a 50m pixel. These data formed part of the original Pennine Data Set[8]. Thereafter, the processing followed the route outlined for the lineament density data.

Bouguer Gravity

The data were produced by digitising BGS Bouguer gravity contour maps at 1:250,000 scale at the intersection of contour lines with east-west lines spaced at 1.25 km. The resulting data were gridded at 250m cell spacing. The raster data form part of the original Pennine Data Set[8].

TABLE III   LEAD PRODUCTION RANKED BY SUB-AREA

Dunham Sub-area[5,6]	Production of lead concentrate tonnes x 10^3	No. of Mines Selected			
5. Weardale	920	2	4	6	9
2. Alston Moor	817	1	3	5	8
11. North Swaledale	555	1	2	4	5
7. Teesdale	377	1	1	2	3
4. East Allendale	306	1	1	2	3
3. West Allendale	209		1	1	2
14. Wharfedale, Grassington	179		1	1	2
15. Niddersdale, Greenhow, Settle	146		1	1	1
8. Heydon Bridge	144		1	1	1
13. South Swaledale - Wensleydale	130			1	1
6. Derwent	51				
1. Escarpment zone	26				
10. Stainmore - Mallerstang	4				
12. Richmond	2				
Total	**3,866**	**6**	**15**	**25**	**35**

11

Gravity, First Horizontal Derivative

A Sobel filter which generates the root sum of squares gradient in the samples and lines direction between a pixel and its immediate neighbours was passed across the gravity image described above. The result was filtered in the frequency domain using a low pass filter with a threshold close to the Nyquist frequency in order to remove artifacts caused by the byte nature of the input and consequent stepping.

Structure

The locations of the Weardale and Wensleydale Granites and their cupolas were taken from published sources[6,7,9,10]. The margin of the Northumberland Trough was taken as the line of the Stublick and Ninety Fathom fault systems. The margins of the Craven Basin were taken as the lines of the Catlow and Sykes Anticlines, the Craven Faults and, very approximately, the Rossendale Anticline and its continuation to the north. The limits of the Stainmore Trough are based partly on Figure 4 of Dunham and Wilson[6].

These outlines were digitised and a raster representation produced using a 250m pixel. Proximity zones were generated around each of the granites, cupolas and basins using ERDAS software.

References

1. Green, P.M. Digital image processing of integrated geochemical and geological information. Journal of the Geological Society, London, vol. 141, 1984, p941-949.

2. Haydon, R., Kholhammer, G. and Volk, P. Application of a modified geographic information system to the use of remote sensing and other thematic data for a mineral and groundwater resource case study. Geological Institute of Ludwig Maximilian University, Munich, FRG.

3. Harding, A.E. and Forrest, M.D. Analysis of multiple geological datasets from the English Lake District. IEEE Transactions on Geoscience and Remote Sensing, vol. 27, no. 6, 1989, p732-739.

4. Bonham-Carter, G.F., Agterberg, F.P. and Wright, D.F. Integration of geological datasets for gold exploration in Nova Scotia. Photogrammetric Engineering and Remote Sensing, vol. 54, no. 11, 1988, p1585-1592.

5. Dunham, K.C. Geology of the Northern Pennine Orefield: 1, Tyne to Stainmore. Memoir of the Geological Survey of Great Britain, 1948, 357pp.

6. Dunham, K.C. and Wilson, A.A. Geology of the Northern Pennine Orefield: 2, Stainmore to Craven. Economic Memoir, British Geological Survey, 1985, 247pp.

7. Plant, J.A. and Jones, D.G. (editors). Metallogenic models and exploration criteria for buried carbonate-hosted ore deposits - a multidisciplinary study in eastern England. (Keyworth, Nottingham; British Geological Survey: London; the Institution of Mining and Metallurgy), 1989.

8. Hunting Geology and Geophysics Limited. Computer Correlation of Geological, Geochemical and Geophysical Prospecting Data with enhanced Satellite Imagery. 1983, unpublished report.

9. Bott, M.H.P. and Masson-Smith, D. The geological interpretation of a gravity survey of the Alston Block and the Durham coalfield. Quarterly Journal of the Geological Society of London, vol. 113, 1957, p93-117.

10. Bott, M.H.P. Geophysical investigations of the northern Pennine basement rocks. Proceedings of the Yorkshire Geological Society, vol. 36, 1967, p39-168.

Fig. 2. Comparison of Bouguer gravity field with lead occurrences (representative selection); a, occurrences in white overlaid on gravity coded from dark blue (low) to red (high) and displayed as if illuminated from the northwest; b, with low gravity threshold; c, with high gravity threshold; d, with optimum gravity threshold.

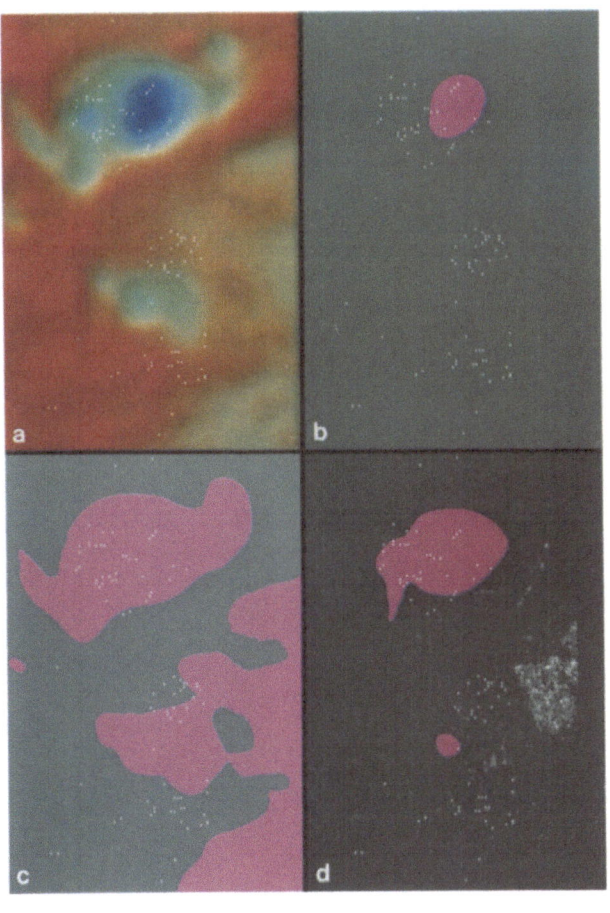

b, first horizontal derivative of Bouguer gravity as if illuminated from northwest;

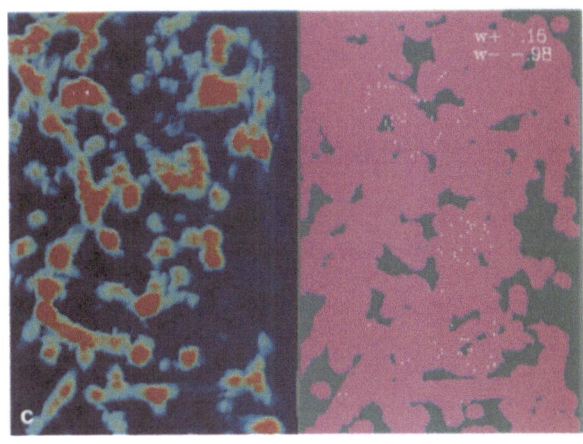

c, Landsat lineament density;

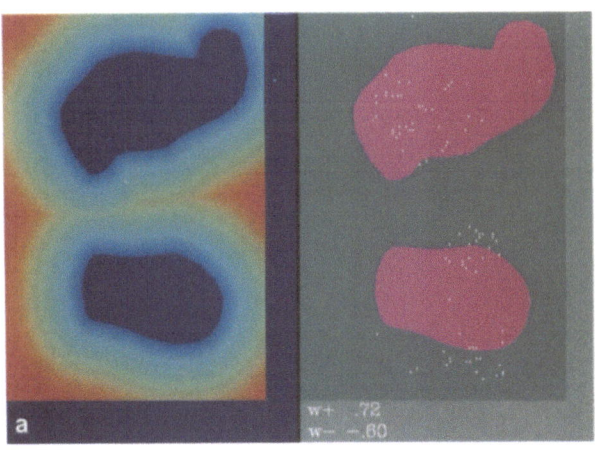

Fig. 3. Derived datasets and optimum thresholds. a; proximity to buried granite;

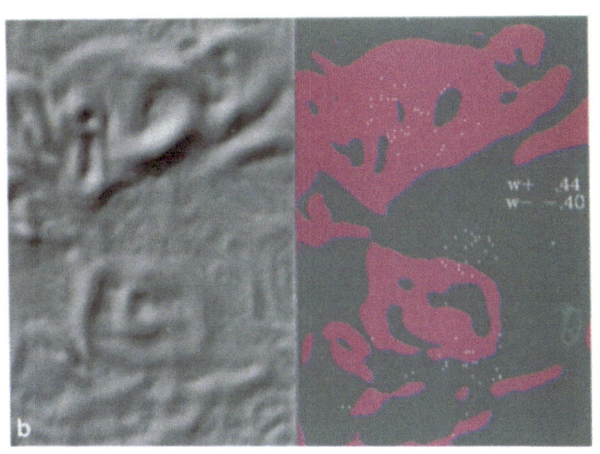

d, stratigraphy; dark red and brown, igneous rocks, lower Palaoezoic and older; blues and flesh, Dinantian; ochre and green, Namurian; dark and medium grey, Westphalian; orange, red and light grey, Permo-Triassic;

13

Fig. 3 Continued...
e, proximity to Dinantian - Namuran Shale basins.

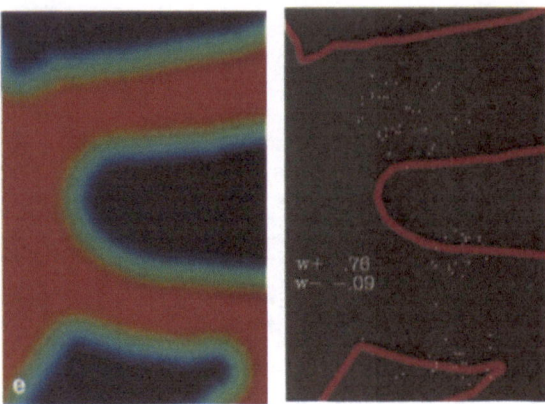

Fig. 5. Likelihood ratio maps compared with lead distribution. a, based on representative selection of lead occurrences; b, based on 6 occurrences biassed towards historical production; c, kilometre squares with lead production (red) and other occurrence (green) over geological map; d, location of 6 mines used in the biassed selection over monochrome Landsat MSS imagery.

14

New interface approach to complex data archive systems for remote sensing and mineral exploration applications

David Cleden
Richard Proud
Nicola Hills
Earth Observation Sciences, Ltd., Fleet, Hampshire, England

SYNOPSIS

The volume of remotely sensed information (both image and non-image based data) required to undertake geophysical processing is constantly increasing as processing techniques become more sophisticated. Often such data analyses depend upon the cross-correlation of elements from different data sets obtained from a variety of sources. It is therefore vital that the data is maintained in a suitable data archive and can be accessed easily and efficiently. The problems of data archive management are well known. However, powerful but flexible interfaces are becoming increasingly important as such systems grow in size and complexity.

ADIS (Advanced Data Interface System) is currently being developed by Earth Observation Sciences Ltd utilizing hypertext concepts to interface to large datasets of image, textual, areal and point-based data. Hypertext techniques can offer the following advantages which are directly relevant to the geophysical processing requirements of many mineral exploration activities:

- incorporation of graphics and icon-based control directives

- flexible data processing functions and data browsing

- handling of multi-format data (e.g. raster imagery, text, along-track data)

- multi-level menu access to all levels of database functions

- distributed database access

- remote, "intelligent terminal" access to mainframe data stores via communication networks

In addition, the application of CD-ROM and optical disk technology greatly increases the scope for fast access to large centralized data stores. The application of these and other techniques has been illustrated in a demonstration system which combines the main features of an extensive relational database with those of an interactive information resource.

KEYWORDS: Catalogue interoperability, hypertext, hypermedia, structured data archive, man/machine interface.

1. INTRODUCTION

Recent developments in digital storage hardware now put more storage capacity than ever at the fingertips of networked computer users. This ever-increasing storage capacity is threatening to exceed the abilities of most data users to process and interact with the data in meaningful ways. The problem is particularly acute in the field of earth observation. ERS-1, a satellite due to be launched by the European Space Agency in the next twelve months, will transmit up to 20 gigabytes of data every day. With a planned mission life of three years, this represents 22 terabytes of data collected during the mission. ERS-1 is just one of many earth observation satellites which will be operational during the 1990s, adding a further order of magnitude to these difficulties.

The obvious problems of transcribing such data on to suitable storage media for subsequent archiving are well understood. In the past, computer compatible tapes (CCT) and high density digital tape (HDDT) have been the standard solutions to this problem. In many quarters, these are now being superseded by optical media which offer the twin benefits of much greater storage density and improved durability.

However, an equally important part of data management is the subsequent retrieval and analysis of the data. Surprisingly, many earth observing missions do not address this problem directly - consequently much valuable data is overlooked or never fully utilized. Standard archive procedures and database technologies are well-established, but relatively little work has been done on the interface between the data storage media and the user. Typically, the user of this data is not a software or database specialist and wishes only to extract the data of interest with the minimum of fuss. The scope and complexity of a data archive makes no difference to such a user; the value of the archive is only as good as its ability to provide him with the information he requires.

A number of concepts and "interface philosophies" have appeared in recent years. Many of these ideas can be adapted to help solve the problems described above, although no single clear solution to the problem of complex data archives has yet arisen.

The principal aim of the work undertaken by Earth Observation Sciences has been to develop a *prototype interface to a complex data archive*. The term *complex* refers to both the physical extent of the data and the nature of individual data items comprising the archive. By carefully choosing the most appropriate aspects of current thinking on man/machine interfacing, the work aims to optimize the interaction between a user and a typical complex data archive. At the same time, the proposed interface solution remains a prototype, so that it can form a testbed for various new ideas as they emerge.

Much of the work in this study relates directly to the authors' experience with remote sensing data sets and their application to the geophysical interpretation of terrestrial surface features and mineral exploration in general. However, the philosophy can be applied equally well to many diverse fields where the primary requirement is for fast and efficient access to large, multi-parameter data archives.

2. NEW CONCEPTS IN INTERFACING AND DATA STRUCTURING

Solutions to this problem can be broadly divided into two categories:

- data structuring
- improved interfacing

Both categories have an equally important role to play. Without a sensible organization of data within an archive, extraction of selected parts of the archive will be time-consuming, impractical or even impossible. Data structuring will obviously have a major impact on the functionality of the archive system, but the dictates of the system should not be allowed to impose constraints upon the user.

Equally, the user must be able to fully utilize the functions of the system to obtain the information he requires. This effectiveness is wholly dependent upon the user interface. In the past, the interface has typically been geared towards obtaining a basic set of information required by the system rather than allowing the user flexibility and creativity in the definition of his requirements from the system.

Data Structuring
The principal aim of data structuring is to allow an item of data to be located and extracted efficiently. In most cases, the primary requirement is for flexibility. An optimization of the system for a particular type of data extraction should not be to the detriment of other data extraction techniques. However, data structuring can - and should - play a much wider role. One exciting prospect is the ability to link discrete, heterogeneous data archives across a network.

Such a system of interconnected data archives is already under development. The concept is known as *catalogue interoperability*. The power of catalogue interoperability lies in the fact that individual data archive managers need only adhere to a very basic set of rules in order to ensure full connectivity with a network of other data archives. The data archives need only be connected electronically to other systems and so may be dispersed around the world. There is no requirement for the data archives to share similarities in the type or organization of data within the archive, other than at a high level "catalogue" level.

The Committee on Earth Observation Satellites (CEOS) has formulated a number of guidelines for the development of interoperable catalogue systems (*reference 1*). The basic structure of such a system is shown in Figure 1.

It consists of three principle components:

Directory:
a set of descriptions of a large number of data sets containing high level information suitable for making an initial determination of the potential usefulness of a dataset for a given application. Information on the location of more detailed descriptions and/or the data set itself will be found at the directory level.

Guide: information necessary to enable the correct usage of the data, e.g. information describing the characteristics of a data set including format, source instrumentation, calibration, processing algorithms, etc.

Inventory:
information about individual data granules (i.e. individual units of archived data). This information is required to locate specific data granules matching the requirements of the user requesting the data. An inventory might contain a summary of data extracted from the granule, together with information required to extract or transfer data from the archive to a suitable workspace.

Inter-archive searching and extraction is enabled by the use of Directory Interchange Formats (DIFs). A DIF contains a standardized description of the contents of a data archive. The contents of a DIF are "meta-data" (i.e. data about data sets) required by users of other data archives to perform basic search functions on remote catalogues.

Each level of the system can operate independently of the others. This not only allows flexibility for the user, making it possible to interface directly to the inventory or archive levels, but ensures that each level receives a set of instructions complete enough for the level to carry out its functions, irrespective of where those instructions originated. This technique of *context passing* is an essential part of the interoperability of data archives.

The benefits of designing catalogue interoperability into a system from the start are clear. On a global scale, access to a large number of data sets may be vital to individual researchers previously only exposed to locally available data archives. The problems of extracting and importing data from external sources must still be overcome, but very often the major hurdle is simply to identify what data is available and establish its usefulness with respect to the task in question. This, in turn, provides an opening into the realm of data set collocation, whereby data from different sensors is collocated and can be used to obtain new insights previously unobtainable from single-sensor data analysis. Practical

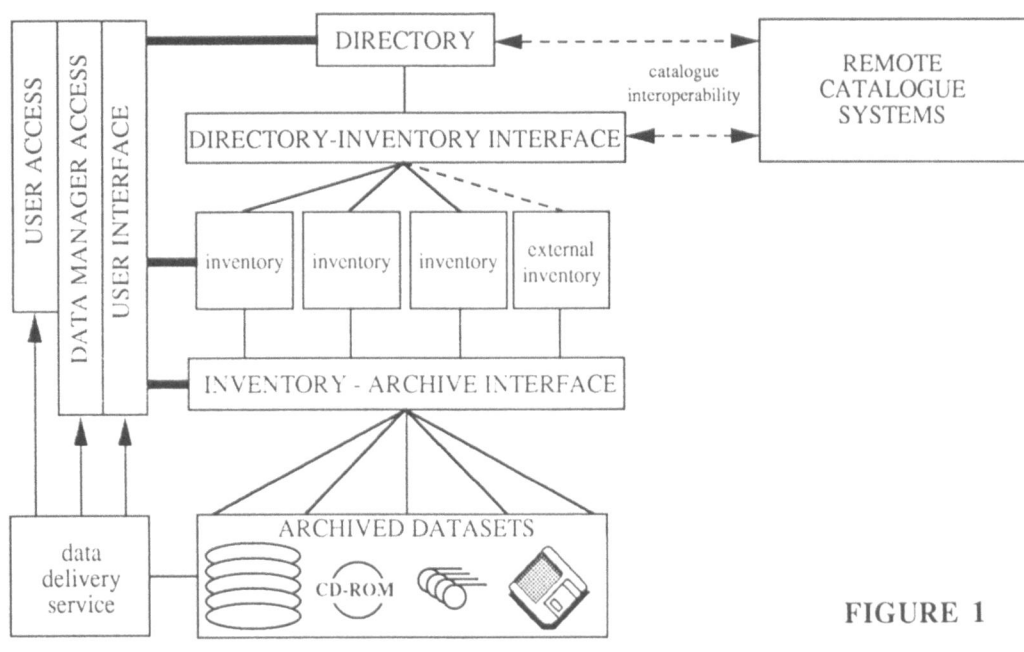

FIGURE 1

17

examples of such collocation include overlaying altimeter data on thermal and infrared imagery. This is an area which has yet to be fully exploited.

New approaches to interfacing

The interface forms a user's principal means of contact with a given system. As such, it is vital that it should assist rather than hinder the user in specifying precisely the information that is required. Various structured query languages (SQL) have been evolved to enable complex logical conditions to be formulated and applied to the archived data. However, for relatively straightforward search criteria, a higher-level graphical interface containing embedded SQL commands hidden from the user, can provide a more effective interface.

For remote sensing applications, search queries can be broken down into four principal components: parameter, coverage, time, and quality. The specification of geographical coverage presents the greatest challenge to the user interface. Three different approaches are illustrated in Figure 2.

In the first example (Figure 2a), the search criteria are specified in terms of the latitude and longitude coordinates representing the corners of a box which defines the area of interest. Two coordinates (e.g. top left and bottom right corners) are sufficient to define the area of interest, which can then be overlaid with further criteria of date, time and individual parameter values. The *search solution* is the result of the search which satisfies the criteria supplied by the user. This is not usually the data itself, but a file containing references, locations and pointers to the required data.

This approach has its limitations however:

- The specified area must be rectangular. This may not be appropriate for the type of data the user requires, causing too much or too little data to be extracted from the archive.

- The search criteria do not take into account how much data match these criteria. A relatively small change to the search area may have a large impact on the quantity and quality of data appearing in the search solution. (See Figure 3 below).

- Visualization of the search criteria is relatively difficult. The interface makes no provision for mapping the search area to a geographical map. This must be done off-line by the user, assuming it is done at all.

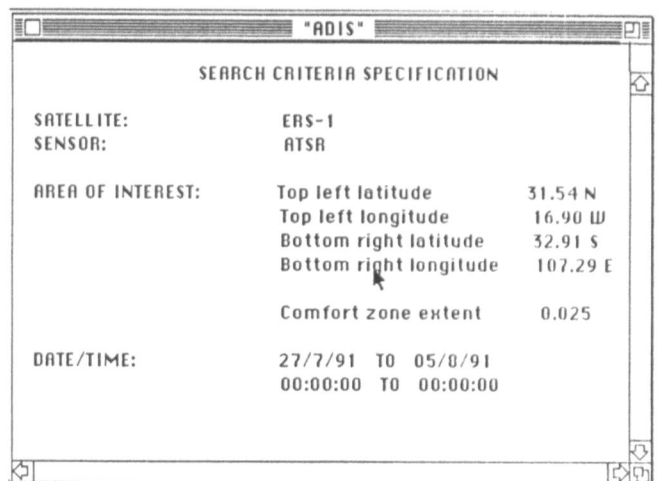

FIGURE 2a:
Search criteria specified by text-based form.

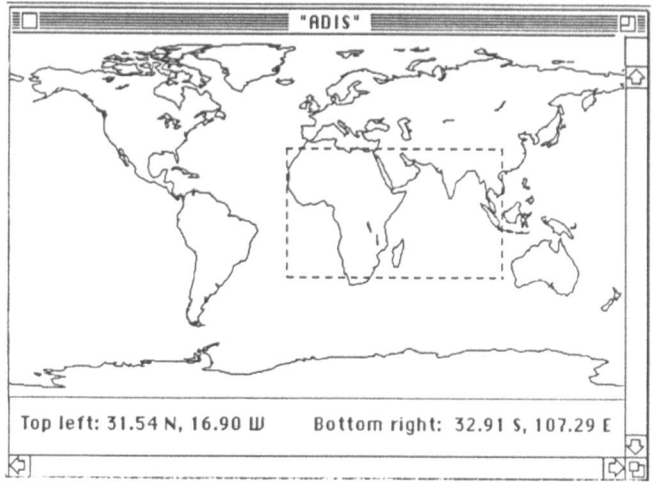

FIGURE 2b:
Search area is superimposed on geographical map of suitable resolution.

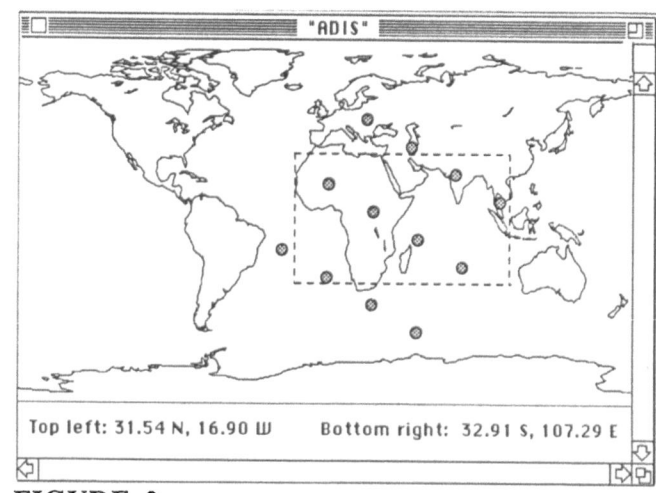

FIGURE 2c:
Search area is superimposed on geographical map of suitable resolution, together with details of available data segments. Using this information, the user may choose to modify his search area accordingly.

18

A better solution to the problem is illustrated in Figures 2b and 2c.

In Figure 2b, the geographical search area is superimposed on a world or regional map. A choice of drawing tools allows an exact region to be defined in as much detail as is required by the user. Ideally, the system is capable of displaying maps of increasing resolution to allow the particular area of interest to be precisely defined.

In Figure 2c, this concept is taken a stage further in an interface which returns information on the data available. The screen displays "product centres" (typically the coordinates of the geographical centre of an image), highlighting which data items will appear in the search solution. This technique has the advantage of allowing the user to modify his search criteria on-line to maximize the results of the search.

The problem can be complicated further by the type of data stored in the archive. Image-based data are relatively simple to deal with since any images overlapping the specified search area can be selected to appear in the "results" of the search. Track-based data composed of discrete, sampled points across a surface are more difficult to deal with.

In Figure 3a, the search solution contains an image partially or fully overlapping the user's search area. However, in Figure 3b, several items of point data lie outside the search area but may well be of interest to the user. In Track 1, although the track itself runs through the search area, the data points

lie outside. Similarly, Track 4 lies wholly outside the search area, but a given data point lies very close to the search area. Many types of point data (e.g. radar altimeter measurements) have a significant footprint size which might well mean that useful data is contained in the track. In both cases, it is quite possible that the user would be interested in this data, although it would not satisfy the search criteria. One solution to this problem is to slightly expand the search area by including a "comfort zone". This ensures that all potentially useful data is included in the solution set but places a requirement on the user to filter out unwanted data.

4. HYPERTEXT AND HYPERMEDIA TECHNOLOGY

HyperText (*reference 2*) is a new approach to the presentation of information. Although the concept of HyperText systems has been in existence since the 1960s, it is only in the last few years that the development has taken off. It differs from more traditional data management systems in that data is linked by context, rather than predefined field structures. The power of the system lies in the ability to define unique links between any individual items of data, which include drawings, pictures and text as well as purely numerical fields. The user is therefore able to use the Hypertext system as a learning tool, extracting various kinds of information in as much detail as required.

HyperText information is structured as a network of nodes connected by links. Nodes can contain

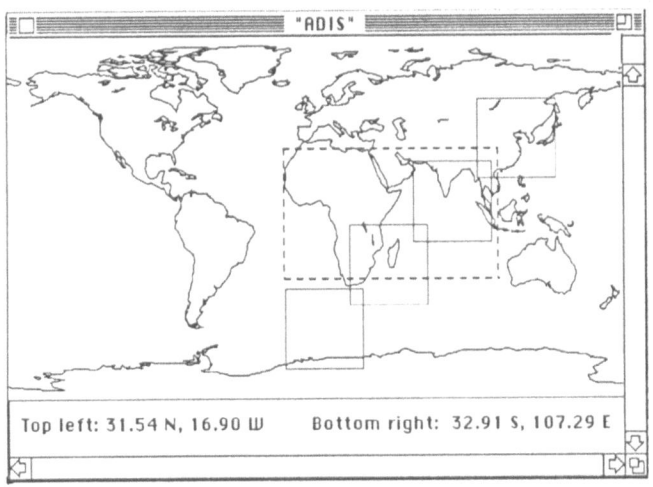

FIGURE 3a:
The search area is superimposed on a geographical map of the region of interest. Data items are selected if the search area fully or partially overlaps them.

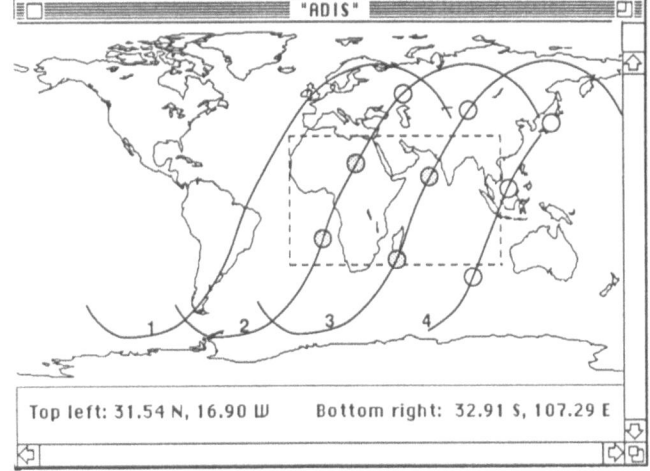

FIGURE 3b:
Only tracks 2 and 3 are selected by the search criteria. Tracks 1 and 4 may also contain useful information.

text, graphics, audio, video as well as executable programs or other forms of data. The nodes (and sometimes the network itself) are viewed through an interactive browse facility and manipulated through a structured editor. This concept of data held in context-linked nodes is fundamental to the power and success of HyperText systems. This will be demonstrated in the next section.

Whilst the organizational structure of standard information database systems and paper catalogues are fixed without major investments in reformatting and restructuring, HyperText links and nodes can be changed dynamically. Information in individual nodes can be updated, new nodes can be linked into the overall HyperText structure and new links added to show new relationships.

HyperText systems can be seen as producing a qualitative change in the way system designers can conceptualize information resources. It is this shift in perspective that has created considerable excitement and such a wealth of new possibilities for the management of essentially unstructured information. These techniques are admirably suited to providing an extremely flexible interface to a data archive.

HyperMedia is a much newer concept that has come from the development of HyperText systems. HyperMedia involves the connection of a wide range of storage-media devices to HyperText nodes. Such devices can include CD-ROM, WORMS, erasable optical disks, laser disks, video recorders, etc. The benefit of including such objects within a HyperText framework is two-fold. It allows the mixture of a much greater range of data within a single system, e.g. text display on a VDU accompanied by speech or a sequence of video frames that visually help to explain the text. It also allows much greater volumes of information to be made available to a user. With CD-ROM holding 600 megabytes of data on a single 5 inch disk, the facilities for creating large and dynamic information resources are greatly increased.

The growth of HyperText and HyperMedia has been greatly facilitated by the development of WIMP (Window-Icon-Mouse-Pointers) computer operating system interfaces. These have become more widely accepted in recent years and have done much to improve the ease of operation of software systems. Many of the developments and improvements to user interfaces rely directly on the WIMP environment and are built firmly on this foundation. These concepts of ease-of-access, efficiency and intuition are fundamental to the

design of *ADIS* and are explored in depth in the next section.

Like most new fields of technology, HyperText and HyperMedia have generated their own set of jargon words. The most important ones have been summarized in Table 1. The list was compiled from the proprietary HyperText system *SuperCard* available for Apple Macintosh computers.

The advantages of HyperText and HyperMedia for man-machine interfaces over more standard approaches can be summarized as follows:

- fast access to the required information
- "intuitive" ease of use
- applicable to both experienced and new users
- inclusion of a wide range of peripheral data to support a user
- efficient integration of many types of data
- can be used for a wide range of tasks - storing, organizing, tracking, communicating, cataloguing, cross-referencing, presenting and retrieving many types of information

5. *ADIS* - ADVANCED DATA INTERFACE SYSTEM

Information on remote sensing data is held in catalogues. Sections 1 to 3 dealt with the major concepts concerning modern computer catalogues. This section will discuss the problems of interfacing to such catalogues and proposes a solution in the form of a new generation catalogue interface - *ADIS*. Although designed with remote sensing catalogues in mind, the generic approach used when designing *ADIS* can be extrapolated to cover many types of data.

A number of requirements are addressed by the *ADIS* system. One obvious requirement is that the system is easy to use, both for experienced and inexperienced users. This has been achieved by providing context-sensitive help, both for specific functions (e.g. defining a search criteria) and for more general information (e.g. descriptions of sensors or processing algorithms). New users inexperienced in the use of particular types of remote sensing data can be given the opportunity to examine example products. A future enhancement to the system will be the addition of a tutorial mode to lead a new user through the user interface by extensive use of examples.

TABLE 1:
Glossary of common Hypertext terms.

SuperCard *A relatively simple Hypertext system developed by Silicon Beach Software and based on Apple's HyperCard. Whilst not containing all of the features of sophisticated hypertext systems, its widespread use makes it an ideal choice for an information system. Its current features are entirely adequate for most information needs and offer exciting possibilities for linking the information to a wide range of audio and visual devices.*

Window/Stack
A name used in the SuperCard system to describe a set of cards (or nodes) based on the same theme. It can be regarded as a SuperCard "document".

Card *SuperCard's basic entity - one screen of information. This is a 'node' in a classic HyperText system.*

Background
A fundamental template shared by a number of cards. The background is composed of a combination of graphics, fields and buttons.

Field *Holds text information. SuperCard has two kinds of field – card fields in which text is only associated with a single card, and background fields in which the field is associated with all cards containing the same background.*

Button *A device for initiating a SuperCard action. Typically, this might be the display of a new item of information or a visual effect.*

SuperTalk *SuperCard's built-in high level programming language that allows the creation of scripts to customize control of the stack. Scripts can be associated with any SuperCard object (cards, buttons etc) and are usually hidden from a user.*

Graphics *A picture is worth a thousand words and this is frequently applicable to man-machine interfaces. The inclusion of graphics performs two necessary functions. It is frequently the most efficient way to convey information that cannot adequately be expressed in words or numerically. Graphics are also important psychologically, providing better means for data visualization.*

Icons *Icons are screen representations of structures or items in the data system, eg. a file of numbers, software program, or section of text.*

Pointers *A mouse-driven pointer allows a user to position a cursor on a section of the screen. "Clicking" the pointer carries out a function related to the pointer's screen position.*

Menus *Menus are a common feature of any user interface. Single selections from multiple layers of menus can be avoided by hiding portions of menus which are not immediately relevant. These menus are termed 'pull-down menus' and usually consist of a list of options that remain hidden from the user until activated by clicking a mouse on an icon or label. At no stage is the user required to leave the application to select the desired mode or option.*

XCMDs/XFCNs
An interface is provided to control other media and run external programs from within SuperCard. The interface consists of commands (XCMDs) and functions (XFCNs) which greatly extend the capabilities of SuperTalk.

Particular consideration must given to the flexibility of the system to allow for the incorporation of new data sets at a future date. This must be done without making assumptions about the format or content of the new data sets.

Taking these requirements and combining them with the major components of a typical catalogue system, the overall structure of *ADIS* and its relationship to the other elements has been designed. It consists of the following components:

- a visual database and browse facility
- a 'Guide' component consisting of general technical information on the satellites and sensors whose data is included in the catalogue
- on-line Help and Utilities
- definition of search criteria
- display of search results
- data ordering facility
- data tools
- mission simulation tools

The linkage of these components with the rest of the catalogue is shown diagrammatically in Figure 4.

Structure of the *ADIS* Demonstrator

Since *ADIS* is a prototype system and is therefore not connected to any operational remote catalogue and archive, it has been necessary to create a simulated catalogue resident within *ADIS*.

The *ADIS* demonstrator is based on an Apple Macintosh IIci work station, with a standard 13" 8-bit Apple colour monitor and four megabytes of RAM. The concepts discussed in Section 4 lend themselves admirably to the implementation of the interface requirements. *ADIS* is therefore written using the SuperCard HyperText application, with external code resources written in Lightspeed C. The simulated data catalogue has been created using ORACLE. Interrogation of the catalogue is performed using standard SQL commands. The visual database consists of a laser disk player (LDP) and an archive of visual products and background information stored on video disk.

The inventory database structure is a simulation of the Earthnet ERS-1 Central Facility's (EECF) Central User Services (CUS) catalogue (*reference 3*). The directory database structure is based on

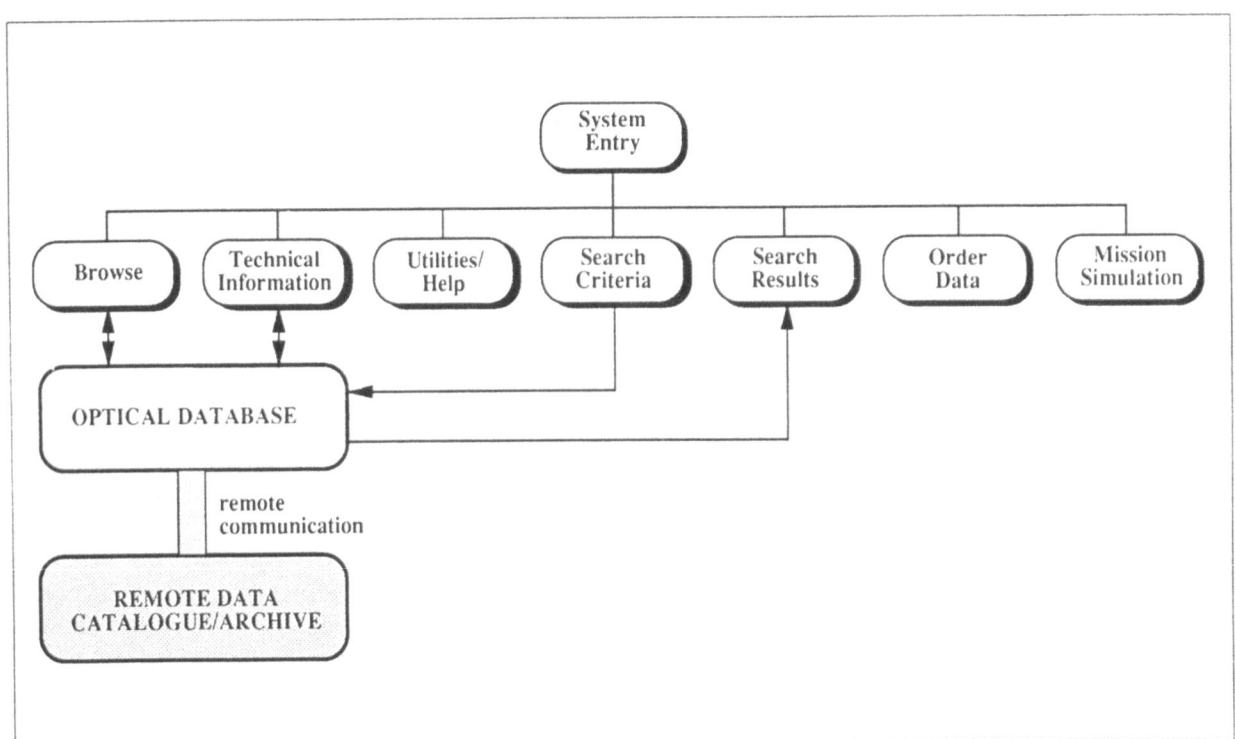

FIGURE 4:
ADIS structure and remote links.

the NASA Master Directory Interchange Format (*reference 4*), in accordance with the requirements of catalogue interoperability, as discussed in Section 2. A definition of inventories and directories has been given in Section 1 of this paper. The *ADIS* inventory data consists of a range of simulated remote sensing satellite products, covering most of the past and present missions. It also contains surface data products, e.g. digital elevation models (DEMS) and publications relevant to remote sensing applications.

The simulated *ADIS* directory data consist of high level dataset descriptions, including datasets from particular satellites and sensors, geophysical and geographical surface data sets, reports describing the technical details of sensors, reports describing the applications of particular remote sensing data etc.

The optical database consists of a videodisk containing example products and quick-looks as single frames plus frame sequences of experts discussing various satellites, their sensors, processing requirements and applications. For the demonstration system, the video frames are displayed on a separate television screen. However, in the next phase of *ADIS*, the analogue video signal will be passed back to the work station and displayed either on a separate high resolution screen or superimposed in a window on the interface terminal.

Search Criteria

The search function allows a user to either define a search query on the database inventories, or search for directory level information to determine what datasets are held in the catalogue. If the latter option is specified, information on the type, extent, source and subject are displayed on the screen. A further option is made available when a summary of all the inventory data specific to that directory can be displayed. Use of this option gives an inexperienced or infrequent user an up-to-date view of what the catalogue contains without potentially wasting time by executing a detailed inventory search for data that may not exist.

To define the search criteria at the inventory level, the user can use a combination of the following information:

- **satellite**
- **sensor**
- **area of interest**
- **date/time**
- quality
- **geographical zone**
- supporting data
- **specific product ID**
- product type (image, track)
- generic sensor type (e.g. altimeter)
- generic satellite type (e.g. geostationary)
- World Reference Grid cell number
- **parameter measured** (e.g. sea-surface temperature)

(essential search criteria are shown in bold)

For most of the criteria, default values are preset. Each user is given a preference set-up file containing these defaults. These defaults can be modified.

Choice of satellite, sensor, quality and geographical coverage is presented either in graphical or text form. Multi-selection of a range of sensors per satellite is also permitted. Date and time information can be specified by numerical definition or by accessing a screen calendar. Supporting data can be selected by icons or from lists. These data types include surface data sets and associated documentation.

High-level search criteria also include the option to define a search on generic parameters - e.g. all radiometric sensors flown in a specified year, all measurements of infra-red reflectance for a given area.

Search Criteria Collation

As the user defines each "layer" of the search, the search definition is built up in structured query language (SQL). For increased speed, the World Reference System (WRS) grid cell number associated with each product centre in the catalogue is calculated for the required geographical area and an index of cell numbers constructed. The WRS is based on the system being developed at Earthnet for the Customer User Services Catalogue. Cell boundaries are defined by lines of longitude/latitude. One cell has the dimensions of 1 degree latitude by 1 degree longitude, except for latitudes greater than 89 degrees, where only one cell will exist. The numbering system for the cells is based on the South-East longitude/latitude of the cell. Six digits are used, split into two fields. The first field uses three digits to represent the latitude of the SE corner of the cell (090 to -90) and the second field uses three digits to represent the longitude of the SE corner of the cell (000 to 360).

Search Execution
The SQL command is passed to the ORACLE database containing the product inventories. Those records matching first the grid cell numbers required and then the remaining search parameters are flagged. For each product identified, a further selection is performed by checking the coordinate boundaries of the products with the actual search coordinates (plus a comfort zone). This method checks that the products flagged do actually lie within the user's search area. It works well for both image and non-image type products, but not for orbit products. These require preprocessing to a suitable form and has not yet been addressed. Once the appropriate records have been located, they are extracted and passed back to the HyperText interface.

Display of Search Results
The results are returned in tabular form from ORACLE to the Search Results stack. The results may be displayed in tabular form or graphically, e.g. Landsat path/row maps or altimeter tracks overlaid on a world map.

Order Data
From the results of a search, the user then has the option to order any products. This can be achieved by selecting either from the results tables or maps. If quick-looks are contained in the visual database and match any products found in the search, these are identified on the screen. These can be displayed on-line immediately.

If the data is requested by the user, an order is passed to the data archive. Data is extracted and passed back to the user's work space. In reality, this operation is likely to involve remote communications with a central archive and has not yet been simulated in *ADIS*.

On-line Help and Utilities
This feature consists of a SuperCard stack giving full on-line help on how to use the catalogue system, plus telemail facilities, contact names and addresses and a graphical display of the structure of the catalogue system. This latter item allows users to understand their present location within the catalogue and the nature of links between other components.

Technical Information
Potential customers of the remote sensing catalogues are frequently unfamiliar with many aspects of earth observation. Therefore an important function of the interface is to supply detailed technical information on satellites, sensors and data contained within the catalogue and archive. Use has been freely made of a modified *EOStacks* (*reference 5*) system. *EOStacks* is an information resource developed for the Royal Aerospace Establishment and contains technical details on products, applications and the space and ground segments for twenty earth observation satellites.

Mission Simulator
The final part of the interface is a mission simulator. This is a model that, for a range of satellites, displays the expected orbit tracks and instrument coverages for a specified time and geographical area.

General Considerations
An important consideration in interfacing to remote catalogues is to ensure that the user interface translates search criteria to a standard text-based format prior to transmission over communication links. Most remote links are not capable of supporting graphics and this must be handled by individual user terminals. In *ADIS*, a user can specify a search area by drawing a box on a map. This is then translated into standard SQL before any connection is made to the data archive.

ADIS also contains an additional help feature - the 'why', 'what' and 'how' concept. At any stage in the system where a decision or action is required, information on **what** is to be done (e.g. select a geographical zone), **why** it has to be done (e.g. otherwise the whole world is selected by default) and **how** it is to be done (e.g. click on the zone of interest shown in the world map) is displayed. This, combined with the context-sensitive help, ensures that the user is presented with all the information necessary to fully exploit the system. Further examples of defining criteria, ordering products, etc. are contained in an *ADIS* tutorial.

6. SUMMARY

We have attempted to show that an efficient and flexible user interface is vital to ensure that effective use is made of data management systems. Nowhere is this more important than in the field of earth observation where large (and frequently diverse) datasets must be manipulated to perform satisfactory geophysical interpretations. Using a

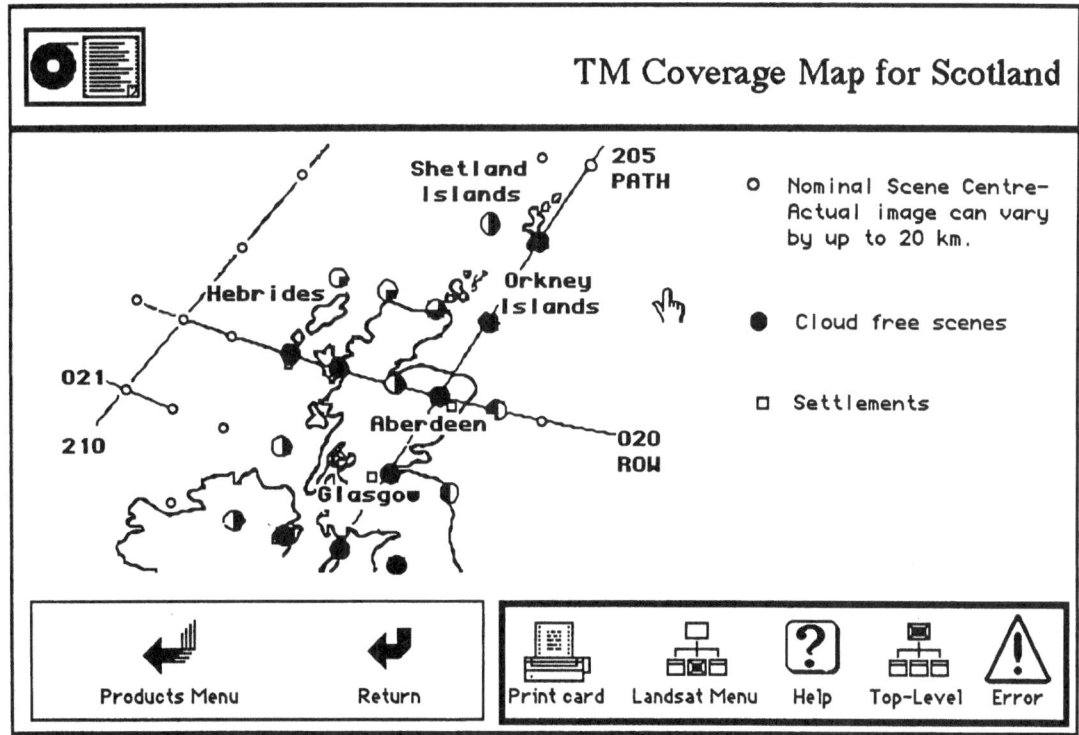

FIGURE 5:
Ordering products from maps (example taken from *EOStacks*).

combination of widely available hardware and software technologies, we have attempted to build a demonstrator system that goes some way towards addressing the problems of effective interaction with complex data archives. Such an approach can prove to be a very powerful tool, placing in the hands of the user instant access to global and regional datasets, with sophisticated mechanisms for the extraction and analysis of this data.

The goal of this work has been to produce a prototype data archive system incorporating the latest developments in interface technology. *ADIS* is not intended to be an operational system although it will form the basis of a number of planned developments in data management. The next stage in its evolution will be the incorporation of CD-ROM, write-once-read-many-times (WORM) optical disk and other mass-storage devices. This branch into multi-media will increase the scope and effectiveness of data archive systems as well as providing the tools for addressing the problems of data manipulation and visualization.

7. ACKNOWLEDGEMENTS

The authors would like to acknowledge the support and cooperation of Dr Mark Elkington (Earth Observation Sciences) during the preparation of this paper. Parts of the work described in this paper were carried out in conjunction with the development of *EOStacks* and the *Algorithm Development Facility* under contract to the Space Department, Royal Aerospace Establishment, Farnborough, UK.

8. REFERENCES

1. *CEOS Catalogue Subgroup, 1990: Guidelines for the Development of Internationally Interoperable Catalogue Systems. Elkington M. D. (ed.), draft*

2. *Hypertext, Byte, October 1988. Fiderio, J., et al, McGraw-Hill*

3. *CU-PG-MDA-SY-0005 CUS Phase B Final Report, 1987. McDonald Detwiler Associates*

4. *Quick Reference Guide: NASA Master Directory, December 1988. NASA*

5. *EOStacks - Earth Observation Information Resource Future Development, December 1988. Elkington, M. D.*

Undocumented faults revealed in multisensor image analysis, Mojave Desert, California, U.S.A.

J. P. Ford B.Sc. (Hons.), Ph.D.
Jet Propulsion Laboratory, California Institute of Technology, Pasadena, California, U.S.A.
R. K. Dokka B.S., M.S., Ph.D.
Department of Geology and Geophysics, Louisiana State University, Baton Rouge, Louisiana, U.S.A.
R. G. Blom B.S., M.S., Ph.D.
Jet Propulsion Laboratory, California Institute of Technology, Pasadena, California, U.S.A.

SUMMARY

Undocumented late Miocene and younger strike-slip and normal faults that extend up to 25 km in the eastern, and northeastern Mojave Desert block have been detected and mapped using Landsat thematic mapper (TM), SPOT panchromatic, and synthetic-aperture radar (SAR) images. The faults are located in the Bristol Mountains, Cady Mountains, Alvord Mountain, and Granite Mountains (Fort Irwin military reservation) areas. Some faults are perceived on the TM images because of spectral contrasts primarily at infrared wavelengths. Other faults are highlighted on SPOT panchromatic and radar images because of topographic contrasts. Corregistration of enhanced TM/SPOT panchromatic data maximizes both spectral and topographic information in the images. The newly detected faults form part of a complex regional network of right shear in eastern California that connects faults in the southern Death Valley region with the San Andreas fault system. Structural relations along the faults suggest at least two intervals of movement and a westward migration of the locus of strain in the southern Mojave. In the Bristol Mountains east of Broadwell Lake, the faults are overlain by undisturbed Pleistocene alluvial fan deposits and are probably inactive. In contrast, the area to the west (Cady Mountains) and south contains faults that cut all deposits and is seismically active. Some faults in the Cady Mountains bound blocks that have experienced differing rotational histories since middle Miocene time. Newly identified faults in the northeastern Mojave suggest that right-slip may be more important to the overall deformation than previously reported.

INTRODUCTION

The Mojave Desert block (MDB) is a fault-bounded tectonic unit in southern California at the western margin of the North American plate[1].

Situated between the San Andreas fault system (SAFS) to the southwest, the southern Death Valley fault zone (DVFZ) to the northeast, and extensional terranes to the east and southeast, the MDB has a complex history of late Cenozoic deformation. Extensional tectonism in the MDB was marked in early Miocene time (ca. 23-19 Ma) by detachment faulting and uplift [2-4]. This was followed after 13 Ma by northwest-oriented strike-slip faulting, distributed mostly across the central and eastern portion of the region (Fig. 1).

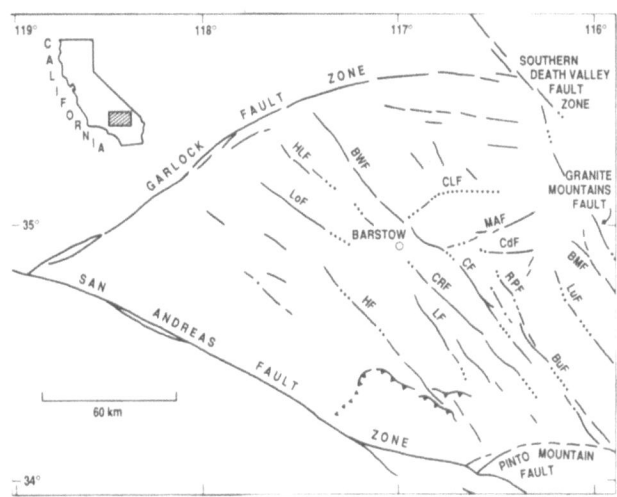

Fig. 1. Major faults of the Mojave Desert Block (MDB). Only the Blackwater-Calico fault is continuous across the MDB. Abbreviations: BuF - Bullion fault, BMF - Bristol Mountains fault, BWF - Blackwater fault, CF - Calico fault, CdF - Cady fault, CLF - Coyote Lake fault, CRF - Camp Rock fault, HF - Helendale fault, HLF - Harper Lake fault, LF - Lenwood fault, LoF - Lockhart fault, LuF- Ludlow fault, MAF - Manix-Afton fault, RPF - Rodman-Pisgah fault.

The right-slip nature of the northwest faults was first recognized by Dibblee[5] who thought they were probably related to the SAFS. Paleomagnetic evidence has indicated that deformation in the MDB also includes block rotations of variable magnitude and direction around vertical axes[6-9]. The northwest faults are important, however, because they dominate the structure and physiography of the area and for the seismic hazard they present.

The MDB consists of rugged mountains that are about equally interspersed with alluviated basins and dry lake beds[10]. Although the rocks are widely exposed in the mountains, the geology of the area is poorly known and largely generalized on published maps. One reason is that much of the MDB is difficult of access and large areas are closed to the public. Also, alluvium and dry lake deposits cover the bedrock geology in many key areas. Under the circumstances, remote sensing provides an ideal means of locating and investigating tectonic features in the area. Images obtained by airborne and spaceborne systems through a range of wavelengths from optical and infrared to microwave (radar) provide map-scale data that are well suited for these purposes. The images were used to study tectonic features in

GEOLOGIC SETTING

Late Cenozoic, northwest-oriented, right-slip faults are distributed across the MDB from the Helendale fault eastward to the Granite Mountains fault[1] (Fig. 1). The faults are spaced ~10 to 20 km apart and, with the exception of the Calico-Blackwater fault, they lack continuity across the block from northwest to southeast. These faults terminate near the latitude of Barstow. Major differences in displacement have been documented on individual faults to the south and on their projected counterparts to the north[12]. In the northeast (FIMR area), the MDB is cut also by east-oriented faults for which left slip is widely inferred[12, 13].

The strike-slip faults are expressed mostly by prominent fault-line scarps and aligned truncated spurs. Such features form topographic lineaments that are clearly visible on the remote sensing images. All the strike-slip faults in the area are considered to be younger than 13 Ma[12]. Bedrock consists mainly of Mesozoic granitoids, Oligocene (?) through early Miocene volcanic

Table I. Image data characteristics

Sensor	Band	Wavelength		Spatial Resolution (m)
Landsat TM	1 (optical)	0.45 - 0.52	μm	~30
"	3 "	0.63 - 0.69	"	"
"	4 (IR)	0.76 - 0.90	"	"
"	5 "	1.55 - 1.75	"	"
"	7 "	2.08 - 2.35	"	"
SPOT	Panchromatic	0.51 - 0.73	μm	10
Airborne SAR	XHH*	2.8	cm	~12
Seasat SAR	LHH*	23.5	cm	~25

*HH denotes horizontal transmit/receive polarization

portions of the eastern and northeastern MDB, where little detailed mapping is available.

The study area extends from Goldstone Lake and the Fort Irwin military reservation (FIMR) in the north to the Argos Valley and Bullion Mountains in the south. In the Cady Mountains and the northern and central Bristol Mountains (Fig. 2), enhanced color-composite Landsat thematic mapper (TM) images have revealed undocumented faults that were previously overlooked[11]. Here, we compare enhanced TM and other types of images obtained in a range of optical and radar wavelengths for detecting and identifying tectonic features in the MDB.

sequences, and younger continental sedimentary deposits. Late Tertiary and Quaternary basaltic lavas cover some areas. Unconsolidated sediments and eolian sand deposits are widespread in the basins.

IMAGE TYPES AND DATA PROCESSING

Multisensor images obtained by the Landsat thematic mapper (TM), Système Probatoire pour l'Observation de la Terre (SPOT), airborne synthetic-aperture radar (SAR), and spaceborne Seasat SAR imaging systems were used in this study. Table I lists the wavelength (band) and spatial resolution of the image data sets used here.

The Landsat TM data were processed to form color-composite images that enhance surface spectral responses while retaining topographic information. TM band ratios that emphasize the compositional differences of rocks at visible and infrared (IR) wavelengths (bands 3/1, 5/4, 5/7) were selected[14, 15]. Each band-ratio channel was then modulated by the band-average reflectance of five bands (1, 3-5, 7) and assigned a primary color. Thus, variations in image chromaticity depict variations in ferric iron (blue), ferrous iron (green), or hydroxyl-bearing minerals and/or carbonates (red). The achromatic band-average data were encoded as intensity.

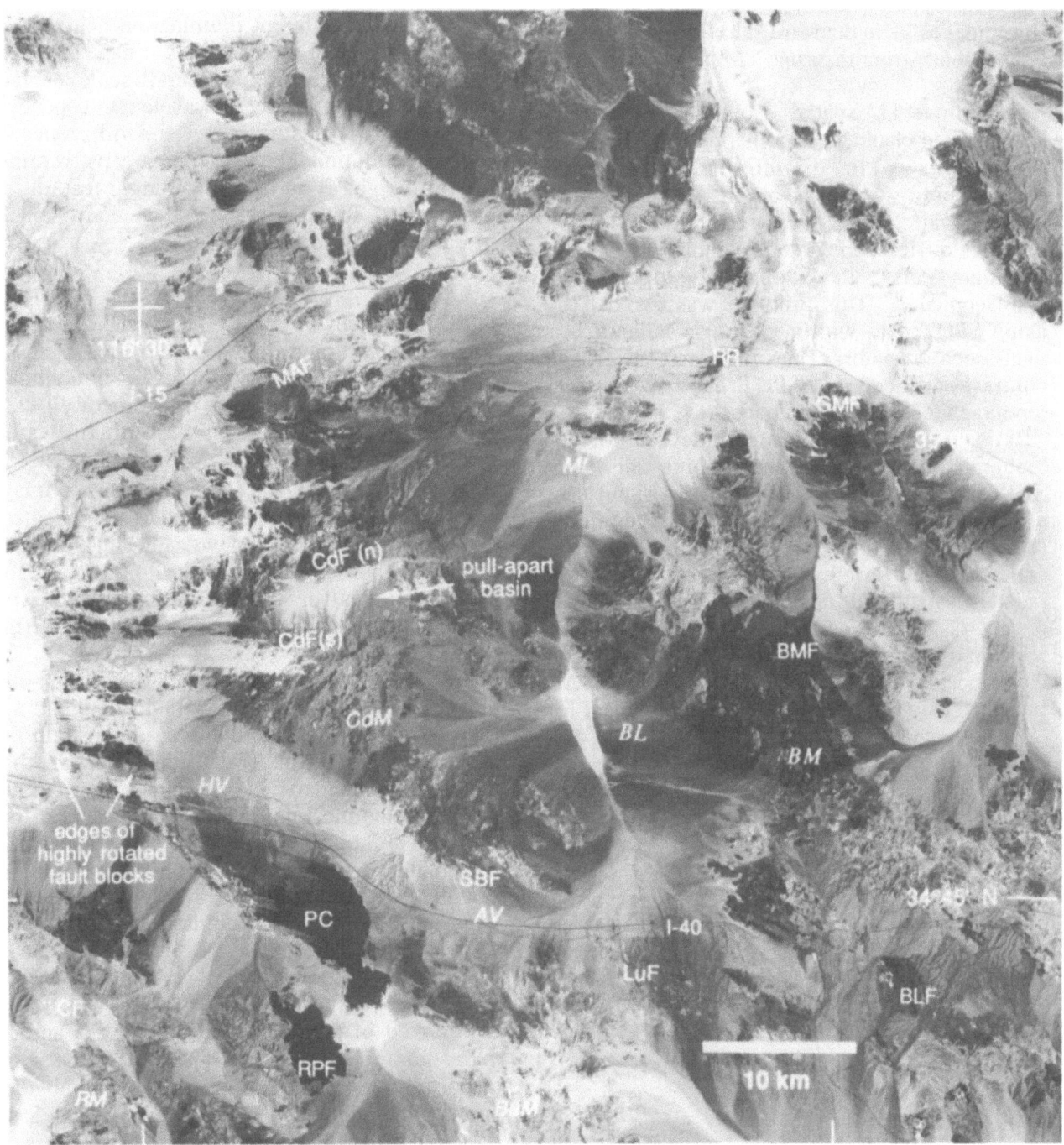

Fig. 2. Landsat TM image (bw) of eastern Mojave Desert. Color variations on the enhanced color image reveal structures that were overlooked in previous mapping. Newly discovered faults: BLF - Broadwell Lake fault, CdF (n) - Cady fault north strand, SBF - Sleeping Beauty fault. Previously documented faults: BMF - Bristol Mountains fault, CF - Calico fault, CdF(s) - Cady fault south strand, GMF - Granite Mountains fault, LuF - Ludlow fault, MAF - Manix-Afton fault, RPF - Rodman-Pisgah fault. Prominent features: PC - Pisgah Crater, BL - Broadwell Lake, ML - Mesquite Lake, BM - Bristol Mountains, BuM - Bullion Mountains, CdM - Cady Mountains, RM - Rodman Mountains, AV - Argos Valley, HV - Hector Valley.

Primarily, this enhances topographic features on images[16]. The potential for confusion of the four dimensions of data in three-dimensional color space is minimized because the spatial differences between topographic and reflectance information are commonly distinct. These enhancement procedures reveal important geologic information that is not apparent in single-band black-and-white (bw) images or in standard false-color-composite images made from the same TM data set (e.g. Fig. 2).

To improve TM spatial resolution (~30m), the band-ratioed channels (chromatic components) of the images were digitally resampled and corregistered to corresponding SPOT 10-m panchromatic data. Image corregistration becomes difficult when corresponding features in the data sets have differing topographic displacements. This problem was avoided by using SPOT data with a viewing geometry less than 3 deg off nadir. Thus, the corregistered and contrast-enhanced TM/SPOT data reveal finer topographic detail on the images (e.g. Fig. 3).

Radar images also provide important topographic information. The airborne and Seasat SAR data sets listed in Table I have significant differences in antenna viewing geometry (depression angle), illumination direction (azimuth), and swath width. Because imaging radar is sensitive to large-scale slope variability, the contrasted SAR system viewing geometries are responsible primarily for the pronounced differences in feature definition seen on the SAR images. This condition is particularly evident for

Fig. 3. Band-ratioed Landsat TM data digitally corregistered with corresponding SPOT panchromatic data (bw) improve spatial resolution in central Cady Mountains. The color composite image discriminates spectral information together with finer topographic detail.

sloping terrain of moderate to high relief. At the low depression angle (14-27 deg) of the airborne SAR, topography is directionally enhanced on the images by shadowing; at the high depression angle (67-73 deg) of Seasat SAR, topography is strongly distorted on the images by geometric compression and layover of radar-facing slopes (e.g. Fig. 4). The directional sensitivity of radar is also clearly evident in the different illumination azimuths of the airborne and Seasat SAR images. The contrasting perception of small-scale surface roughness on the images is wavelength dependent (Table I). In this instance, the difference in spatial resolution plays a comparatively minor role in the ability to detect geologic features on the images. Both SAR data sets were available as mosaics in map format.

IMAGE ANALYSES AND FIELD OBSERVATIONS

Image analyses and directed field investigations have led to the discovery of previously unknown faults and related structures. The Broadwell Lake fault was detected from prominent color alignments first observed on the enhanced TM image[11]. The fault trace identified on the image was observed in the field to be an alignment of discontinuous outcrops of volcanic tuff and tuff breccia in fault contact with granitoid rocks. The fault continues for about 25 km in the Bristol Mountains, through Black Ridge to the Lava Hills (Fig. 2).

The fault is located in a broad zone of right-slip movement; however, field work to date has not revealed piercing points that would constrain the direction and amount of displacement that occurred. Kinematic indicators suggest dip-slip motion at the most northwesterly exposure of the fault and strike-slip motion at several places along its length. As undeformed Pleistocene alluvial gravels overlie faulted bedrock at several exposures, the fault does not appear to have been recently active.

The Broadwell Lake fault does not have obvious topographic expression. For this reason, it was overlooked in previous field mapping[17] and on aerial photographs. It was also not detected on airborne or Seasat SAR images. A number of subparallel ridges and slopes are highlighted on the SAR images (Fig. 4) because the terrain is oriented transverse to the illumination in both cases. However, the ridges do not indicate the position of the fault on the SAR images.

In the Cady Mountains, a number of strike-slip and normal faults from 3-14 km in length appear as aligned topographic features that coincide with abrupt textural and color discontinuities on the enhanced TM image. The topographic and

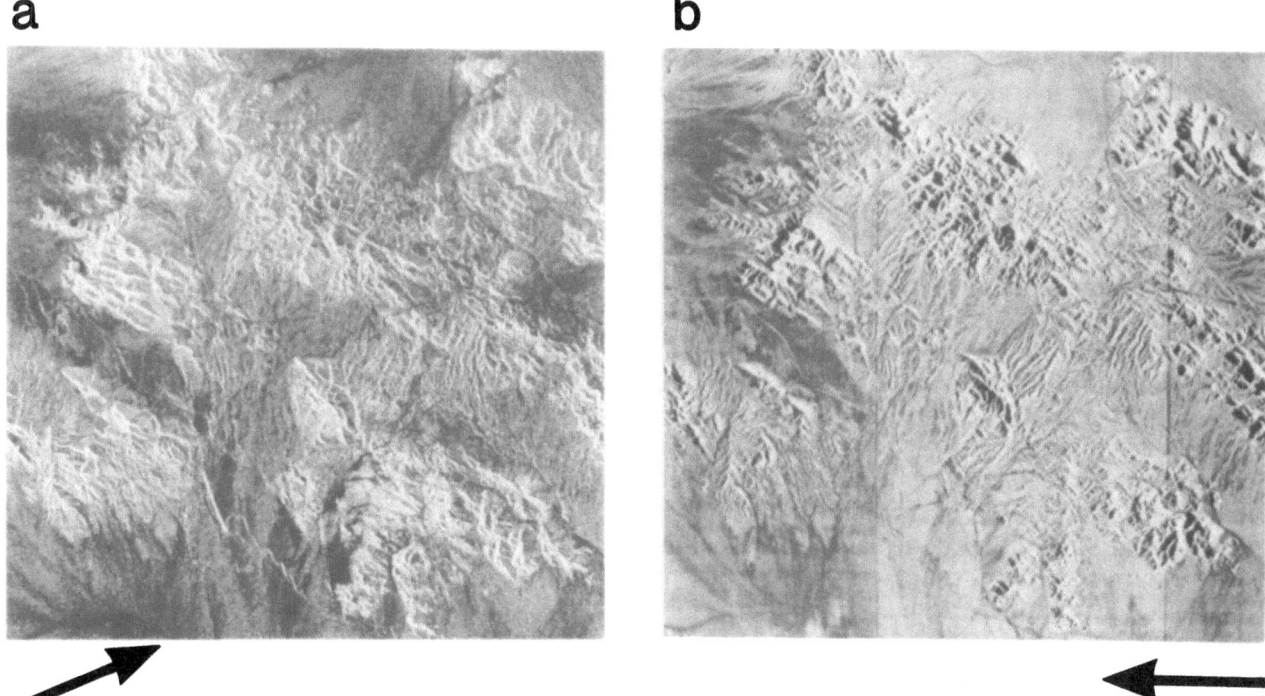

Fig. 4. Broadwell Lake fault in the Bristol Mountains-Lava Hills extends from upper left to lower right across images: (a) Seasat SAR obtained at 23.5 cm wavelength and 67-73 deg depression angle looking northeast (August, 1978); (b) airborne (Aeroservice/U.S.Geol. Surv.) SAR obtained at 2.8 cm wavelength and 14-27 deg depression angle looking due west (November, 1985). Note effects of differing viewing directions, shown by arrows, and depression angles on terrain perception and feature identification and compare corresponding coverage on TM image (Fig. 2).

compositional information depicted on the image have led to the field verification of faults that were overlooked in previous field mapping[18-19]. Examples include the north strand of the Cady fault[11], and linear boundaries of highly rotated fault blocks[9] (see Fig. 2 for locations). A wide depression (Hidden Valley) between the north and south strands of the Cady fault has been interpreted as a pull-apart basin that formed along the left-stepping left-slip Cady fault system[12]. The Cady Mountains area is seismically active and contains faults that cut deposits of Pleistocene and younger age.

The higher resolution of the combined TM/SPOT image (Fig. 3) defines the topography more sharply, which is important for detecting subtle low-relief features. Linear relief is highlighted on the corresponding SAR images, although subtle features oriented within ~15 deg of the illumination azimuth are mostly not distinguishable.

Most of the Goldstone/FIMR area is closed to the public. Available geologic maps of the area present only generalized lithologic and structural information[20]. TM, SPOT, and SAR images provide more detailed coverage than is obtainable from other sources. Preliminary investigations of

airborne SAR images have shown the presence of a number of unmapped northwest-oriented fault scarps and truncated spurs in the area. The features are pronounced on the SAR images because of their transverse orientation relative to the illumination. Although the features remain to be checked in the field, their linearity, extent, orientation, and tectonic setting suggest that they represent unmapped right-slip faults. In this event, right slip may be more important to the overall deformation of this region than has previously been reported.

In addition to the advantages that remote sensing images offer in locating previously unmapped structures in inaccessible terrain, they also enable the verification of existing map data. For example, the Ludlow fault (Figs. 1-2) was mapped in bedrock south of Ludlow[21] and projected northward for over 30 km along the east margin of the Cady Mountains. As shown on published maps, the fault is extrapolated across wide alluviated areas to join short fault segments with differing orientation in disconnected outcrops [17, 22]. This interpretation, however, is not supported by any of the images or by our field observations.

31

DISCUSSION

Both topographic and lithologic information are needed to interpret tectonic features from remote sensing images. The topographic highlighting observed on radar images and on high-resolution wideband optical images of the Mojave Desert enhances the perception of aligned features that denote fault structures. The ability to detect and interpret such features, however, is subject to constraints imposed by the sensor resolution, the scene illumination vector, and the geometric orientation of the features. By comparison, color patterns observed on enhanced multispectral images denote compositional differences among the exposed rocks and sediments. Enhanced images discriminate surface materials at infrared wavelengths that do not display strong visible color contrasts. This has allowed the detection of faults in localities where the topography is not diagnostic.

With suitable processing, multispectral TM data sets have provided color images on which lithologies are discriminated by their contrasting absorption and reflection, and landform depiction is retained. Improved landform depiction has been achieved by the digital corregistration of corresponding higher-resolution SPOT panchromatic data. SAR data have provided an alternative source of images that display pronounced, though strongly directional landform enhancement.

The newly discovered faults[11] are interpreted as part of a regional network of right shear that connects the Death Valley region with the San Andreas fault system[12, 23]. Structural relationships along the faults suggest at least two intervals of movement. Faults located east of Broadwell Lake are overlain by unconsolidated alluvial fan debris and are probably inactive. In contrast, the area to the west (Cady Mountains) contains faults that cut all deposits and is seismically active. The present pattern of faulting and strain indicates that regional shear is now concentrated in the central and southern parts of the MDB[24].

Recently, Dokka and Travis[12,23] have suggested that strain across the MDB since middle Miocene time is regionally heterogeneous and is partitioned into domains that have independent histories and styles of late Cenozoic deformation. This model is based on field mapping, which shows that most northwest-striking faults in the MDB lack continuity across the block and that there is a major discrepancy in the displacement on individual faults to the south and on their projected counterparts to the north. This model differs substantially from earlier models[13, 25] of regionally distributed simple shear that require continuity and uniform displacement of the faults across the entire MDB.

The newly observed faults provide supporting evidence for the heterogeneous strain model. For example, the Broadwell Lake fault is limited in its extent, in common with other right-slip faults in the area to the northeast (Bristol Mountains and Granite Mountains faults) and west (Ludlow fault). Multisensor images have allowed the recognition of important faults in the Mojave Desert that were previously overlooked by field geologists; thus, the images have facilitated tectonic analysis of the region. The results clearly demonstrate the utility of such images in support of local and regional tectonic studies in well exposed terrain.

References

1. Dokka, R.K. Displacements on late Cenozoic strike-slip faults of the central Mojave Desert. Geology, v. 11, (1983), p. 305-308.
2. Dokka, R.K. Patterns and modes of early Miocene extension of the central Mojave Desert, California. In, Continental Extension Processes, L. Mayer, ed., Geol. Soc. Amer. Spec. Paper 208, (1986), p. 75-95.
3. Dokka, R.K. The Mojave extensional belt of southern California. Tectonics, v. 8, (1989), p. 363-390.
4. Dokka, R.K., and M.O.Woodburne. Mid-Tertiary extensional tectonics and sedimentation, central Mojave Desert, California. Louisiana State Univ. Publ. Geol. Geophys. Tecton. Sediment., v. 1, (1986), 55 pp.
5. Dibblee, T. W. Jr. Evidence of strike-slip faulting along northwest-trending faults in the Mojave Desert. U. S. Geol. Survey Prof. Pap., 424-B, (1961), p. B197-B199.
6. Golombek, M., and L.Brown. Clockwise rotation of the western Mojave Desert. Geology, v. 16, (1988), p. 126-130.
7. MacFadden, B.J., M.O.Woodburne, and N.D.Opdyke. Paleomagnetism and Neogene clockwise rotation of the Northern Cady Mountains, Mojave Desert of Southern California. Jour. Geophys. Res., v. 95, no. B4, (1990), p. 4597-4608.
8. Ross, T.M., B.P.Luyendyk, and R.B.Haston. Paleomagnetic evidence for Neogene clockwise rotations in the central Mojave Desert, California. Geology, v. 17, (1989), p. 470-473.
9. Ross, T.M. and R.K.Dokka. Structure of a highly rotated crustal block in the southwest Cady Mountains, California: Implications for the kinematic development of the Mojave extensional belt. Geol. Soc. Am. Abstr. Progr., v. 22, no. 3, (1990), p. 79.

10. Jennings, C.W. (compiler). Geologic Map of California. Sacramento, California: Calif. Div. Mines and Geol., Geologic Data Map No. 2 (1977), scale 1:750,000.

11. Ford, J.P., R.K.Dokka, R.E.Crippen, and R.G.Blom. Faults in the Mojave Desert, California, as revealed on enhanced Landsat images. Science, v.248, (1990), p. 1000-1003.

12. Dokka, R.K., and C.J.Travis. Late Cenozoic strike-slip faulting in the Mojave Desert, California. Tectonics v. 9, (1990), 311-340.

13. Garfunkel, Z. Model for the late Cenozoic tectonic history of the Mojave Desert and its relation to adjacent areas. Geol. Soc. Am. Bull., v. 85, (1974), p. 1931-1944.

14. Podwysocki, M.H., M. S. Powers, and O. D. Jones. Preliminary evaluation of the Landsat-4 Thematic mapper data for mineral exploration. Adv. Space Res. v. 5 (no. 5), (1985), 13-20.

15. Crippen, R.E., E.J.Hajic, J.E.Estes, and R.G.Blom. Statistical band and band-ratio selection to maximize spectral information in color composite displays. Remote Sensing of Environment, (submitted).

16. Crippen, R.E. Image display of four components of spectral data. Remote Sensing of Environment, (submitted).

17. Dibblee, T.W., Jr. Geologic map of the Broadwell Lake quadrangle, San Bernardino County, California. U.S. Geol. Survey Misc. Geol. Inv. Map I-478 (1967), scale 1:62,500.

18. Dibblee, T.W., Jr., and A.M.Bassett. Geologic map of the Cady Mountains quadrangle, San Bernardino County, California. U.S. Geol. Survey Misc. Geol. Inv. Map I-467, (1966) scale 1:62,500.

19. Dibblee, T.W., Jr., and A.M.Bassett. Geologic map of the Newberry quadrangle, San Bernardino County, California. U.S. Geol. Survey Misc. Geol. Inv. Map I-461, (1966) scale 1:62,500.

20. Jennings, C.W., J.L.Burnettt, and B.W.Troxel (compilers). Geologic Atlas of California. Sacramento, California: Calif. Div. Mines and Geol., Trona Sheet (1962), scale 1:250,000.

21. Dibblee, T.W., Jr. Geologic map of the Ludlow quadrangle, San Bernardino County, California. U.S. Geol. Survey Misc. Geol. Inv. Map I-477 (1967), scale 1:62,500.

22. Bortugno, E.J., and T.E.Spittler (compilers). Geologic Map of the San Bernardino Quadrangle, California. Sacramento, California: Calif. Div. Mines and Geol., Regional Geologic Map Series No. 3A, (1986), scale 1:250,000.

23. Dokka, R.K., and C.J.Travis. The role of the eastern California shear zone in accomodating Pacific-North American transform motion. Geophys. Res. Lett., (in press).

24. Sauber, J., W.Thatcher, and S.Solomon. Geodetic measurement of deformation in the central Mojave Desert, California. Jour. Geophys. Res., v. 91, (1986), p. 12,683-12,694.

25. Carter, J.N., B.P.Luyendyk and R.R.Terres. Neogene clockwise rotation of the eastern Transverse Ranges, California, suggested by paleomagnetic vectors. Geol. Soc. Am. Bull., v. 98, (1987), p. 199-206.

Acknowledgments

This research was performed at the Jet Propulsion Laboratory, Pasadena, California under contract to the National Aeronautical and Space Administration, Washington, DC. Additional support was provided from a Summer Faculty Fellowship award to R.K.Dokka under the JPL/Caltech NASA/ASEE Program. Field investigations by R.K.Dokka were supported by grants from the National Science Foundation. Landsat data copyrighted by EOSAT. SPOT data copyrighted by CNES.

Case study on metal-stressed vegetation from a copper orebody, Kiruna region, northern Sweden

H. Isaksson M.Sc.
Swedish Geological Company, Luleå, Sweden
L. C. Andersson M.Sc., T.Lic.
Swedish Space Corporation, Solna, Sweden

SYNOPSIS

Soils with high content of copper in the Kiruna region, northern Sweden, are known to correspond to open grass areas with the flower "Viscaria Alpina". Such locations also occur at two different copper-occurences, Viscaria and central Pahtohavare. A study of metal effect on vegetation has been performed using satellite data, a digital elevation model and vegetation mapping from airphotos. The result from the study is integrated with other geological information to provide a priority of the targets.

INTRODUCTION

During the last five years improved satellite based image sensors and new methods for analysis of both image data and traditional geological measurements have been developed. Modern information technology has partly modified and enhanced the technique for geological interpretation. The visualisation of geophysical and geochemical measurements as images and the combination of this information with remote sensing data in a digital image processing system have become valuable new tools for the exploration community.

A case study has been carried out to test the applicability of remote sensing for exploration in vegetated and glaciated terrain and to develop methods for co-analysis of satellite, airborne and ground measurements. Soil with high content of copper in the Kiruna region, northern Sweden, are known to correspond to open grass areas with the flower "Viscaria Alpina". Such locations occur at two different copper-occurences, Viscaria and central Pahtohavare. The study examined the possibility to detect such metal effects on vegetation using Landsat TM and elevation data, and furthermore tried to improve the efficiency in targeting these areas by integration of additional geological information.

This study has partly been performed within the framework of a project called GEOVISION, and partly run for the Swedish State Mining Property Commission (NSG). The GEOVISION project is a joint effort between the Swedish Geological Co (SGAB) and Swedish Space Corp (SSC) to develop a geologically adapted and raster-based GIS.

OBJECTIVES

The aims of the case study have been the following:

- to develop and test a geobotanical remote sensing method applicable in glaciated terrain,

- to develop integration techniques for co-analysis of the special types of data used in exploration work (Fig 1),

- to provide an example and demonstrate the practical application of this new technology.

Spatial distribution / Physical characteristics	Rectangular grid		Scattered points
	Satellite Aerosurvey	Ground survey	
Potential field	magnetic fields electromagnetic fields electrial fields		gravity
Partly continuous characteristics	natural emission	geochemical analysis	
Discontinuous characteristics	radiance topography	natural emission susceptibillity density geochemical analysis geological observations	

Figure 1: Different types of measured data, regarding spatial distribution and physical characteristics.

APPROACH

The fundamental ideas for the case study have been; 1) to <u>correct</u> the remote sensing data as much as possible for inhomogenities such as sun shadows and different vegetation; 2) to <u>complement</u> the surface sensitive remote sensing method with deep sensitive geological information, such as geophysical measurements; and 3) to <u>visualize</u> the analysis with a flexible and interactive image processing system.

With this approach remote sensing of vegetation has been shown to be a promising exploration tool even in glaciated terrain.

STUDY AREA

The Pahtohavare deposits are situated 9 km SSW of Kiruna in northern Sweden (Fig 2). The geology around Pahtohavare is described in a generalized geological map (Fig 3). For further details reference should be made to information published by the NSG (1988).

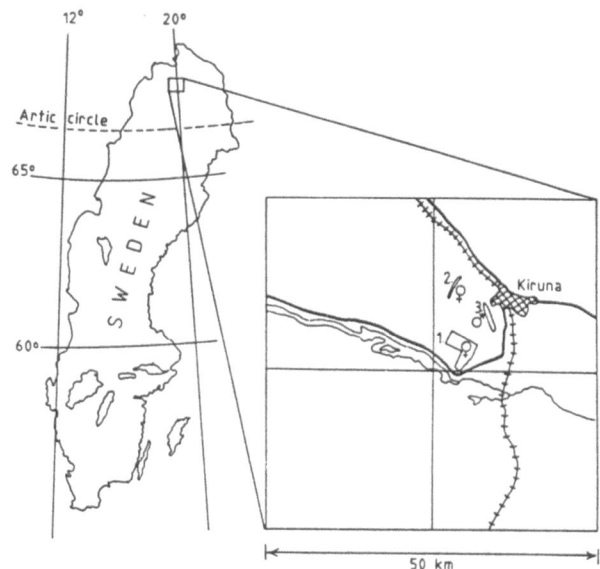

Figure 2: Location map, Map sheet 29 J Kiruna.
1. Pahtohavare deposits,
2. Viscaria, copper ore,
3. Kirunavaara, iron ore

Satellite data describe the reflectance and emittance of vegetation and the ground. Small changes in the reflectance of vegetation can occur where there is natural or artificial metal stress. However, geobotanical remote sensing is a complex subject involving many different factors. Descriptions of the current position in this subject include that by Hodcroft and Moore (1988).

The areal around Kiruna offers a number of advantages as a test site for geobotany. The vegetation is often uniform with large areas covered by mountain birch forests. As a result the effects of man, such as forestry, are few. There are also earlier indications that geobotany can be used in this terrain. Bölviken et al. (1977) indicated the possibility of using Landsat MSS to detect natural copper poisoning on Finnmarksvidda-Norway. Also the Viscaria copper mine, just west of Kirunavaara iron ore, takes its name from the flower "Viscaria Alpina", which was found growing on top of the ore. At Pahtohavare, a mineral occurence was discovered in 1986 whose exposure coincided partly with Viscaria Alpina growing in a small opening at a spring. The Pahtohavare deposit lies within the Kiruna greenstone group, which consists mainly of mafic lavas and tuffs. The position is somewhat reminiscent of the Viscaria deposit.

Excavations exposed two sulphide bodies, one 17.5 m wide, containing 8.9% copper and 2.2 g/t of gold and the other 23.5 m wide, containing 4.2% copper and 0.8 g/t of gold. Large contents of copper were also indicated in the moraine. The working theory was therefore that these relatively high copper contents should effect the closest mountain birch forest and subsequently provide anomalous spectral signatures. It was hoped that similar signatures could be identified at other locations and thereby result in further investigations.

The study was based mainly on the following sources of information:

- Landsat 5 TM precision corrected
- Spot Panchromatic precision corrected
- Vegetation mapping, from infrared air-photos
- Geochemical, peat bog samples
- Aerogeophysical measurements, EM, 50x50 m grid, 30 m ground clearance
- Digital elevation model, 50x50 m grid

METHODS

The first study, in spring 1987, was based solely on Landsat Thematic Mapper data. Anomalous spectral signatures for birch and aspen (Fig 4) could be identified in the mountain birch forest where the ore was exposed. The anomalies were normalized by dividing the difference (T-B) by $B\sigma$ where T is the mean reflectance at the training site, B is the mean reflectance and $B\sigma$ is the standard deviation for mountain birch in the surrounding region.

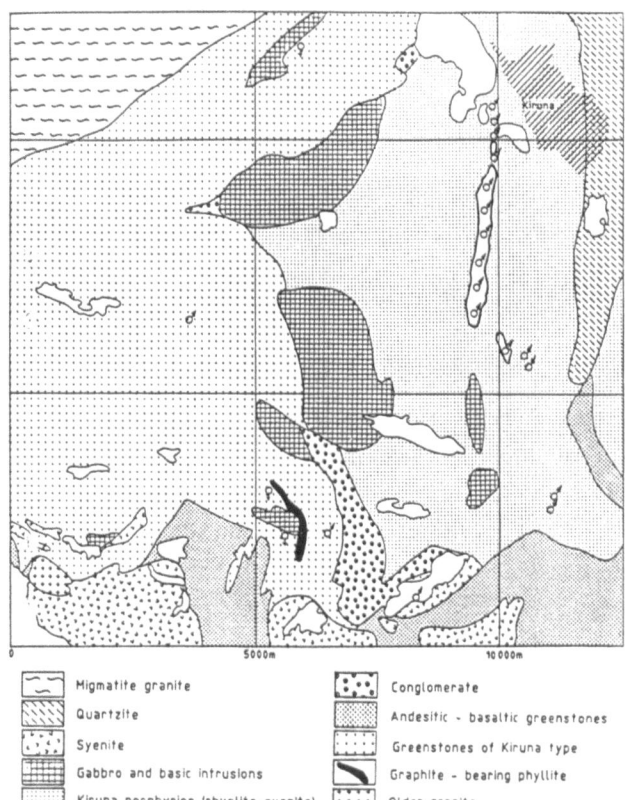

~~ Migmatite granite	Conglomerate
Quartzite	Andesitic - basaltic greenstones
Syenite	Greenstones of Kiruna type
Gabbro and basic intrusions	Graphite - bearing phyllite
Kiruna porphyries (rhyolite-syenite)	Older granite
Iron ore	Cu deposite

Figure 3: Generalized geology, after SGU Ser. Af nr 2, 1967.

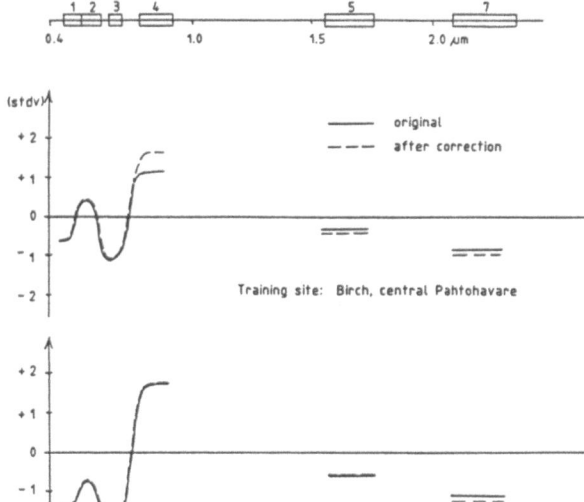

SPECTRAL SIGNATURES

Training site: Birch, central Pahtohavare

Training site; Aspen, SE Pahtohavare

Figure 4: Spectral anomaly at the training sites. The anomaly is normalized by dividing the difference (T-B) by $B\sigma$ where T is the mean reflectance at the training site, B is the mean reflectance and $B\sigma$ is the standard deviation for mountain birch in the surrounding region.

The maximum likelihood classification which followed showed, however, that these types of signatures were very widespread on southern slopes. The topographical slope and aspect are very significant for the growth of the vegetation. This is especially evident at these northern latitudes. The results were therefore difficult to interpret and very difficult to follow up in the field.

Subsequent stages of the project included vegetation mapping from infrared air-photos (9200 m, equivalent to 1:20 000 scale), with a smallest map unit of 1 ha. Work was also carried out within the framework of the GEOVISION project using test data, 25x25 km from the mapsheet 29J Kiruna NE.

Geometrically precision corrected data from the SPOT Panchromatic sensor was

used to locate the excavations made between 18 June 1986 and 17 July 1987, thereby permitting the identification of the training area in the Landsat TM data with greater precision than earlier.

A terrain location model (TLM) was calculated from the digital elevation model to permit correction of the Thematic Mapper data for terrain attitude (Fig 5). The model is based on shaded relief images with illuminations from NE, SE, SW, NW and the inclination 45 degrees. The idea was to use the variables to describe the attitude of the terrain, aspect and slope, to permit subsequent corrections of reflectance values from the satellite data.

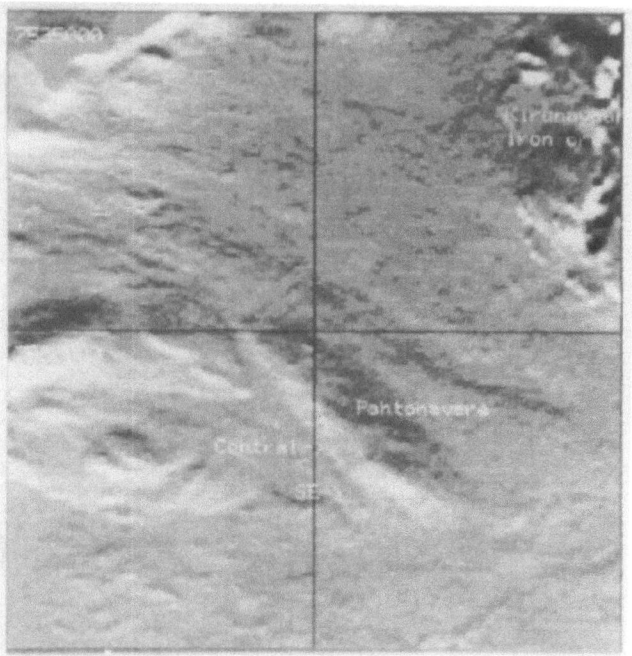

Figure 5: Terrain Location Model, TLM. Shaded relief image, added illumination from NE, SE, SW and NW with the inclination 45 degrees, calculated from a digital elevation grid, 50x50 m. One square in the mesh corresponds to 5x5 km.

Corrections between reflectance and terrain location were made for mountain birch by comparing XY-plots of the TLM images and each satellite channel. Linear regression between reflectance and two terrain location images was used to create new functions (look-up tables) for compensation. The corrected reflectance was obtained by updating the original data with the new functions and secondly, subtracting the result from the original data.

Landsat TM channel 4 provided the clearest relationship as about 70% of the spectral variation within the mountain birch forest could be explained by its attitude in the terrain. Some channels permitted no identification of relationships, mainly because of an overall small variance. Fig 6 shows the plot from channel 4 before compensation and Fig 7 after compensation. Fig 4 shows the change in spectral signature before and after compensation.

The technique used results in an overall reduction in the variance which, in turn, reduces the classification error (maximum likelihood). The anomalous class now showed a more coherent pattern with less correlation to the terrain attitude. Furthermore it was also easier to differentiate between birch forest and aspen. The number of targets that emerged in the classification also dropped, but not sufficiently to be manageable.

To reduce the number of targets for follow up, additional geological data consisting of copper analysis on peat bog samples were used. Greater weight was attached to areas of vegetation close to sample locations with anomalous copper content. In addition, geophysical measurements (airborne EM) indicating graphite-bearing phyllites similar to Pahtohavare have been assigned greater weight. The weighting scheme used was as tabel 1 below.

The final product was a thematic vege-tation map indicating metal stress, associated with different weights (Fig 8). This permits ranking of the anomalous areas, selecting those with the greatest weight for checking in the field.

The selection criteria used below are experimental, not necessarily optimal, and simply illustrate the potential of this technique. However, a first preliminary comparison with other geological information show interesting correlation. A more careful follow-up in the field of the new targets will hope-fully be performed during the summer 1990.

Table 1

Base	(value)	Higher weight	(add value)
Birch, central Pahtohavare	1	d:o within 3 pixels 90 m from each other	1
Aspen, SE Pahtohavare	1	Cu >2.5 stdv, within 90 m	1
		Cu >3.0 stdv, within 90 m	2
		Moderate conductor, within 90 m	1
		Good conductor, within 90 m	2

Maximum possible weight = 6

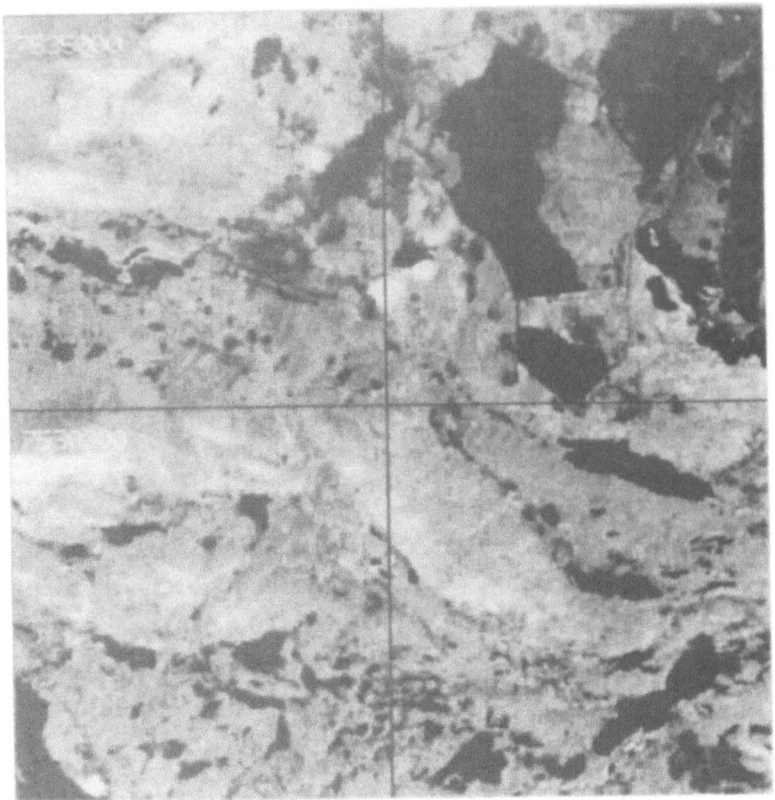

Figure 6: Channel 4, original Landsat 5 TM, part of 196/12 860618, precision corrected, by SSC.

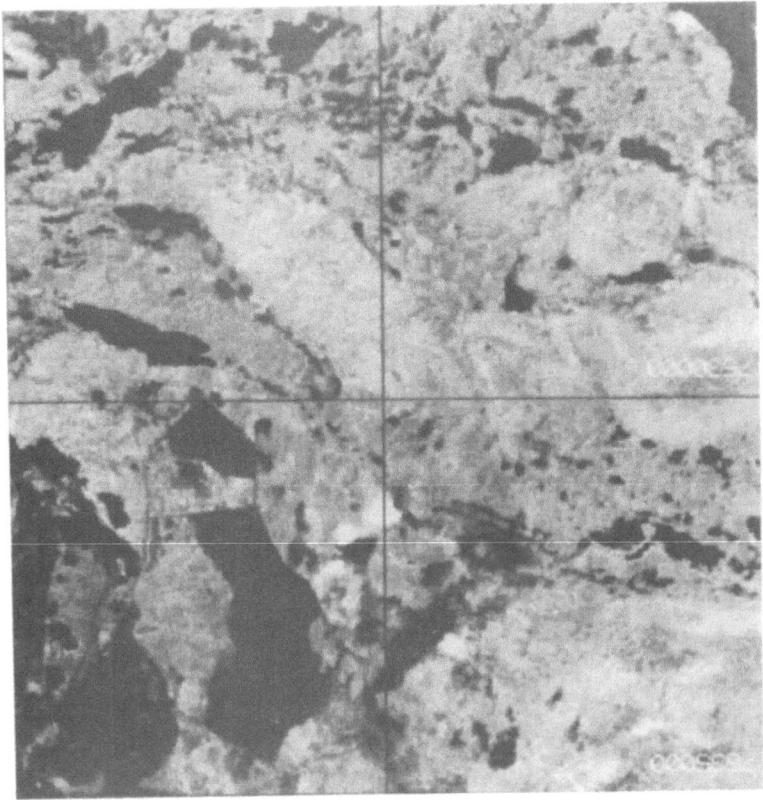

Figure 7: Channel 4, corrected for the effects of slope and aspect.

40

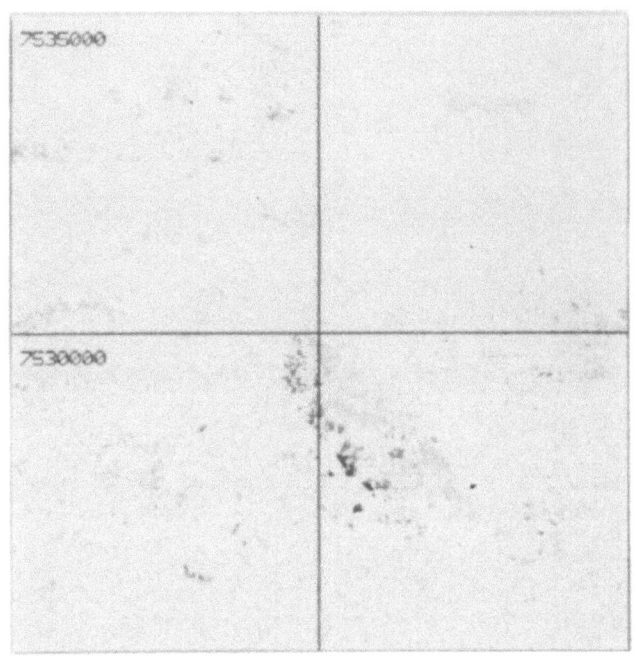

Figure 8: Classification of vegetation with possible metal stress. The result is weighted according to high copper in peat bog samples and favourable airborne EM patterns. The areas with highest weight are shown as the darkest classes.

SUMMARY

The case study from the Pahtohavare copper-occurence showed promising results:

-   it was possible to succesfully detect known geobotanically anomalous areas and to classify potentially new interesting metal stressed vegetation with Landsat TM data,

-   to present geological data as images in combination with the image processing system was shown to be a flexible tool for co- processing and co-analysis of multilayered geological informa- tion,

-   the integration of remote sensing data with traditional geological measurements was an efficient

approach for the ranking of exploration targets.

We belive that the techniques of inter- active processing, visualization and analysis of remote sensing data and geological information are only in their infancy. Development will take place even faster and therefore it will be necessary for the geological community in the future to have a good knowledge of the methodology and a flexible basic system.

ACKNOWLEDGMENTS

We thank all members of the GEOVISION project for their contributions to this paper. We also thank the State Mining Property Commissions for permission to present the results of the "Pahto- havare - Metal stress" project.

REFERENCES

Bölviken, B Honey F., Levine S.R., Lyon R.J.P. and Prelat, A.:
1977:   Detection of naturally heavy- metal-poisoned areas by Landsat-1 digital data, Journal of Geochemical Exploration, 8 p. 457-471.

Hodcroft A.J.T. and Moore J.McM.:
1988:   Remote Sensing of vegetation - a promising exploration tool, Mining Magazine, October 1988, p. 274-279

The State Mining Property Commission (NSG):
1988:   Pahtohavare - gold and copper in Kiruna, A Prospectus, 1988.

Isaksson, H.:
1987:   Pahtohavare - Digital bild- analys, satellitdata - Metod- utveckling. NSG internal

report PRAP 87030

1988:   Pahtohavare - Metallstress II.
           NSG internal report PRAP 88043

Albertsson, J.:
1988:   "Specialkarta - Pahtohavare"
           NSG-commission, LMV 1988.

Rock identification by advanced data processing methods

T. Kawakami
Earth Resources Satellite Data Analysis Center, Tokyo, Japan
Y. Shibata
Bishimetal Exploration Co., Ltd., Tokyo, Japan
T. Yamakawa
Bishimetal Exploration Co., Ltd., Tokyo, Japan
H. Takizawa
Mitsubishi Petroleum Development Co., Ltd., Tokyo, Japan

ABSTRACT

Rock identification using advanced data processing methods has been carried out in the Lower Indus Basin, Pakistan. Data processing methods applied to TM VNIR-SWIR data are; ① False color composite, ② Principal component analysis, ③ Band ratio and ④ Principal component compression. Data processing methods applied to LANDSAT TM thermal infrared data are; ① False color composite, ② ΔT processing and ③ ATI processing. Among these processings, Principal component compression provided sufficient information for the geological interpretation. The processing using thermal infrared data confirmed the applicability to geological interpretation.

INTRODUCTION

The target areas for hydrocarbon exploration of today tend to be remote or hazardous. Remote sensing technology makes the preliminary geological study for such areas possible quickly and economically.

At this time, the identification of rock units by advanced data processing methods using not only the conventional day-time TM data, but also night-time TM data has been carried out for the northern part of the Lower Indus Basin in the Islamic Republic of Pakistan.

STUDY AREA

The East-Sulaiman area is situated in the northern part of the Lower Indus basin and 50 km to the NE of the Sui gas field which is the largest gas field of Pakistan. The area is in mountainous terrain and free from vegetation.

GEOLOGICAL SETTING

Pakistan is tectonically situated on the boundary between the Eurasian and Indian plates, and the Lower Indus basin spreads over the margin of the Indian plate.

The Lower Indus basin is filled with a thick accumulation of Mesozoic to Cenozoic sedimentary rocks. The Mesozoic sedimentary rocks are composed of limestone accompanied with intercalations of shale and sandstone. In the East-Sulaiman area, the Cenozoic sedimentary rocks including the Sui Main Limestone of the Laki Formation, which is an important gas reservoir, are widely exposed, and the area is characterized by well developed fold structures.

Fig.1 Location Map of the Studied Area

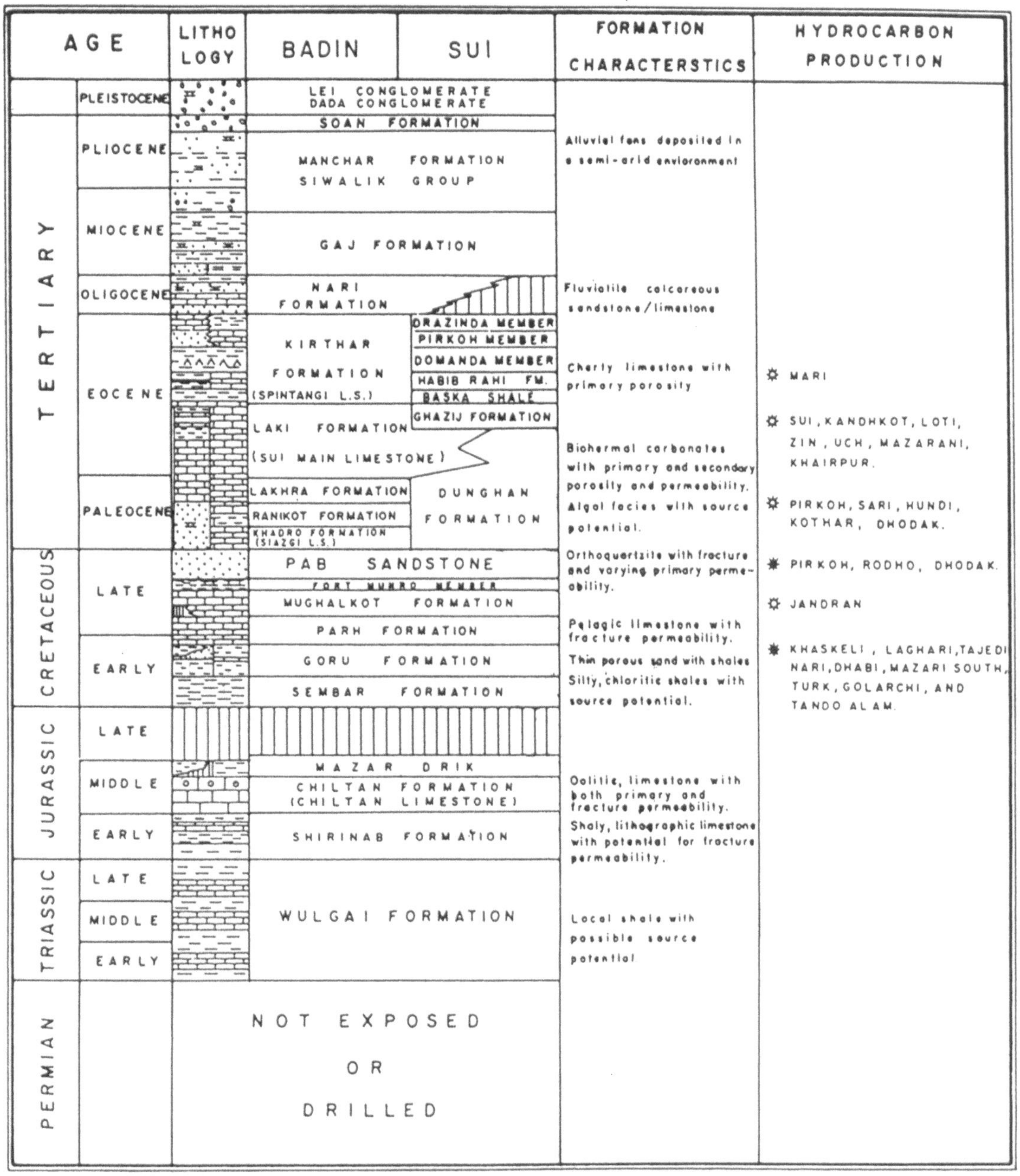

AGE		LITHOLOGY	BADIN	SUI	FORMATION CHARACTERSTICS	HYDROCARBON PRODUCTION
TERTIARY	PLEISTOCENE		LEI CONGLOMERATE / DADA CONGLOMERATE			
	PLIOCENE		SOAN FORMATION		Alluvial fans deposited in a semi-arid enviornment	
			MANCHAR FORMATION / SIWALIK GROUP			
	MIOCENE		GAJ FORMATION			
	OLIGOCENE		NARI FORMATION		Fluviatile calcareous sandstone/limestone	
	EOCENE		KIRTHAR FORMATION (SPINTANGI L.S.)	DRAZINDA MEMBER / PIRKOH MEMBER / DOMANDA MEMBER / HABIB RAHI FM. / BASKA SHALE	Cherty limestone with primary porosity	☼ MARI
			LAKI FORMATION (SUI MAIN LIMESTONE)	GHAZIJ FORMATION	Biohermal carbonates with primary and secondary porosity and permeability. Algal facies with source potential.	☼ SUI, KANDHKOT, LOTI, ZIN, UCH, MAZARANI, KHAIRPUR.
	PALEOCENE		LAKHRA FORMATION / RANIKOT FORMATION / KHADRO FORMATION (SIAZGI L.S.)	DUNGHAN FORMATION		☼ PIRKOH, SARI, HUNDI, KOTHAR, DHODAK.
CRETACEOUS	LATE		PAB SANDSTONE		Orthoquartzite with fracture and varying primary permeability.	❋ PIRKOH, RODHO, DHODAK.
			FORT MUNRO MEMBER / MUGHALKOT FORMATION			☼ JANDRAN
			PARH FORMATION		Pelagic limestone with fracture permeability.	
	EARLY		GORU FORMATION		Thin porous sand with shales	❋ KHASKELI, LAGHARI, TAJEDI NARI, DHABI, MAZARI SOUTH, TURK, GOLARCHI, AND TANDO ALAM.
			SEMBAR FORMATION		Silty, chloritic shales with source potential.	
JURASSIC	LATE					
	MIDDLE		MAZAR DRIK / CHILTAN FORMATION (CHILTAN LIMESTONE)		Oolitic, limestone with both primary and fracture permeability.	
	EARLY		SHIRINAB FORMATION		Shaly, lithographic limestone with potential for fracture permeability.	
TRIASSIC	LATE		WULGAI FORMATION		Local shale with possible source potential	
	MIDDLE					
	EARLY					
PERMIAN			NOT EXPOSED OR DRILLED			

(After Kadri, 1986)

Fig.2 Stratigraphic Column of Lower Indus Basin

DATA USED

LANDSAT TM day-time and night-time data were used for this study. Data descriptions are in Table I.

Path	Row	Date of Acq.	Cloud	Remarks
151	40	17 / 01 / 87	00%	Day-time
018	204	21 / 04 / 88	00%	Night-time

Table I   Data description

DATA PROCESSING AND INTERPRETATION OF VNIR AND SWIR DATA

Data processing

The following methods were applied to TM VNIR (Visible Near Infra-Red) and SWIR (Short Wavelength Infra-Red) data.

(a) False color composite
(b) Principal component analysis
(c) Band ratio
(d) Principal component compression

Principal component compression was calculated by the following formula;

$$CPC(X,Y) = PC1(X,Y) * PCn(X,Y)   n = 2,3 \cdots,6$$

where

$CPC(X,Y)$: Principal component compression data at (X,Y)

$PC1(X,Y)$: 1st P.C data to be convoluted at (X,Y)

$PCn(X,Y)$: n - th P.C data at (X,Y)

Interpretation

False color composite image

The sedimentary rocks distributed in the East-Sulaiman area were divided into 9 geological units.

Principal component analysis image

This image provided information of the main geological framework. As the first principal component represents the albedo of the surface, this image emphasized geomorphological expression.

Band ratio image

4 lithological groups were recognized on this image; they were, sandstone unit, shale unit, limestone unit and alternation of sandstone and shale.

Principal component compressed image

This image accomplished by the convolution of the first principal component to other components, contains more then 99% of TM data information, and provides more effective information both on geomorphological and geological features.

As a result of these data processings, the number of rock units identified were 14 instead of 9 by the false color image.

DATA PROCESSING AND INTERPRETATION OF TM THERMAL INFRARED DATA

Data processing

The following methods were applied to the TM thermal infrared data.

(a) False color composite
(b) ΔT processing (change in temperature)

ΔT shows a cycle of thermal variation in a day and can be calculated by the following formula;

$$\Delta T = N(T\ day - T\ night)$$

T day : DN values of band 6 day-time data

T night : DN values of band 6 night-time data

N : scaling factor

Fig.3　False Color Composite Image
　　　　(BGR=1・4・5)

Fig.4　Principal Component Compressed Image

Fig.5　TM Band 6 Night-time Image

Fig.6　ΔT Image

(c) ATI processing

ATI means apparent thermal inertia. It can be calculated by the following formula;

$$ATI = NC(1-a) / \Delta T$$

where

N : scaling factor

C : $\sin\theta \sin\phi (1-\tan^2\theta \tan^2\phi)^{1/2} +$
$\cos\theta \cos\phi \ arc \ cos(-\tan\theta \tan\phi)$
(θ = Earth latitude,
ϕ = solar declination)

a : apparent albedo

ΔT: N(T day - T night)

Interpretation of TM thermal infrared data

False color composite image

Differences in lithological expression were recognized in the same rock unit interpreted by the conventional false color image.

ΔT image

On this image, four geological unit were recognized based on grey level indicating thermal variation in a day cycle. The distribution of sandstones surrounded by shales was clearly indicated.

ATI image

The ATI image is similar to the image reversed ΔT image, but some differences in lithological expression were observed.

This image suffers from the effect of shadow, and is interpreted with caution.

Though the application of TM thermal band 6 to geological analysis is limited because of the broad wavelength interval, the possibility of rock identification using the petrophysical characteristics (density, thermal conductivity, heat capacity) is suggested.

CONCLUSION

Rock identification was done separately in the VNIR-SWIR and the thermal infrared data.

Advanced data processing methods, particularly principal component compression provided sufficient information for geological interpretation.

Furthermore, the data processing methods using thermal infrared data both in the day-time and at night images confirmed the applicability of the TM thermal band 6 data to geological interpetation.

REFERENCES

1. Geological Survey of Pakistan, 1964, Geological Map of Pakistan, 1:2,000,000

2. Kadri, I.B and Abid, M.S., 1986, Geology of Hydrocarbon Accumulations in Pakistan, 6th Offshore South East Asia Conference Preprints, p226-235

3. Quadri,V.N. and Shuaib,S.M, 1986, Hydrocarbon prospects of Southern Indus Basin, Pakistan, American Association of Petroleum Geologists Vol.70, No.6, p.730-747

4. Short,N.M.and Stuart, JR,L.M, 1982, The Heat Capacity Mapping Mission(HCMM) Anthology, NASA

Image enhancement of epithermal gold deposit alteration zones in southeast Spain

John McMahon Moore
Liu Jian Guo
Department of Geology and Centre for Remote Sensing, Imperial College of Science, Technology and Medicine, London, England

Abstract

This paper describes the application of digitally enhanced Landsat Thematic Mapper imagery to spectral discrimination of terrain types in a semi-arid area containing a variety of rocks and hydrated mineral deposits. Single band contrast stretched TM images, colour composites, band difference, principal component and hue-composite images have been used for clay mineral and gypsum identification and general lithological mapping.

A number of new image processing techniques have been used in this study. These include Balance Contrast Enhancement Technique (BCET), statistical assessment for colour composite band selection (IOBS) and Hue Image Colour Composition (HRGB). Hue colour composite images have been classified to produce simple combined solid and drift geological maps of parts of the Cabo Gata Volcanic Belt.

Using the Rodalquilar Mine area as a 'control site' the enhanced images have been successfully used to identify areas of horticultural crops and relatively dense natural vegetation. The colour composite images with statistically lowest interband correlations is a TM band 5/4/1 combination. After appropriate contrast enhancement, this image is the most useful for rock and vegetation mapping. Individual specialized enhancement methods including principal components and hue colour composite images identify kaolinite-illite hydrothermal alteration, bentonite clays, and gypsum. Groups of mafic (andesitic) and less mafic (dacitic) volcanics, Miocene limestones and other Tertiary strata and superficial deposits have been separated in classified hue images.

INTRODUCTION

The Miocene volcanic rocks of the Cabo de Gata peninsula in Almeria Province, Spain contain several epithermal mineral deposits including an active gold mine in Rodalquilar (Fig.1). Other areas of hydrothermal alteration have attracted prospecting interest. The peninsula is one of the most arid areas in Europe, and natural vegetation is sparse. The volcanic rocks and hydrothermal alteration zones associated with the ore deposits are relatively well-exposed.

This paper describes a series of image enhancement techniques applied to Landsat Thematic Mapper (TM) multi-spectral imagery for mapping of hydrothermal alteration of the type associated with the epithermal ore-bodies. Among the objectives was discrimination of silicic/argillic epithermal alteration from other rock types including gypsum and bentonite clay deposits.

REGIONAL GEOLOGY AND ORE DEPOSITS

The Cabo de Gata lies in the Betic Cordillera, part of the Alpine Orogenic Belt which crosses southern Spain. The area has 'basin and range' topography consisting of mountainous metamorphic basement blocks, separated by Miocene and Pliocene molasse sediment-filled basins containing commercially important gypsum deposits.

Andesitic and dacitic eruptions occurred in later Miocene times to form an arcuate volcanic belt extending from the Cabo de Gata north-eastwards to Murcia.

The Miocene volcanics contain swarms of epithermal gold and silver-lead-zinc veins and associated alteration. Most important of these is the Rodalquilar gold deposit. This orebody supports an open-pit and heap leach mining operation [1]. At Rodalquilar, gold occurs in quartz veins and silicified (chalcedonic) wall-rocks with associated alunite and jarosite. Surrounding a quartz-alunite zone with small areas of propylitic alteration, are concentric kaolinitic, illitic and vermiculite zones [2]. The silicic and kaolinite-illite alteration covers an area of some 15 square kilometres around the mine site and extends eastward beneath a cover of younger Tertiary limestones towards the coast.

Surrounding the hydrothermal alteration zones are relatively unaltered and weathered dacite and andesitic pyroclastic volcanics. The volcanoclastic strata west and south of the mine area contain altered tuff horizons which have been worked for bentonite (montmorillonite).

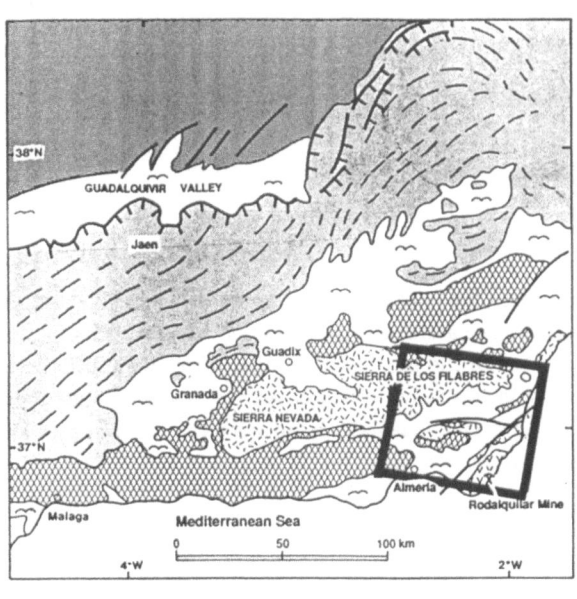

Fig.1 Simplified geological map of SE Spain showing location of the Landsat TM quarter scene used in this study.

TERRAIN SPECTRAL CHARACTER

The volcanic terrain and adjacent Tertiary inter-mountain basins (which will be familiar to readers who watch 'spaghetti western' movies), comprise approximately 60% regolith, dominated by lithic clast debris, 20% outcropping bedrock and 20% vegetation as scattered bushes and thin undergrowth, particularly in dry valleys and on north and west facing hill slopes. Rock outcrops are commonly obscured by weathering veneers and, locally, by lichen.

The spectral character of picture elements (IFOV) in the imagery used for this study is dominated by information from rock fragments and soil with contributions from vegetation and outcropping bedrock. Some of the most distinctive pixels represent unsurfaced roads in the mine area and the abandoned mill tailings.

The only pixels dominated by the spectral signature of vegetation are those representing areas of small-scale, irrigated horticulture.

THEMATIC MAPPER IMAGERY

Thematic Mapper (TM) is a seven channel, multispectral imaging system carried by Landsat 4 and 5. Picture elements each represent approximately $30m \times 30m$ on the ground.

Six of the seven TM bands (band 1-5 and 7) represent terrain in terms of reflected radiation in wavelength ranges between 0.45 and 2.35 micrometres. Images of TM band 6 represent emitted thermal radiation in the wavelength range 10.4-12.5 micrometres. Picture elements (IFOV) for band 6 are approximately $120m \times 120m$.

For general geological applications the band TM 5 (1.55-1.75μm) and 'clay band' TM 7 (2.08-2.35μm) have been proven most useful, but TM 1 (0.45-0.52μm) and TM 3 (0.63-0.69μm) are used for mapping mafic rock and iron rich minerals and soils. TM 4 (0.76-0.90μm) images assist in locating vegetation.

In theory, the thermal band (TM 6) could be used to identify minerals, with characteristic thermal properties, notably unusual thermal inertia, e.g. quartz. In reality this phenomenon is commonly obscured by the thermal effects of shadowing in uneven terrain [3]. McGann (pers. comm.) found some evidence of anomalously low emissivity in the area of maximum silicification at Rodalquilar. Combined with other TM bands to improve apparent spatial resolution in display, TM band 6 has some potential for crude discrimination of quartz and carbonaceous metamorphic rocks in the study area.

IMAGE ENHANCEMENT PROCESSING

Two Landsat TM sub-scenes consisting of 512 lines, each with 512 pixels, chosen from a TM quarter scene (Fig.2A), were selected for detailed enhancement. The geography of Rodalquilar mining area is shown more clearly by SPOT panchromatic imagery (Fig.2B) with much higher spatial resolution ($10m \times 10m$).

The digital image processing programme began with simple enhancement techniques and was extended progressively to more sophisticated methods to highlight subtle spectral differences.

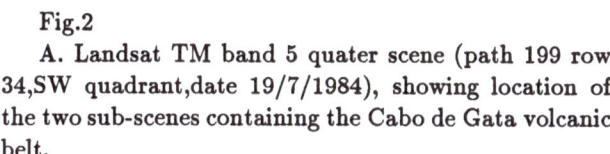

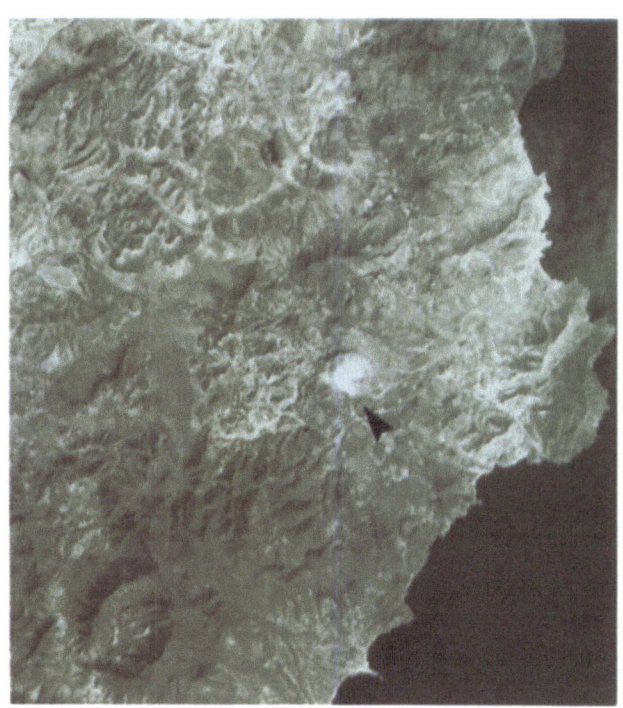

Fig.2

A. Landsat TM band 5 quater scene (path 199 row 34,SW quadrant,date 19/7/1984), showing location of the two sub-scenes containing the Cabo de Gata volcanic belt.

B. SPOT panchromatic image sub-scene (approx 16×16km) showing topographical features in the Rodalquilar mine area. Note abandoned mill tailings dump (arrowed).

Single band contrast enhancement

Single band monochrome images can be used for location and applications in structural geology,but they are not satisfactory for rock type or vegetation mapping [4].

Interactive contrast enhancement is applied individually to single band images as preparation for colour composition. A colour composite will show strong colour bias if the means and DN value ranges of the three bands are significantly different. This problem can be solved by stretching or compressing each band to an equal mean and value range using the balance contrast enhancement technique (BCET) which is defined by a parabolic or cubic function [5]. BCET is a monotonically increasing function and its shape is quite straight;thus the basic shapes of the histograms of the images and the image information are not changed. BCET is therefore a valuable prerequisite to single band image study and to colour composition. An primary atmospheric correction can also be easily applied by setting 0 as the minimum of the value range of the output image when performing BCET.

Colour composites

Three band colour composite images are the most useful products for general soil and rock interpretation. Colour composite images are produced by simultaneous display of three contrast enhanced, registered single band

images, represented in three primary colours (red, green and blue). Multiple three band colour combinations can be produced from combinations of the six TM reflective radiation bands. The number of permutations is increased significantly by interchange of the colour allocations for display.

The factors which influence the value of the colour composite images for lithological interpretation are: the choice of image bands for the colour composites; and the colour allocation for presentation.

Choice of image bands for the colour composite can be made qualitatively on the basis of known spectral reflectance properties of the rocks and soils in relation to TM bands or using statistical methods based on correlation between bands. The Index of Optimal Band-triplet Selection (IOBS) has been developed to produce the most effective band combinations for colour composition [6].

The most satisfactory colour allocation for presentation and production of aesthetically interesting and psychologically appropriate composite images is made by judicious of display colour allocation. (Note: With normal colour vision the eye can discriminate more than 600 combinations of hue, saturation and intensity but only 30-40 grey tones. Also, the eye reacts selectively to colour, with the strongest response to intensity and saturation variations in reds and lowest response to greens.)

A variety of infrared band image combinations have been used to display subtle spectral differences among rock types which appear similar in the visible wave length range.

Our study has shown that for the Cabo de Gata volcanic terrain the most useful colour composite images are combinations of the following bands after Balance Contrast Enhancement:

TM 4/3/2 (RGB) - Standard false colour composite. The image displays vegetation effectively in reds, iron rich soils (including the outer areas of the alteration zones) in greens and andesitic volcanics in blues. There is poor discrimination of Tertiary and Quaternary deposits including limestone strata. Apart from the simulated true colour imagery (bands 3/2/1 as RGB) which is very unsatisfactory in this case, the standard false colour composite image is the most familiar colour display for experienced photo-interpreters.

TM 3/4/2 (RGB) - This image differs from the previous combination only in the display colour allocations. It displays iron mineral weathering in rocks and soils as red brown, vegetation in green and mafic volcanics in deep blue. This image content is the same as 4/3/2 (RGB) but re-allocation of display colour shifts the eye catching red-brown hues from vegetation to iron-rich, red soils and rocks.

TM 5/4/1 (RGB) - Quantitative band selection. This is the least correlated three band combination from the six TM bands of reflected radiation according to the statistical analysis of TM data for the Rodalquilar TM subscene area using the IOBS method. This composite is the most suitable general image for terrain discrimination of rock, soil and vegetation.

TM 7/5/1 (RGB) - Displays andesitic (mafic) volcanics as red/brown and dacite (iron stained) volcanics as olive green. Areas with limited hydrothermal alteration and red soils appear green. The area of intense alteration silicification/ kaolinization zone is represented in yellow-green. Limestone terrain, Tertiary age sediments and vegetation are poorly discriminated.

TM 7/5/3 (RGB) - Effective in rock type discrimination, particularly for red soils which are displayed more clearly as red-brown.

TM 5/3/1 (RGB) - This band combination was selected by Crosta and Moore [7] as the most appropriate for mapping of sedimentary rock types in the Tabernas Tertiary Basin.

The colour composite images are standard products. The colour allocations are easily understood and interpreted. Iron minerals (red soils and ferruginous mineral weathering in rocks), dark (andesitic) volcanics and vegetation are clearly displayed. Apart from 5/4/1 combination which was selected on inter-band correlation criteria, colour composite band choice is a matter of qualitative judgement.

Colour composite images do not discriminate satisfactorily among silicic/argillic alteration minerals, bentonite (montmorillonite clays), limestone and gypsum deposits in this area.

Band difference and ratio images

Differential reflectance of soil/rock and vegetation between the TM bands can be used to highlight vegetation and certain rock and mineral types, including alteration mineral assemblages [8,9]. This is achieved by performing subtraction or division between pixel DN values of two images (bands).

Image differences or ratios display as monochrome images, after appropriate contrast enhancement, show reflectance variations in the original data. Band difference is the simpler and, according to Cañas (pers. comm.), the more effective of the two techniques.

Band subtraction was used in this study specifically to show: hydrated minerals (clays and gypsum); iron (red) minerals; and vegetation.

TM5 minus TM7: TM band 5 represents the wave length range of maximum rock and mineral reflectance. TM band 7 coincides with an absorbtion feature which anomalously reduces the reflectance of hydrated minerals, notably clays and to a lesser extent gypsum. The arithmetical difference image of TM5 minus TM7 therefore highlights clays and gypsum. In the Rodalquilar image these include kaolinitic and montmorillonitic (bentonite) clays and gypsum (Fig.3A).

TM3 minus TM1: This image is the 'red minus blue' difference and displays the red soil and rocks effectively (Fig.3B).

Experiments with colour composite images made by combinations of three band difference images have been used to emphasize combinations of features e.g. FCC TM3-TM1, TM4-TM3, TM5-TM7 (RGB) to show iron colouration, vegetation and clay minerals respectively in red, green and blue hues.

Difference and ratio images suppress topographic shadow effects. Band difference images do not separate kaolinitic clays from bentonite or gypsum.

PRINCIPAL COMPONENT (PC) IMAGES

Principal component transformations applied to multiband data sets compress information from several bands and can produce very effective separations of terrain types which appear similar in single band and band difference images. The first principal component image usually incorporates more than 90% of the image information, succeeding PC images represent progressively smaller percentages of data but emphasize data values spread along the shorter axes in feature space. It is therefore in the intermediate principal component images that otherwise obscure spectral information may be revealed. Generally, the image signal/noise ratio worsens progressively in the last members of a PC image series.

In this study a series of PC transformations were carried out using various band combinations. The most successful results were achieved from combinations of TM bands without utilizing band 6 (whose low spatial resolution degrades results). From some PC combinations TM2 was also excluded because of poor raw data quality.

PC images for alteration zone mapping

PC5 derived from TM 1,3,4,5,7

This monochrome image (Fig.3C) effectively shows the areas of hydrothermal alteration dominated by clays. The image can be displayed in both positive and negative form and annotated by allocation of pseudo colours by 'density slicing' of pixel DN values. PC5 images effectively highlight the clay alteration zones associated with Rodalquilar and other epithermal mineralization elsewhere in the volcanic belt (Fig.4). Pixels representing gypsum and clays are also highlighted in PC5 images.

B. Band difference image TM3-TM1. Note areas of red (ferruginous soils and rocks) in light tones.

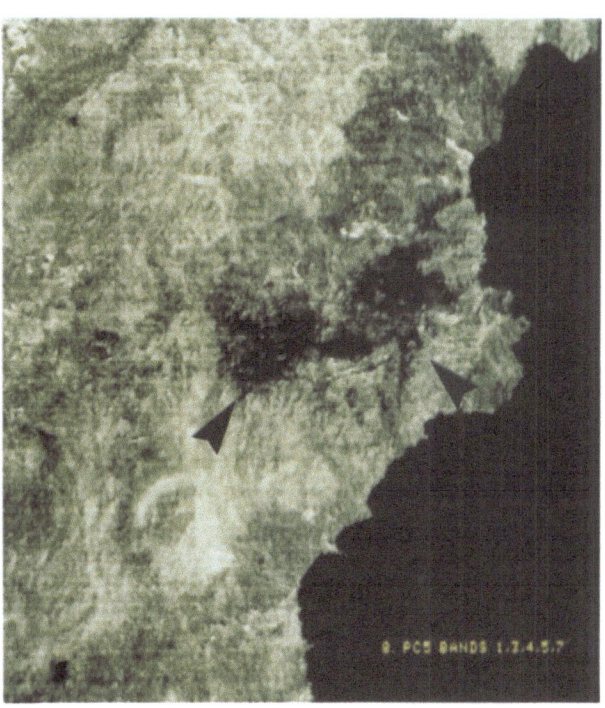

C. Principal component 5 image (derived from TM bands 1,3,4,5,7). Areas of argillic alteration are in dark tones (indicated by arrows).

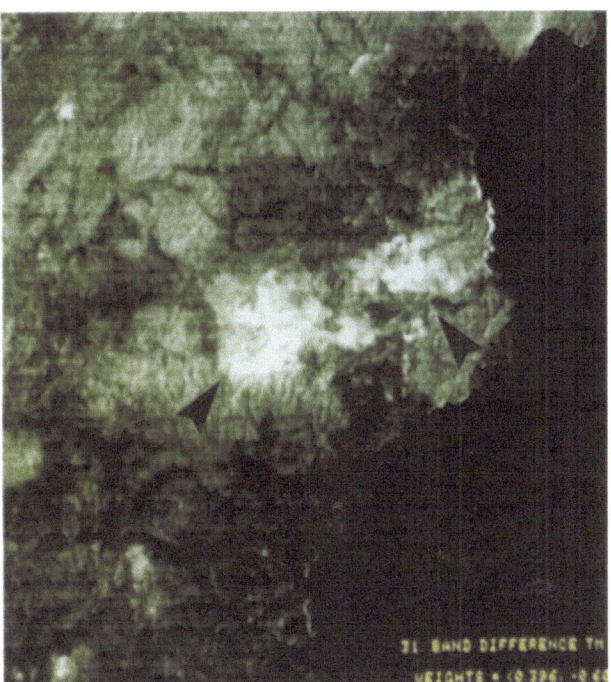

A. Band difference image TM5-TM7. Note the highlighting areas of argillic alteration in light tones.

Fig.3 Rodalquilar sub-scene

PC3 derived from TM 1,3,4,5,7

In this PC image gypsum pixels are brighter than those of clays. PC3 is therefore useful for hydrated mineral discrimination.

PC colour composite images

Principal component images can be combined in groups of three as PC colour composites which are spectacularly dominated by intense and saturated colours. PC transformation can achieve very selective and subtle discrimination of terrain types. Unfortunately, there is no simple and direct relationship between image colours and terrain types such as vegetation, soil and rock. Also, rock types which are fundamentally different can be represented similarly in lower order PCs. These images are difficult to interpret except by empirical reference to field observation or geological maps.

PC colour composite PC 2/1/3 (RGB), derived from TM 1,3,4,5,7

The PC 2/1/3 (RGB) colour composite used in this study effectively separate bentonite localities (yellow without green) from the kaolinite-illite clay alteration zones (dominantly green).

PC colour composite PC 5/4/3 (RGB), derived from TM 1,3,4,5,7

This image enhancement achieved the most effective separation of gypsum and kaolinitic alteration mineral assemblages. Gypsum, bright in PC3 and PC5, appears magenta while kaolinite/bentonite which is brighter in PC5 than PC3 appears red.

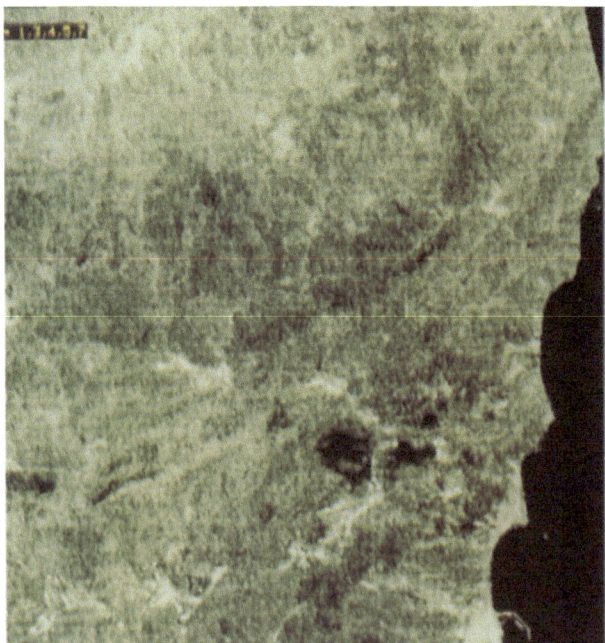

HUE/SATURATION/INTENSITY IMAGES

Various authors have used hue, saturation and intensity transformations to improve colour representation in an image. We have developed a technique to use of the HSI transformation in a different way. 'Hue colour composition' is a method particularly suitable to rock/soil image display in the Cabo de Gata area [6,10].

The method consists of extraction of a hue image from a three band colour composite by HSI transformation. This process is repeated with different TM band combinations to produce three hue images. The three images are combined for display as a three hue image RGB colour composite.

The product is an effective way to display the hue (spectral) information from the original images. This method provides a very colourful product suitable for empirical interpretation. One consequence of hue transformation is elimination of shadow effects which are a function of intensity and saturation.

Hue colour composite image of Rodalquilar TM subscene provides general lithological information and effectively separates Miocene limestones from clay minerals of the hydrothermal alteration zones and bentonite deposits as shown in the geological interpretation map (Fig.5).

HUE IMAGE CLASSIFICATION

The image processing techniques described above, selectively enhance rock and soil types for visual interpretation. Classification of terrain type (including rocks and soils) is an extension of selective enhancement and a step towards automatic mapping.

Hue colour composite images are almost shadow free and show limited variation in colour within the areas of the same rock/soil type. They are therefore a convenient image base for spectral classification.

Fig.4 Carboneras sub-scene: principal component 5 image (derived from TM bands 1,3,4,5,7). Note areas of argillic alteration in darker tones, indicated by arrows. The alteration coincides with the Carboneras gold prospect. Darker tones near the western edge of the subscene are gypsum outcrops.

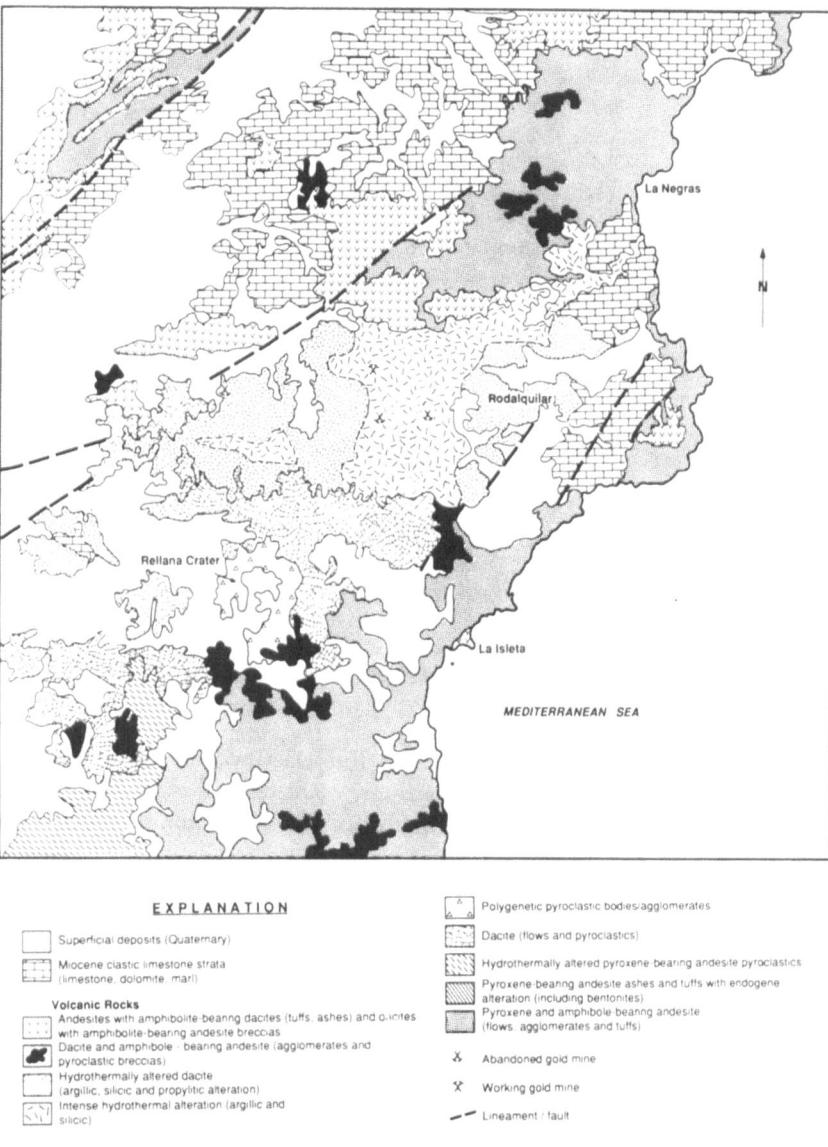

EXPLANATION

| | Superficial deposits (Quaternary) |
| | Miocene clastic limestone strata (limestone, dolomite, marl) |

Volcanic Rocks

	Andesites with amphibolite-bearing dacites (tuffs, ashes) and dacites with amphibolite-bearing andesite breccias
	Dacite and amphibole - bearing andesite (agglomerates and pyroclastic breccias)
	Hydrothermally altered dacite (argillic, silicic and propylitic alteration)
	Intense hydrothermal alteration (argillic and silicic)

	Polygenetic pyroclastic bodies/agglomerates
	Dacite (flows and pyroclastics)
	Hydrothermally altered pyroxene-bearing andesite pyroclastics
	Pyroxene-bearing andesite ashes and tuffs with endogene alteration (including bentonites)
	Pyroxene and amphibole-bearing andesite (flows, agglomerates and tuffs)

丼	Abandoned gold mine
𝑋	Working gold mine
⸺	Lineament / fault

Fig.5 Rodalquilar sub-scene: geological interpretation map based on the hue colour composite image.

A practical method of three dimensional feature space iterative clustering for image classification has been developed [11]. In this method the clustering iteration is performed in three dimensional feature space rather than by image pixel by pixel scanning. This method permits the cluster size and pixel frequency to be taken into account. Thus a more advanced decision rule, the optimal multiple point re-assignment can be used [12].

In the case of Rodalquilar TM sub-scene area, this method produces classes or groups of classes which correspond well to mapped rock/soil and other terrain cover types (Fig.6). The classification image shows high agreement to the geological interpretation of hue colour composite (Fig.5) and existing geological maps (Fig.7).

Enhanced imagery is the starting point for preparation of classification image map. Automatic rock type classification of imagery require further work to incorporate textural information, together with the control by field work and known geological information.

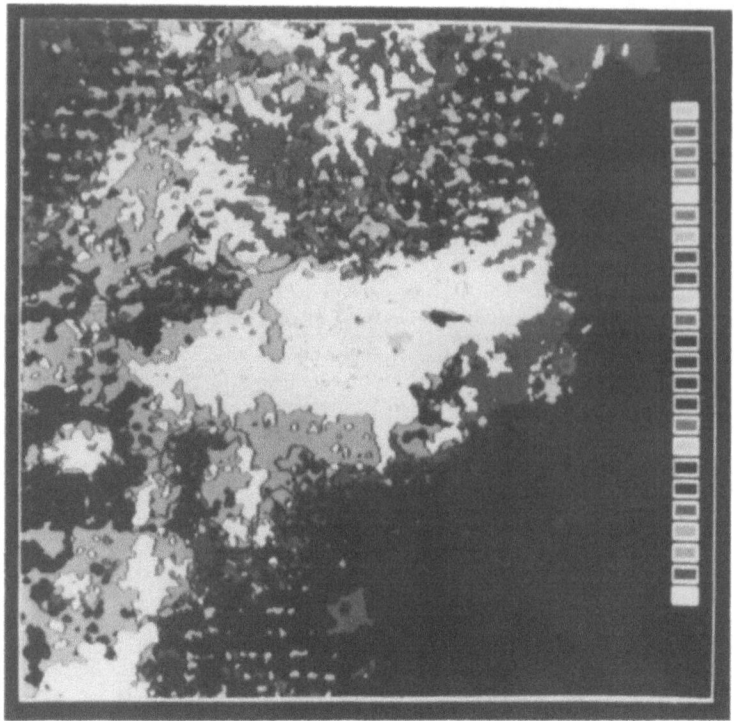

Fig.6 Rodalquilar sub-scene: Hue classification image (original in colour), reproduced in black and white, 24 classes. The 24th class indicates the active mining areas.

CONCLUSIONS

Individual TM bands and SPOT panchromatic images are relatively unsatisfactory for rock type and alteration zone mapping in the Cabo de Gata area.

The most useful and comprehensible images for general lithological mapping are colour composite images produced from combinations of individually contrast enhanced TM bands. The most useful combinations (RGB) are:TM 5/4/1, TM 7/5/1, TM 7/5/3 and TM 5/3/1. These images are readily understood and can be used for vegetation, volcanic rock types, soils and general hydrothermal alteration zone mapping.

Band difference images can be used to locate clay and gypsum (TM5-TM7) and red (iron-mineral) soil and rock outcrops (TM3-TM1).

Colour composite combinations of the principal component images derived from various combinations of TM bands of the study area distinguish clays from gypsum and discriminate between kaolinite-illite clays of the hydrothermal alteration zone and bentonite tuff outcrops.

Colour composite image made from hue image triplets suppress shadow effects and effectively discriminate rock types, including Miocene limestones which are among the most difficult rock type spectra to separate. Classified versions of the three hue image colour composite represent the nearest digital product to a combined solid and draft lithological map.

This study demonstrates a systematic image processing for geological (rock type) mapping and hydrated mineral identification.

ACKNOWLEDGEMENTS

Imagery of the Rodalquilar area has been studied for several years by members of the Imperial College Centre for Remote Sensing and Geology Dept research group. The results presented in this paper are based on the authors' studies but owe much to the work and ideas of colleagues and many many M.Sc. and research students. The image processing contribution of Dr A.A. Cañas of the Applied Optics Group, Physics Dept is particularly acknowledged, as are the works of Dr J. Huckerby and Dr R.W. Magee, Dr Alvaro Crosta and Hugo Forero-Onofre.

Particular thanks go to Caroline McGann, Simon Harvey and many other M.Sc. students of the London University inter-collegiate Remote Sensing course for their stimulating unpublished work on image enhancement of the Rodalquilar area.

Thanks also offered to Empressa Nacional ADARO, St Jo Mining, and its successor Minas Transaccion SA, particularly the company geologist Peré Hernandez, who have generously granted repeated access to the mine sites.

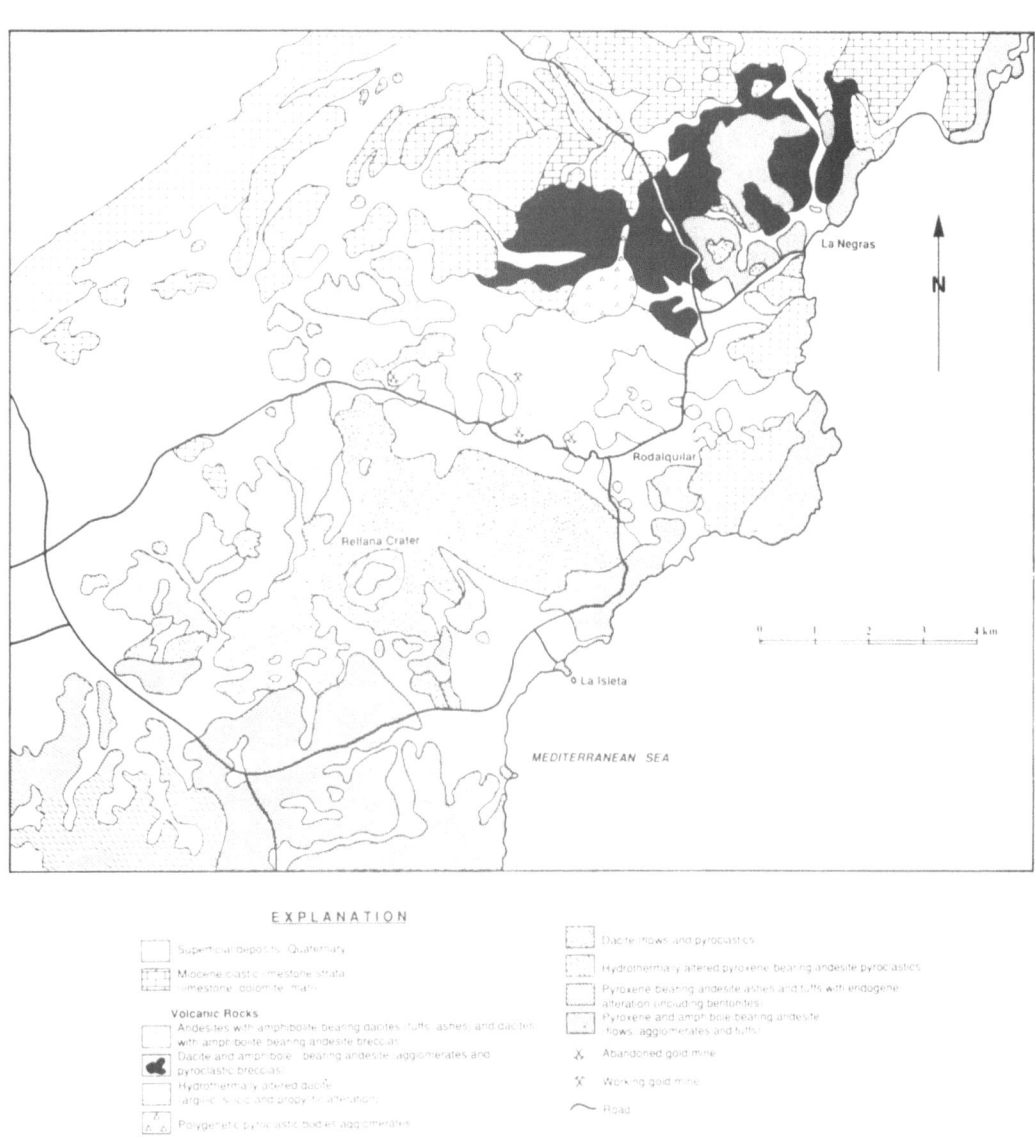

EXPLANATION

Fig.7 Simplified geological map of the Rodalquilar district based on published geological maps (IGME geological maps scale 1:50000 Carboneras and Pozzo Frailes). Compare with Fig.5 and Fig.6.

REFERENCES

1. Arribas A.jr., Rytuba J.J., Rye R.O., Cunningham C.G., Podwysocki M.H., Kelly W.C., Arribas A.sr, McKee E.H. and Smith J.G. 1989, Preliminary study of the ore deposits and alteration of the Rodalquilar caldera complex, southeastern Spain. US Geological Survey, Open file Report 89-327, 39p.

2. Hernandez P.A., Garcia-Estrada P.A. and Cowley P.N. 1989, Geological Setting,alteration and lithogeochemistry of the Transaccion epithermal gold deposit, Rodalquilar mining district,southeast Spain. Transactions of Institute of Mining Metallurgy,London,98. B. 78-80.

3. Anon, Landsat 4 Data Users Handbook. NOAA, US Geological Survey, Washington DC, USA, 1984.

4. Forero-Onofre H. 1989, Mapping of hydrothermal alteration zones in gold-bearing volcanic rocks:a comparison of different products. M.Sc. Thesis (unpublished), University of London, Inter-Collegiate Remcte Sensing M.Sc. Course, 82p.

5. Liu J.G. Balance Contrast Enhancement Technique and its application in image colour composition. 1990 International Journal of Remote Sensing, in press.

6. Liu J.G. and Moore J.McM. Colour enhancement and shadow suppression technique for TM images. In: Proc. Seventh Thematic Conference on Remote Sensing for Exploration Geology, 2 - 6 Oct., 1989, Calgary, Alberta, Canada. ERIM, Ann Arbor, Michigan. Vol.2, p. 901-915.

7. Crosta A.P. and Moore J.McM. Geological mapping using Landsat Thematic Mapper Imagery in Almeria Province, southeast Spain. International Journal of Remote Sensing. 1989, vol.10, No.3. p. 505-514.

8. Huckerby J.A., Magee R.W., Moore J.McM. and Coates D. Thematic Mapper applied to alteration zone mapping for gold exploration in southeast Spain. In: Proc. Fifth Thematic Conference on Remote Sensing for Exploration Geology, 29 Sept.- 2 Oct., 1986, Reno, Nevada, USA. ERIM, Ann Arbor, Michigan. p. 591-599.

9. Magee R.W., Moore J.McM. and Brunner J. Thematic Mapper applied to mapping hydrothermal Alteration in southwest New Mexico. In: Proc. Fifth Thematic Conference on Remote Sensing for Exploration Geology, 29 Sept.-2 Oct., 1986, Reno, Nevada, USA. ERIM, Ann Arbor, Michigan. p. p. 373-382.

10. Liu J.G. and Moore J.McM. Hue image RGB colour composition. A simple technique to suppress shadow and enhance spectral signature. International Journal of Remote Sensing. 1990, Vol.11, in press.

11. Liu J.G. and Haigh J.D. A three dimensional feature space iterative clustering method for multi-spectral image classification. In: Remote Sensing for the Nineties - Proc. 10th Annual International Geoscience and Remote Sensing Symposium, IEEE, 20 - 24 May, 1990, University of Maryland, Washington DC, USA. Vol.3, p. 2367-2370.

12. Kittler J. and Pairman D. Optimality of reassignment rules in dynamic clustering. Pattern Recognition. 1988, Vol.21, No.2, pp.169-174.

Automatic detection of three-dimensional geological features from remotely sensed imagery and digital terrain models

Kevin P. Morris M.Sc.
Remote Sensing Unit, Department of Geography, University College London, London, England

ABSTRACT

Traditionally, the role of remote sensing in geology has been to provide the geologist with visually enhanced imagery offering synoptic coverage of an area for manual interpretation. More recently, automated techniques have been introduced in an attempt to remove the subjectivity of human interpretation. The need for these techniques will grow as a result of the increase in data volumes and the expectations afforded by sensors on board the Polar Platforms, due to be launched in the 1990s. This paper describes a number of automated techniques designed to extract geological information from remotely-sensed imagery and digital terrain models (DTMs). In particular, an edge detection algorithm is applied to the imagery and the DTM to identify lithological boundaries and faults. Various rules and conditions are identified which enable separation of 'geological' edges from other types of edge (e.g. man-made features). The edge images are then thresholded and thinned to produced line entities. The resulting two-dimensional data are combined with the elevation data to determine local dip and strike measurements. A least squares approach is employed to 'fit' these three-dimensional data to a planar surface. The results are compared with measurements of dip and strike obtained in the field and derived from a 1:50,000 scale geology map of the area. In many parts of the study area the results are accurate to within ±10°. Other areas require improved/additional image processing techniques, knowledge-based rules and a structural model if accurate measurements are to be extracted successfully.

KEYWORDS – Image Understanding, Structural Geology, Digital Terrain Models, Edge Detection, Surface Fitting

BACKGROUND

Remote sensing has long been used as a tool in geological applications. Motivated predominantly by the need for oil and mineral exploration, applications include lithologic mapping, geobotanical studies and structural analyses (both local and tectonic). To date, fairly standard image process-ing techniques have been adopted, primarily to enhance the visual appearance of the remotely-sensed imagery; with the emphasis being placed on manual interpretation of these data to extract the desired geological information. The approach is, by definition, a subjective one and results may vary considerably between different expert interpreters[1]. One way of overcoming this is to incorporate various automated techniques into the image analysis procedure. The aim would be to create a knowledge-based system which would include the standard image processing methods, as well as sophisticated algorithms more closely related to human perception and, more importantly, accumulated 'expert knowledge'.

A second, and perhaps more urgent, requirement for an automated geological image understanding system stems from the increases in data likely to be produced by the next generation of remote sensing devices, such as the Earth observing system instruments on board the NASA[2], ESA[3] and NASDA[4] Polar Platforms. These will record data over a wider range of wavelengths and in many more wavebands than are presently available (e.g. the High Resolution Imaging Spectrometer (HIRIS) with 192 wavebands[5]). Data will also be acquired at many different sensor look angles and spatial resolutions. It has been calculated that the NASA Polar Platform, EOS-A, will download approximately 10^{13} bits of data every day[6]. For many applications there is little hope of manually interpreting such large volumes of data in an appropriate timeframe. Although geological applications do not have the same time-critical needs as, for example, agricultural monitoring, they will nevertheless need to make use of the vast amount of information available in order to improve final outputs and decision making.

In addition to the increase in the volume of data available to users, the expectations of the usefulness of such data will also be raised dramatically[7]. New procedures must therefore be developed to automate and, hence, accelerate the image interpretation stages in all remote sensing applications, including those geological. Whilst recognizing possible limitations (or under-use) of these procedures as simply a means to reduce data volumes and effect 'educated guesses' to guide human interpretation of a scene, their value in these areas alone should not be underestimated.

PREVIOUS INVESTIGATIONS

A considerable amount of work has been undertaken in the fields of image understanding and knowledge-based systems (e.g. [7, 8, 9, 10]). The consensus of opinion suggests that a completely automated system should include:-

   i. models of the image formation process[8], e.g. modelling the effects due to topography, the atmosphere, geometric distortion and Sun-target-sensor geometry,

   ii. 3D information[7]; since most features on the Earth's surface are better described using three, as opposed to two, dimensions,

   iii. models of bidirectional reflectance[7],

   iv. previous interpretations or classifications of the area or ancillary data, if available, e.g. maps[10],

   v. contextual geometric reasoning, i.e. recognizing features or shapes as part of the context of the scene[9],

   vi. error handling capabilities which are flexible enough to cope with image noise and image analysis problems, and

   vii. specialist knowledge.

These techniques are commonly incorporated into a computer system consisting of an image processing library, a knowledge-base (including general image formation knowledge and specialist knowledge), a database (containing all image, map and ancillary data), a model of the present understanding of the scene, and a 'manager' which oversees the whole task, directing further analysis of the scene depending on the results of previous image processing and knowledge-based rules.

In the past, structural mapping in remote sensing has been limited to the study of lithological identification and lineament analysis in two dimensions only. More recently, authors such as Thiessen et al.[11] and Sauter et al.[12] have started to use DTMs (i.e. information on the height, slope and aspect of each pixel[13]) to aid geological interpretation and to utilize fully the three-dimensional information about the Earth's surface.

Thiessen et al.[11] have employed a Geologic Spatial Analysis (GSA) system which uses advanced three-dimensional spatial correlation and coplanarity techniques in an attempt to identify lineaments belonging to the same geological structure. The input data to the system can be lineaments identified in the field or manually from imagery, or information derived from seismic profiles. As yet, no method has been developed for automatically detecting lineaments or surfaces in remotely-sensed imagery.

Sauter et al.[12] have used elevation data to classify various geological features. They used edge detection techniques to identify steep slopes, plotting these on rose diagrams to determine directional frequency. In their research, Sauter et al. also employed several ridge and valley detection and Lambertian shading techniques to delineate and enhance geological features. These techniques have only been used to aid manual rather than automatic interpretation.

PROJECT INTRODUCTION

A fully automated system designed *specifically* to extract geological information, should aim to include image processing techniques and knowledge-based rules for the analysis of all the major geological remote sensing problems, from spectral identification of rock and mineral types through to structural mapping. This paper concentrates on a specific aspect of this, namely the task of structural mapping, using an integrated analysis of remotely-sensed imagery and Digital Terrain Models (DTMs). While recognizing the requirements for a complete image understanding system, mentioned above, and for analysis in other disciplines of geological remote sensing, the paper outlines several ways in which both elevation data and remotely-sensed imagery can be used to extract 3D geological information (e.g. dip and strike measurement of bedding and faults). The procedure developed is shown diagrammatically in Figure 1. Firstly, lineaments are detected in both the DTM and the imagery, Several simple rules are applied to suppress those lineaments that are not likely to be related to geology e.g. lake shores and roads. The two-dimensional edge data is then combined with the elevation data to calculate a best-fit plane or curved surface which describes the data. Further rules are applied to include only those orientations most likely to represent geological features. Finally, a structural map is made.

STUDY AREA AND DATA ACQUISITION
Study Area

The site chosen for this investigation includes an area of mountainous terrain located near Capel Curig, Snowdonia National Park, U.K.. The location was selected because of the pronounced relationship between the topography of the area and its geology, and because of the availability of suitable remotely-sensed imagery. The geology consists of Ordovician slates, mudstones and siltstones, interspersed with sandstones (greywackes), tuffs and ash-flow tuffs (predominantly acidic in composition)[14]. This sedimentary sequence is intruded by dolerite sills and dykes, also of Ordovician age. The topographic expression of this geology is partly masked by the effects of glaciation, but is manifested as quite distinct geomorphological features caused by the interlayering of hard and soft rocks. The harder and more resistant rocks include dolerite, ash-flow tuff and sandstone, and are generally located along ridges or prominent features such as cliffs; while the soft rocks comprise slate, mudstone, siltstone and tuff, and are found in more low-lying areas, which are more likely to be covered by vegeta-

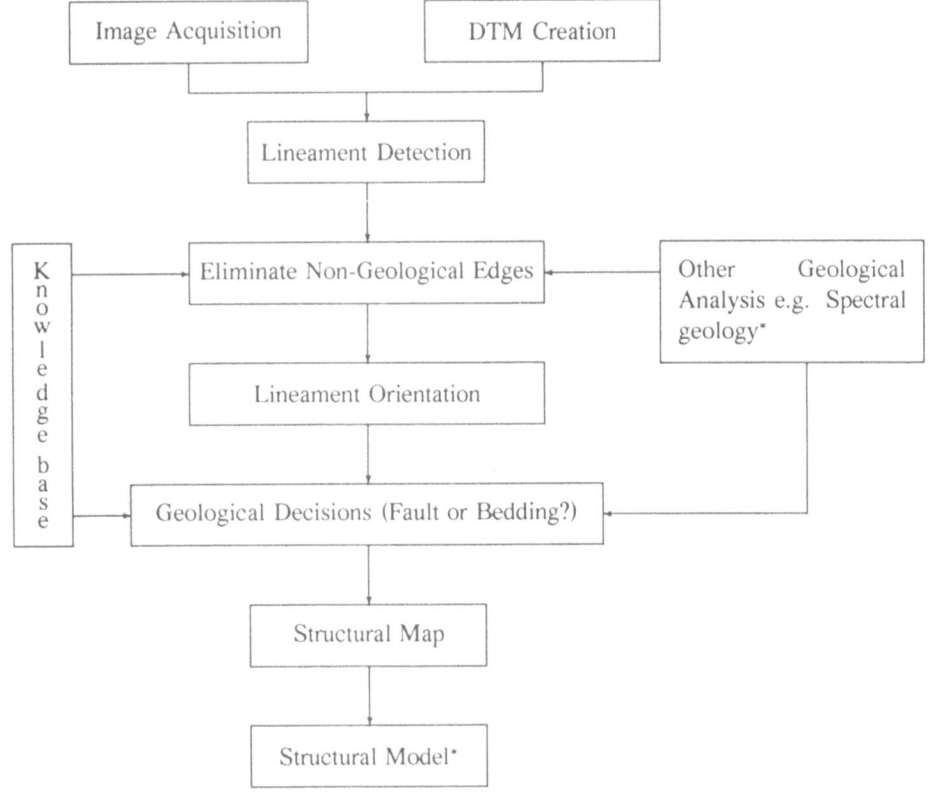

* – *Not yet implemented*

Figure 1, Flow diagram showing the present configuration of the proposed automated system.

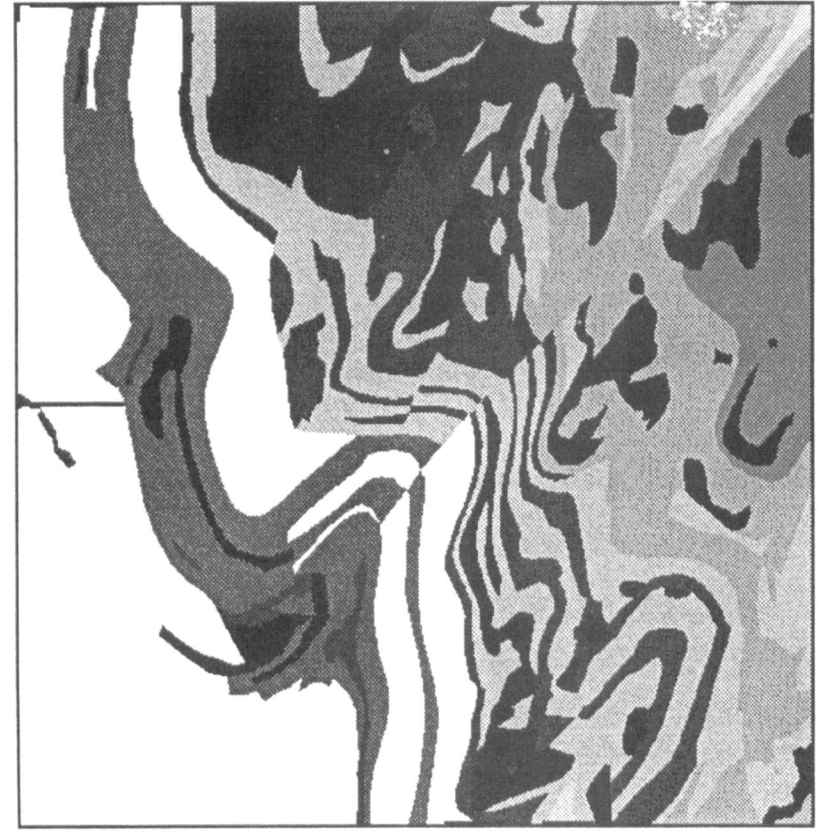

Key:-

Black   Dolerite
  ↓   Acid ash-flow tuff
  ↓   Sandstone
  ↓   Mudstone
  ↓   Tuff
  ↓   Siltstone; mudstone
  ↓   Basic Tuff
White   Silstone

Figure 2, Geology map of Llyn Cowlyd.
Area is 5x5km.

tion. The main deformation phase occured during the early stages of the Caledonian orogeny and caused major folding and faulting along a north-easterly orientation. A second, less pronounced deformation occured along a south-easterly direction. A simplified geological map of the area is shown in Figure 2.

Data Acquisition

The image data used in this investigation were acquired by the NERC Daedalus Airborne Thematic Mapper (ATM) scanner (AADS-1268) which records data in 11 separate wavebands from the visible to the thermal infra-red wavelengths. Images were obtained from three parallel flight lines, spaced at 1km intervals, at 9:15 GMT on 19th July 1989. The altitude of the aircraft was approximately 2000m giving a nominal spatial resolution of 5m in the nadir viewing position, although the rugged terrain and the wide scan angle of the ATM (84°) results in extremely variable spatial resolution throughout an image[15].

The image data were both radiometrically calibrated, using a technique described by Wilson[16] and geometrically corrected to Ordance Survey base maps using a high-order polynomial warping function and nearest neighbour interpolation, to preserve the radiometric integrity of the data[17] (Figure 3).

METHODS
Creation of Digital Terrain Model

A DTM of a 5km² area centred on Llyn Cowlyd has been created at a spatial resolution of 5m, in order to correspond with that of the ATM imagery (Figure 4). The DTM was generated by digitizing contours and spot heights from a 1:10,000 scale Ordnance Survey topographic map (©Crown Copyright, 1976). These data were processed using a digital mapping package, MAPICS. The MAPICS routine 'GRID' generates a Triangulated Irregular Network (TIN), based on the production of a set of Theissen polygons between the data points. For each grid location, weighting factors are determined for points adjacent to the polygon. These are calculated using the proportion of the area of a new Theissen polygon, introduced by the grid location, which is included in each of the original polygons sited in that area[18]. Height values are calculated using the weighting factors and the neighbouring data points. Many different methods exist for the interpolation of irregular data points to a regular grid, including weighted average, or bilinear gridding, and kriging[19]. The TIN method has been chosen for this investigation as preliminary results show that it retains the highest geometric fidelity when compared with the original topographic map data[20]. There are several large lakes in the area, for which no depth information is available. Consequently, large areas of the scene have no detailed height information. To avoid any spurious results during the following investigation, each of the lakes in the area were digitized and a constant height value was assigned to each of these areas.

In future studies, DTMs could be created by automatically stereo-matching a pair of stereo aerial photographs or high resolution digital images, as described by Day and Muller[21]. The stereo-matching procedure can produce high quality Digital Elevation Models (DEMs) from stereo SPOT data[21], but it is considerably more complicated to match pixels in images which may be severely affected by geometric distortions introduced by altitude, roll, pitch and yaw changes of the aircraft.

In addition to providing quantitative information on elevation, slope and aspect, DTMs can be used to aid in the 3D visualization of a scene either by overlaying imagery onto a perspective view of the elevation data, or by illuminating the DTM from a particular angle. Figure 5 shows a perspective view, looking from the South West with the illumination source positioned somewhere to the South East; the DTM has been shaded using a Lambertian model[22].

Lineament Detection

The surface expression of folds, faults, joints, strikes, lithological contacts and other geological features is often in the form of lineaments[23]. A common way to detect lineaments within an image is to use simple $3 * 3$ gradient edge filters, such as the Sobel and Prewitt filters[17, 25, 26]. The resulting 'edge strength' image can be 'thresholded' and 'thinned' using standard image processing techniques, to produce definite edges, one pixel in width. Many other edge detection algorithms exist, varying in complexity and performance, which may provide improved results, particularly in terms of scale dependency. Several detectors which have been used previously in geological application are the Haar transform[28], the Hough transform[29, 30] and the Fourier transform; a comprehensive summary of these and other edge detectors is given by Blicher[27].

Lineament detection is normally performed on spectral data. However, the same techniques can also be applied to a DTM, to obtain additional information on lineaments. In this study two orthogonal Sobel filters have been applied to the data and combined to obtain both edge strength and orientation information. The filters were applied to both the imagery (ATM band 7 (0.76μm–0.90μm), as this image exhibits a particularly high contrast of geological information) and the DTM. Edges derived from spectral data are commonly used and are relatively easily understood. By contrast, edges derived from DTMs have not been widely used. A gradient filter applied to a height image will enhance those areas of the image with the steepest slope and will therefore be virtually identical to the slope image derived from the interpolation procedure (the slope image has been used in further processing). A similar filter applied to a slope image will enhance areas with the largest *change* of slope, and when applied to an aspect image the filter will highlight areas where there is a rapid change in the direction of slope (e.g. along sharp ridges). One problem with the aspect image is that it represents azimuth angles between 0–360°. Therefore, if there is a small change in angle between 359° and 1° a standard Sobel operator will produce

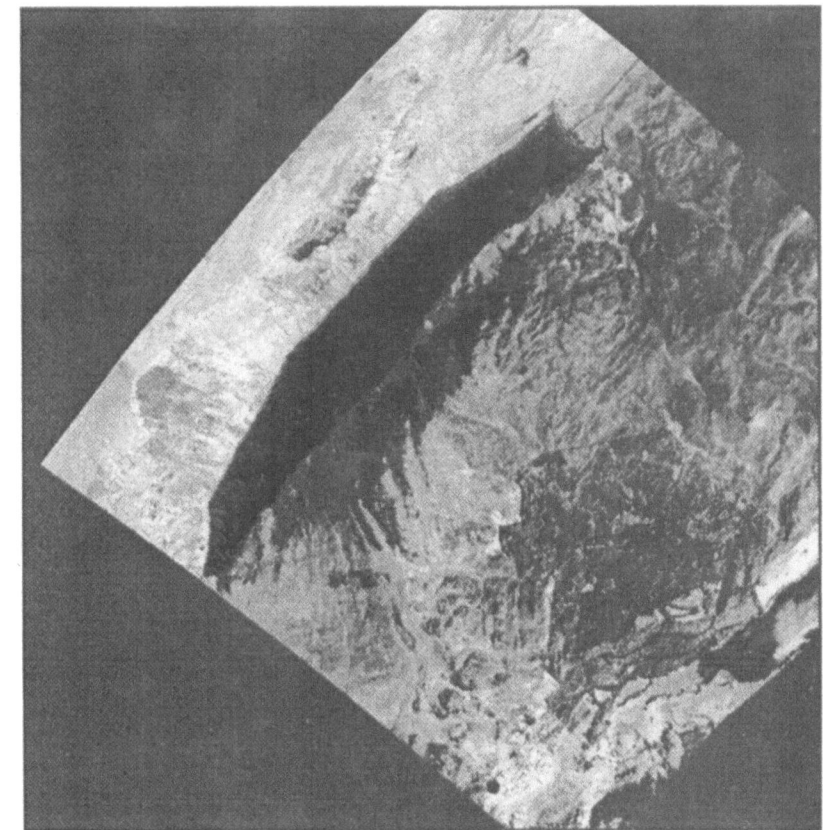

Figure 3, ATM band 7, Llyn Cowlyd, near Capel
Curig, Snowdonia, U.K.

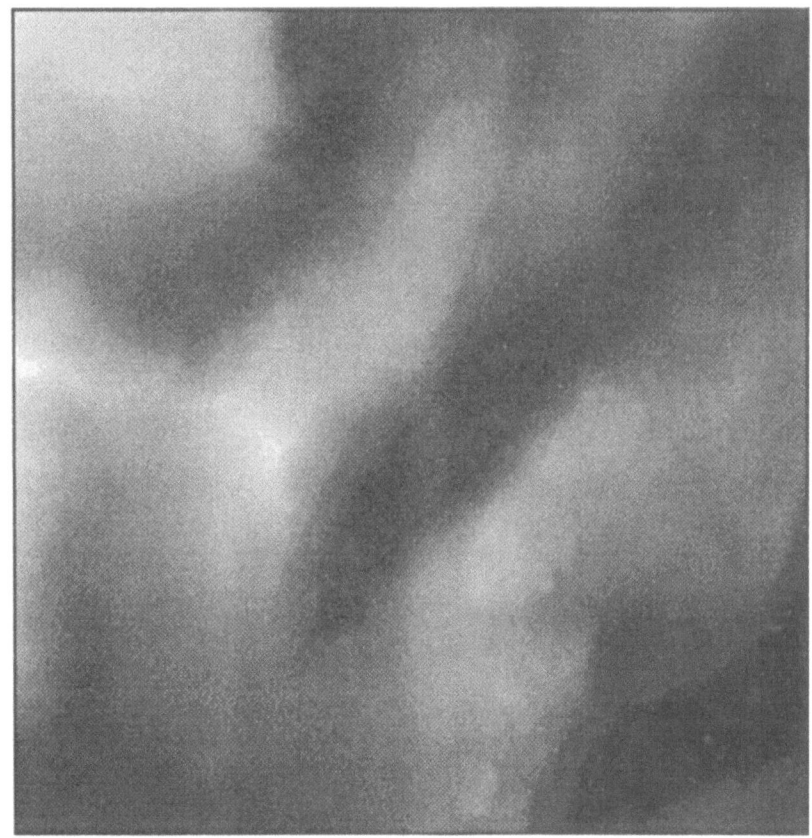

Figure 4a, Height image of Llyn Cowlyd DTM.
Image width = 5 km. White signifies highest
elevations and black signifies lowest.

Figure 4b, Slope image of Llyn Cowlyd DTM, lightest
pixels indicate steepest slopes.

Figure 4c, Aspect image of Llyn Cowlyd DTM.

Figure 5, Perspective view of Llyn Cowlyd, from the SW

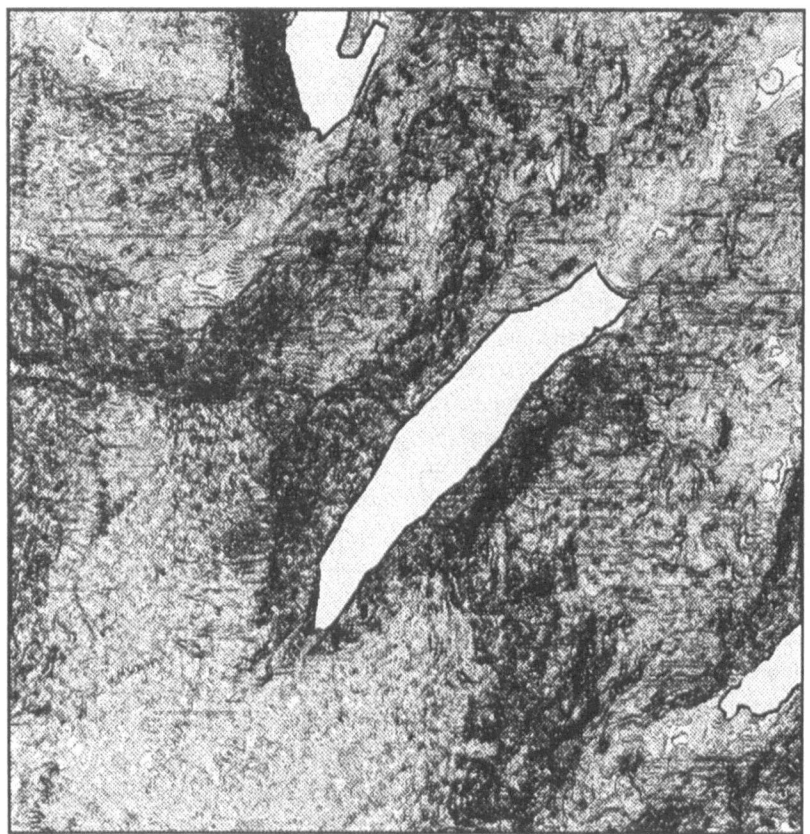

Figure 6a, Sobel edge strengths derived from the slope
image, dark colours indicate strongest edges.

a large edge value. This can be overcome by assigning all angle differences (A) which are $\geq 180°$ to $(360° - A)$, during the edge detection process.

It would be expected that each of the techniques outlined above should give different but complementary results. Each should describe different properties of the geomorphology of the area and therefore of the relationship between the geology and topography. The results of the edge detection are shown in Figures 6 and 7. Application of the edge detection algorithm to the slope image identifies strong edges where there is a pronounced *break* of slope. In many cases this may identify two geological boundaries (e.g. at the top and bottom of a steep slope, rather than the one which would by highlighted in the slope image itself. A further advantage over the slope image is that the slope edges pick out changes in slope in areas of gentler incline, especially at rock outcrops, where a sudden change of slope usually occurs.

The results obtained using the aspect image (Figure 6b) are markedly different from the two previous methods. This method tends to identify ridges and valleys, and is particularly useful in areas with a generally uniform slope. In this respect it is entirely complementary to the other two methods.

Finally, the results of the edge detection of ATM band 7 are shown in Figure 7. Most of the edges shown in Figure 6a are repeated here due to the natural shading of the scene although in some areas the shading is so severe that the image contrast in these parts of the scene is very poor and therefore no edges have been derived. The spectral data contains considerably more information than the DTM and as a result exhibits many different types of edge e.g. lakes, forests, tracks, roads, changes in vegetation and various other man-made features.

The product of the edge detection algorithm is an image which indicates the strengths of the edges for all pixels in the image. However, if edges are to be used as entities in order to calculate such physical parameters as dip and strike, decisions need to be made as to what constitutes a reasonable edge and what is background noise. The simplest method of doing this, and that which has been used here, is to 'threshold' the image. This results in a binary image where all edge strengths greater than or equal to the threshold are assigned a value of 1 and all those below are assigned a 0. A more sophisticated method, which will be considered in future stages of this project, is that of line following. Here a line can be followed through areas of low edge strength provided that certain conditions are met regarding the strength and orientation of neighbouring pixels[31].

After thresholding, the binary images are 'thinned'. This process recursively thins an image until each edge has a width of one pixel, while retaining the integrity of any branching structure in the edge. An algorithm has been developed which creates entities of these thinned edges. The process searches the thinned image for endpoints and then follows

each edge looking for the longest path through the branching structure of the edge (if present). Each edge entity is then stored as a series of 3D co-ordinates representing each pixel along the path, to be used later in the study for the calculation of dip and strike. Measurements are only included if the original edge is greater than a specified length; in this case the threshold has been set to 9 pixels (i.e. the edge must be at least 50 meters in length).

Ridge and Valley Line Extraction

Ridges and valleys are often closely related to lithology, structural geology and geomorphology[32] and the drainage network as a whole is often indicative of geological environments (e.g. sedimentary versus igneous). Parvis[33] and Howard[34] give comprehensive empirical descriptions of some thirty different drainage patterns relating to various forms of geology. Argialas[35], who has reduced this list to eight general patterns, also describes an expert system designed to recognize different drainage patterns on the basis of branching angles, distance between angles and concentration of branches, in a detected drainage network. Furthermore, segments of the drainage network may be used to identify specific geological features, e.g. streams can often follow the strike of sedimentary strata or the line of a fault. Hence, these segments may be used in a similar manner as the edges from the previous section.

A variety of algorithms exist which are designed to extract ridge and valley lines from a DTM, most of these employ a $3 * 3$ kernel which is passed across the image to identify different geometric properties e.g. $\cap$ or $\cup$ shaped, sink holes or saddle points[32, 36]. Riazanoff *et al.*[32] introduce what they define to be a 'structuralist' approach, in which lines are followed through the DTM, starting from isolated points such as saddle points, and then, for ridge line detection, climbing along the steepest slope. A similar approach has been developed as part of this project. In this respect an algorithm has been developed which simulates the progress of a raindrop (falling on each pixel) as it flows in the direction of the steepest slope. As each raindrop passes through a point the raindrop count for that point is incremented, hence the final valley-line image represents the number of raindrops passing through each point (given that each pixel in the DTM has equal run-off properties) and results in a what may be described as a 'stream-order image'. A more detailed description of the algorithm is given in Morris[20].

The results of the 'raindrop' algorithm applied to this area are shown in Figure 8. The algorithm has successfully identified all of the drainage network in the Ordnance Survey 1:10,000 scale topographic map plus many more smaller scale features not shown on the map. Many of these additional features are strongly related to the local geology, in that they circumscribe rock outcrops or follow small ravines through rock exposures (possibly caused by faulting). The raindrop image has the advantage over other valley line detectors in that it may be thresholded to highlight streams and rivers of any specified

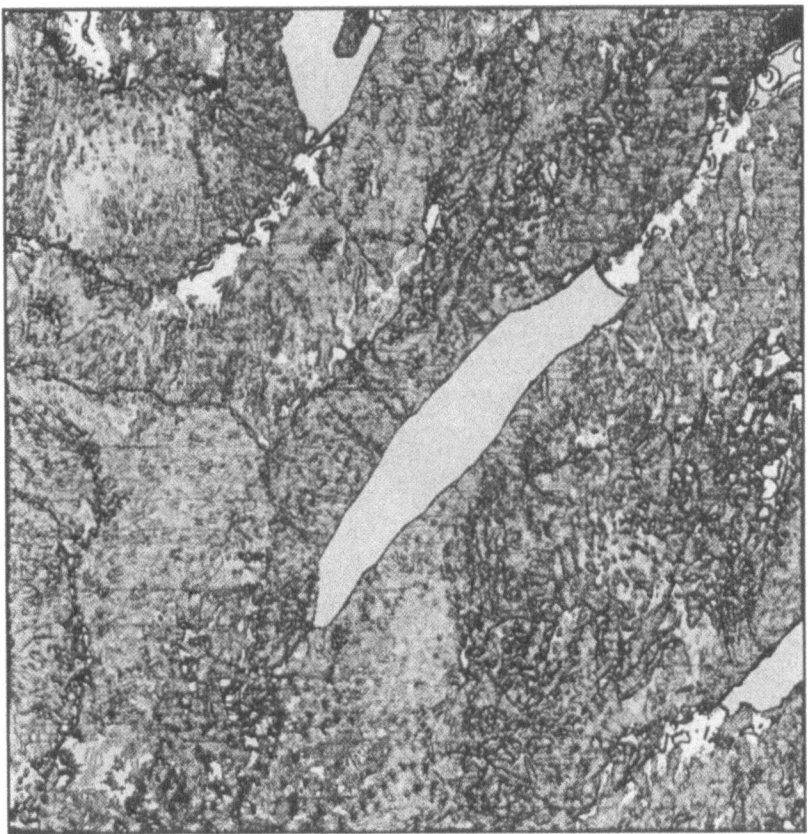

Figure 6b, Sobel edge strengths derived from the aspect
image, dark colours indicate strongest edges.

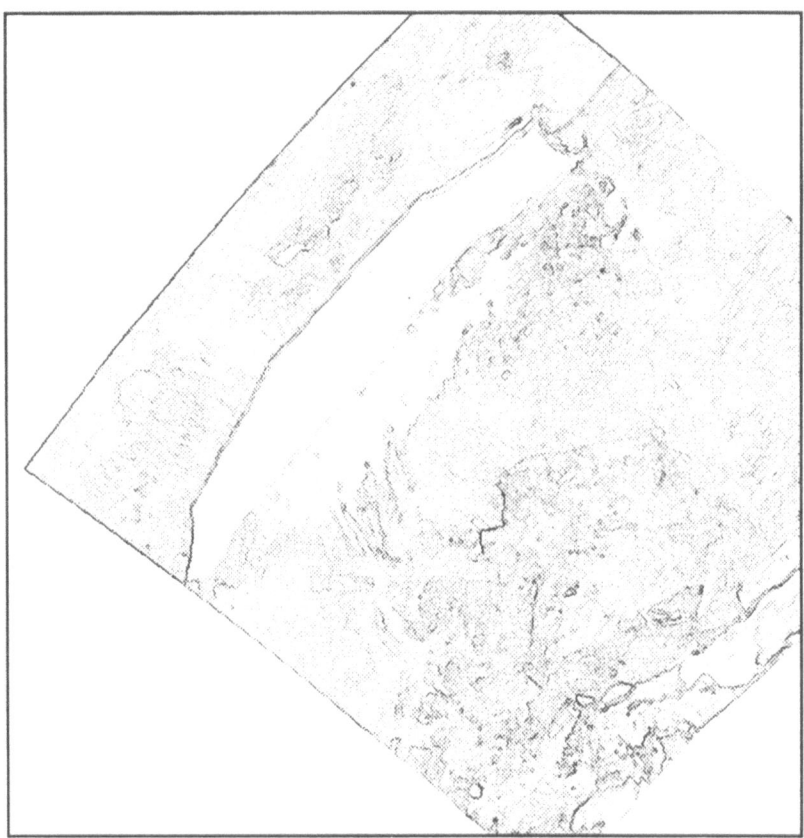

Figure 7, Sobel edge strengths derived from ATM
band 7, dark colours indicate strongest edges.

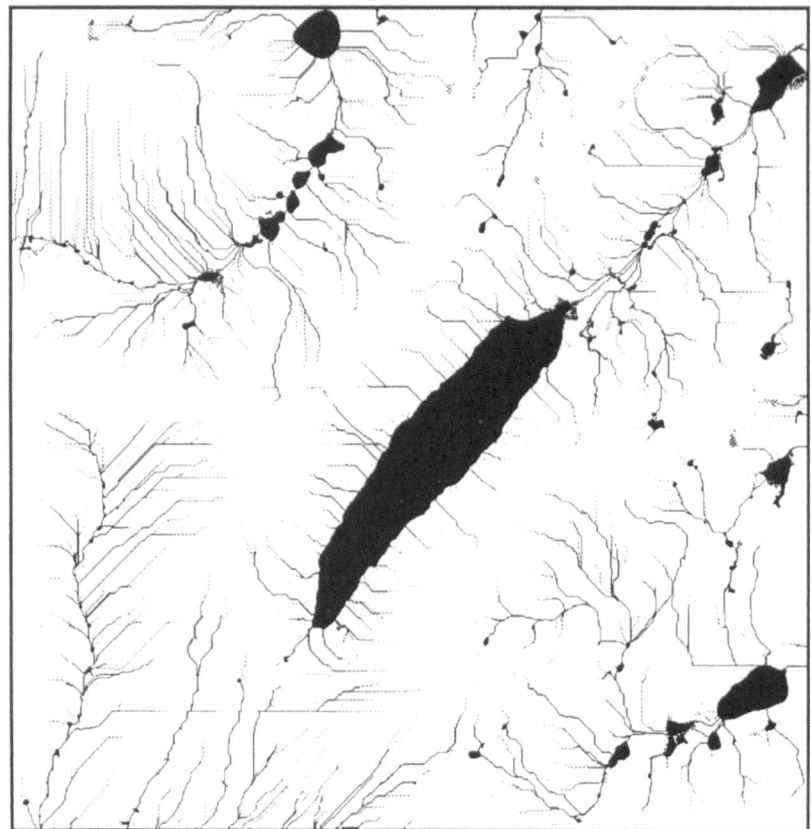

Figure 8, Drainage network of Llyn Cowlyd, using
'raindrop' algorithm (threshold=60)

Figure 9a, Dip and strike measurements for the slope image results.

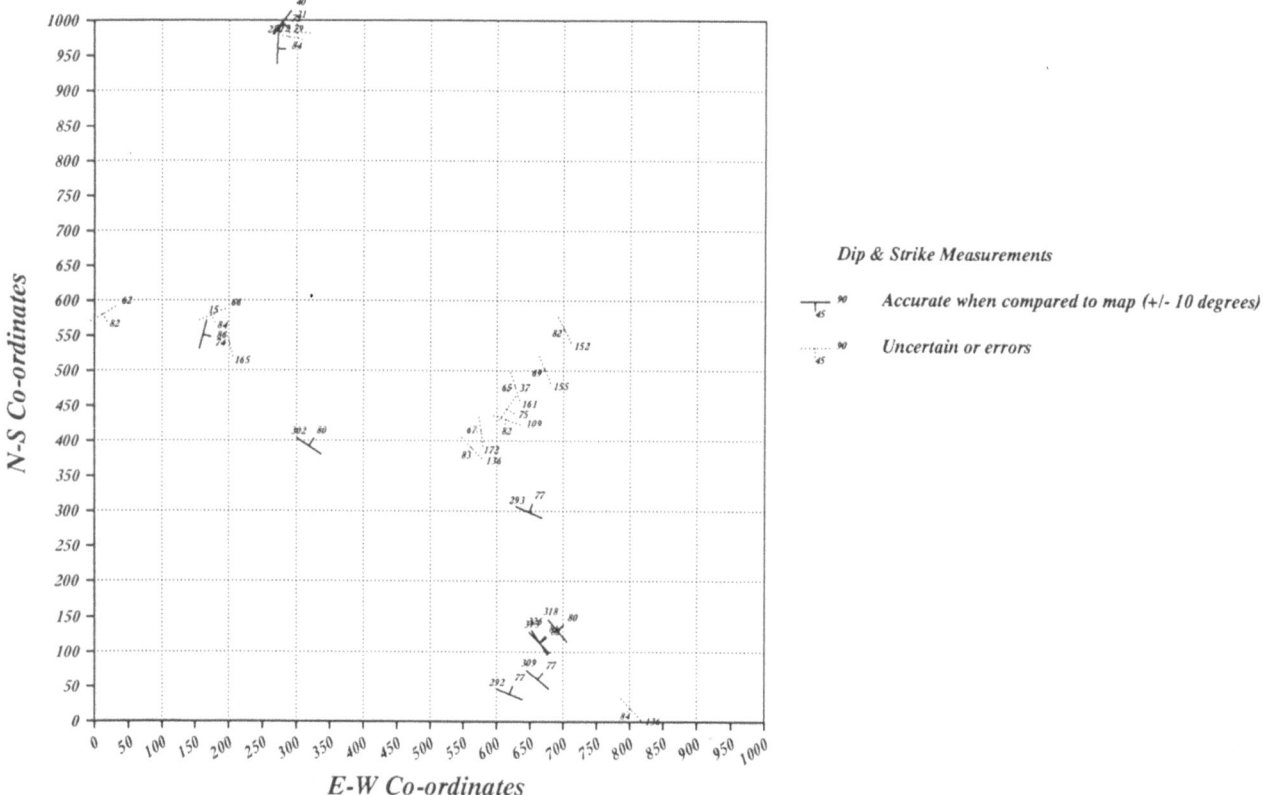

magnitude; therefore, various types of information may be extracted quite readily.

Selecting Geological Lineaments

We now have a plethora of edge information derived from the techniques outlined above. The next problem is to decide which edges to use and to what each edge might relate. Most of the edges derived from a DTM can be thought of as being geologically or geomorphologically relevant, unless any gross man-made features are present in the scene, such as road and railway cuttings, dams, quarries and habitation. Conversely, many of the edges found in remotely-sensed imagery are due to man-made features, water bodies or vegetation/agricultural boundaries.

In an attempt to eliminate unwanted edges, knowledge-based rules can be applied to the edge images. Take the example of the edge detected around Llyn Cowlyd reservoir. Several significant features discriminate this type of edge from any other:-

    i. the edge has a constant elevation along its entire length,

    ii. if the edge is complete it will form a closed shape, and

    iii. one side of the edge will have the distinctive spectral signature of water.

Extreme care must be taken when designing and implementing rules of this kind. They should be general rules which may be applied to any imagery of any area, and should not be made specific to a particular scene. This point can be illustrated with respect to the identification of roads. In this area of Snowdonia the roads are generally low lying (following broad valleys and passes), straight or slightly curved, have a gentle incline, are connected in a road network and are usually bordered by vegetation. However, in other more mountainous areas the roads may follow a more tortuous path obeying few of the rules set above. Similarly, in less-hilly terrain roads might follow higher ground, and so forth.

Similar identification rules may be created for field boundaries, forest plantations, tracks, habitation and any other man-made features. Once the non-geologically related features have been identified to some specified degree of confidence, the edges may be deleted from the database of edges. The only rule of this kind implemented to date is the lake identification rule.

Lineament Orientation

The data points derived from the edge detection techniques represent lineaments which lie on a 3D surface. If the edge represents a lithological boundary or a fault the points may be used to calculate the dip and strike of the bed or fault. A least-squares approach has been used to fit the data to a planar surface, producing an equation of the form $z = ax + by + c$ where x, y and z are the 3D co-ordinates and a, b and c are the coefficients derived from the fit.

A major difficulty with this technique is that if all of the data points lie along a straight line in 3D space then any number of planes could also be fitted through the data where the line belongs to the plane. One test to determine whether this is happening is to investigate and compare the coefficient of determination (R^2) derived from the planar fit with that derived from from a linear fit of the data in a new co-ordinate system defined by the planar surface. If the linear R^2 is $\geq$ the planar R^2 then the edge should not be used for the calculation of dip and strike, as it is more likely to define a line than a plane. In this study lines have been rejected if the R^2 for the planar fit is below a 0.6 threshold and if the linear R^2 is greater than the planar R^2 or greater than 0.9. These threshold limits have been set somewhat arbitrarily, but include two provisos. Firstly, to give lenience to the planar fit, as the edge may delineate a slightly curved surface, and secondly, to be strict on the linear fit because of the gross inaccuracies caused by calculating dip and strike for a straight line.

In many cases it is possible that an edge may represent a lithological boundary which is folded. In such cases a curved surface may need to be fitted to the data, again using a least-squares method and producing an equation of the form $z = ax^2 + by^2 + cxy + dx + ey + f$ where a, b, c, d, e, and f are the derived coefficients.

Once the equations have been determined it is an elementary process to calculate dip and strike measurements for each edge. Figure 9a shows some of the structural measurements calculated from the height information only.

DISCUSSION

From the results shown in Figures 9a–e it is unclear whether the measurements indicate the dip and strike of bedding or of a fault, or indeed whether they represent a geological feature at all. Although it is possible at this stage to interpret the results manually, the automatic continuation of the process requires more geological knowledge and the use of further processing. This constitutes part of the complete image understanding system which was outlined in the introduction.

When related to the geological map[37], some of the results show a very good comparison. Despite this, there are many questionable results, probably due to noise in the data, the problem of fitting planar surfaces to straight lines or curved surfaces, and the fact that, in some cases, edges may cross geological boundaries. For results derived directly from the slope image the most accurate results (Figure 9a) occur where the steepest slopes cross a ridge and therefore describe an arc on the Earth's surface to which a plane may be readily fitted (e.g. in the south of the area). Surprisingly, all of the steepest slopes detected on the eastern slopes of Llyn Cowlyd (where geological features are most visually apparent in Figure 5) produce dips in the opposite direction to those taken from the map, although most of the strike values are accurate to within ±15°. This is partly due to the fact that these lines have a

Figure 9b, Dip and strike measurements for the slope edge results.

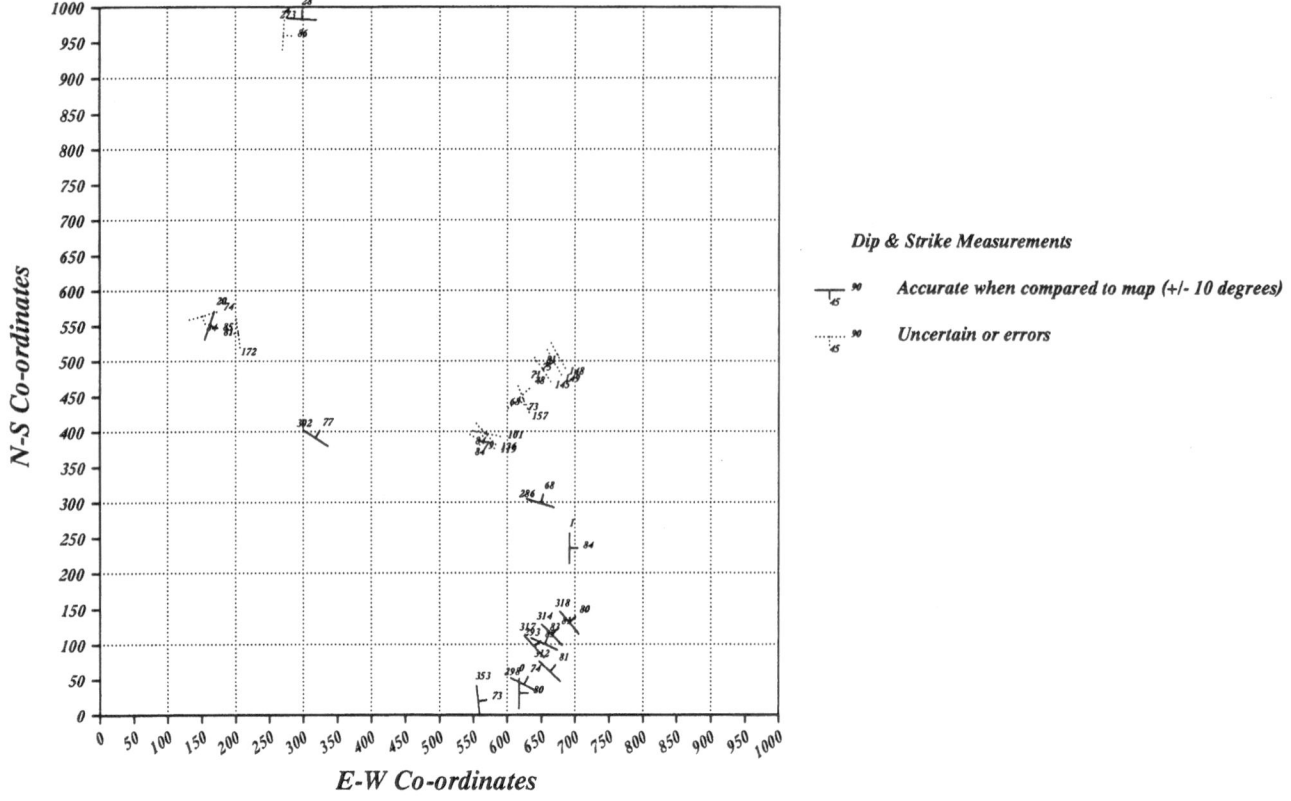

Figure 9c, Dip and strike measurements for the aspect edge results.

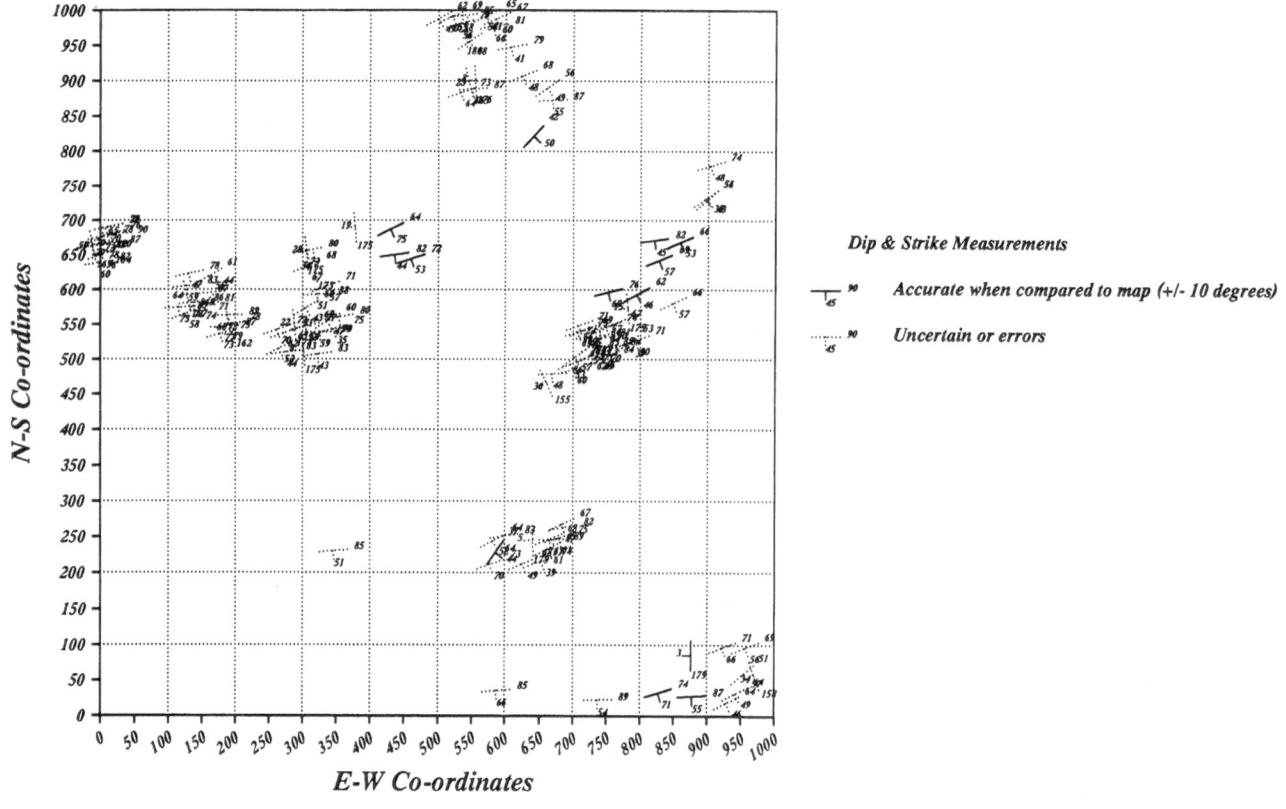

Figure 9d, Dip and strike measurements for the ATM image results.

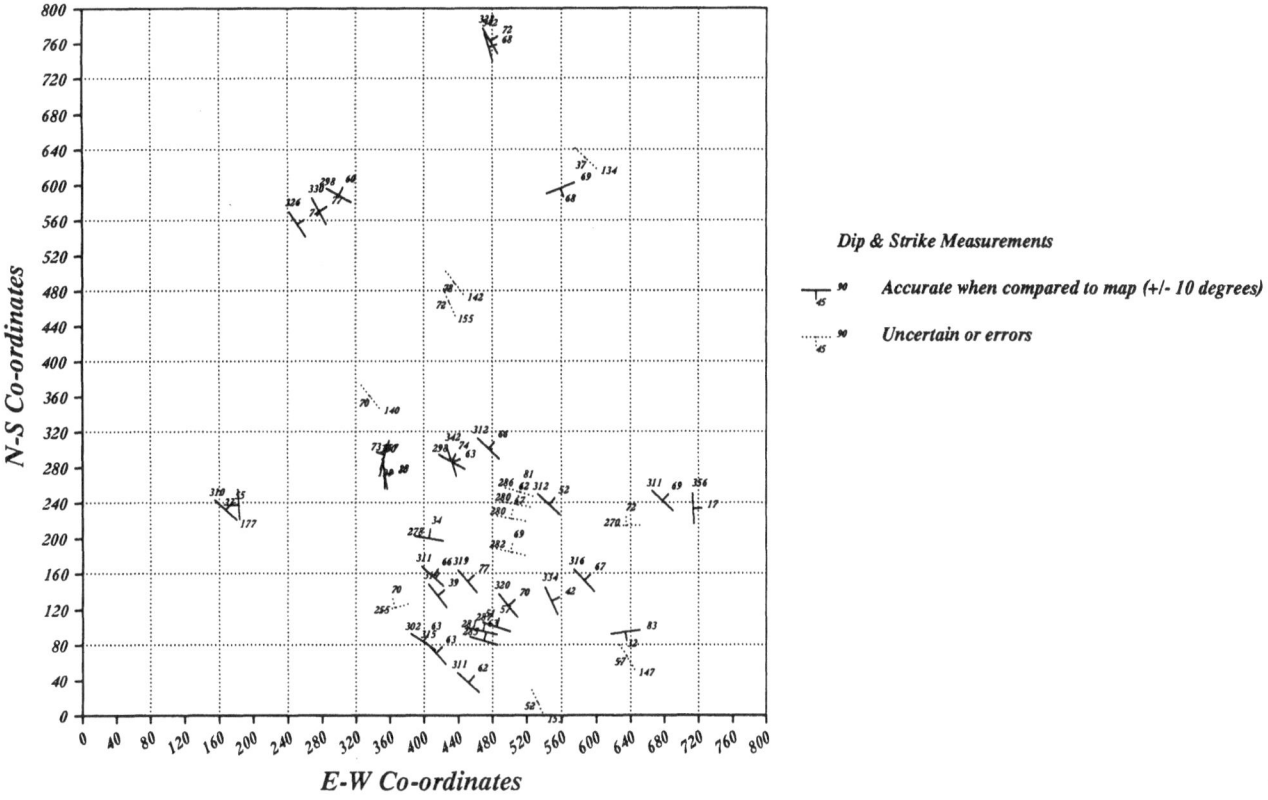

Figure 9e, Dip and strike measurements for the drainage network results.

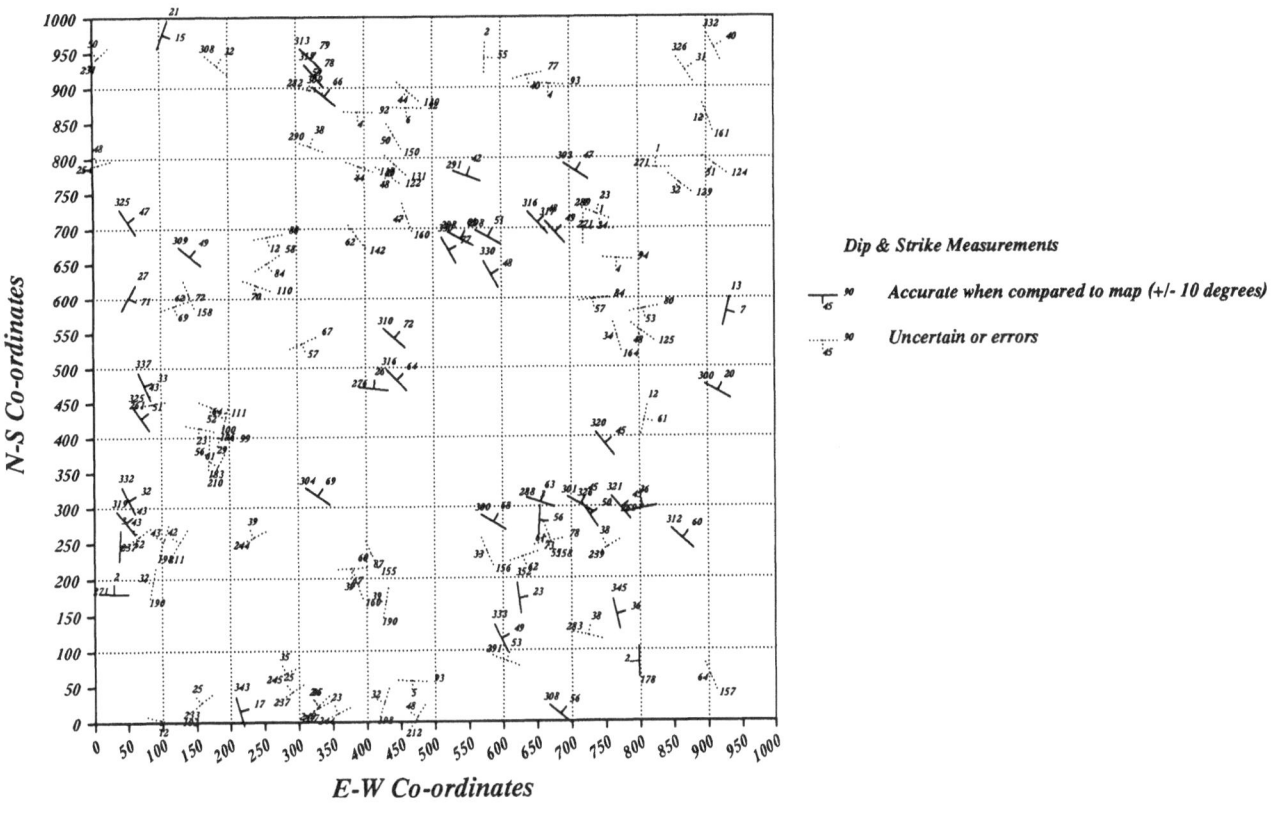

fairly high linear R^2 value and partly due to problems of simple thresholding. In this part of the DTM the slopes approach vertical cliffs in places, and as a result the steepest slopes may cross many geological boundaries. Here, as mentioned earlier, it would be beneficial to know the spectral properties of the lithology that the edge is attempting to represent. This knowledge could then aid a line-following process designed specifically to avoid crossing geological boundaries.

The edges derived from the slope image (Figure 9b) are almost identical to those derived from the height image, providing very little additional information using the thresholds set in this study. However, the slope-edge image holds a considerable amount of useful information at different edge strengths to those selected here, for example, subtle changes of slope can occur in areas of generally uniform slope, indicating the position of rock outcrops; these outcrops produce edge strengths towards the middle of the dynamic range of the edge image. Edges derived from the aspect image tend to enclose surface features (e.g. rock outcrops) and as such do not necessarily follow lithological boundaries. For instance, in this area outcrops of dolerite frequently occur at or near the top of ridges; here only one edge of the outcrop, usually the lowest, describes a geological boundary, while the others peter out into a bordering softer rock unit or into vegetation. As a result, very few of the dip and strike measurements derived from the aspect image correspond to the geological map. Nevertheless, as with the slope image, the aspect image contains a considerable amount of potentially useful information which has not been fully extracted using the techniques described in this paper. The aspect-edges require segmentation into constituent parts before more accurate analyses can be undertaken.

ATM Imagery

As mentioned previously, many of the results obtained from the ATM imagery correspond to those obtained from the height image, due to natural scene shading. Additional features, including man-made objects, are caused by spectral changes in the image. Apart from lake-edge identification, no rules specific to recognizing non-geological features have been implemented as yet. However, a large proportion of these features have been omitted from the final results (Figure 9d) because they either exhibit a low planar R^2 (e.g. forest tracks) or a high linear R^2 (e.g. roads and forest boundaries). A further problem with this set of results is the misregistration of the ATM imagery and the DTM. The average spatial error between the two data sets is approximately 18 meters, which can precipitate a large error in the 3D co-ordinates produced for each line and, hence, in the final calculation of dip and strike. In the future, it is hoped to include a geometric correction technique based on Delauney triangulation, designed specifically for airborne data, which should provide greater accuracy than is presently possible[38]. The results shown in Figure 9d are obtained from ATM band 7 alone. The ATM sensor records in 11 separate wavebands, each of which contains additional information and which should

be used to aid the spectral identification of rock types or lithological boundaries delineated by changes in vegetation. A substantial review of techniques for using and combining spectral data in geological applications is provided by Drury[39]; many of these techniques, including band ratioing and principal component analysis are likely to aid, significantly, the automatic extraction of lithological boundaries.

Drainage Network

The results derived from the drainage network (Figure 9e) have been obtained by thresholding all pixels having a raindrop count of between 150 and 450, which picks out most of the first-order streams. These streams are strongly related to the geology, whereas the larger streams have been greatly affected by glaciation. Accurate results have been obtained near the tops of mountains (e.g. to the southeast of Llyn Cowlyd) where streams often follow the lower edges of rock outcrops. In other areas the streams tend to cut across geological boundaries often following lines of faulting. In these cases either erroneous results are produced or the line is rejected because of its high linear R^2 value. In situations where it is thought that many of the lines are indicative of faulting it would be advantageous to employ the coplanarity techniques of Thiessen et al.[11] to identify the most frequent orientations of the faulting. These, in turn, could be used to provide rules for fault identification, i.e. if an edge with a high linear R^2 value fits one of the known fault orientations, then it is likely to represent a fault.

The techniques developed thus far represent only the first basic steps towards a future 'image understanding system'. The dip and strike measurements, once it has been ascertained what they relate to, could be used to drive further processing. For instance, if a dip and strike is known at some point, the DTM could be used to predict all other places in the scene at which the strata should outcrop, given a simple linearly dipping structural model. A more intensive search could then be conducted in these areas for further evidence to confirm the model. If the strata are identified, with the same dip and strike, then the model is verified; if not, then the hypothesis should be modified, perhaps to include a fault or a fold. Unfortunately, the process is less staightforward than this, since information on more than one dip and strike, is required to fit the proposed model. However, this example serves to show the general philsophy behind the image understanding system approach. This type of approach, which is a mixture of top-down (i.e an expected model drives the system) and bottom-up (i.e. the data drives the system) processing should provide a more complete and human-like method than using either top-down or bottom-up alone[9, 40].

CONCLUSIONS

The work presented in this paper represents the first steps towards a fully automated system for the extraction of geological information from remotely-sensed imagery and digital terrain models and outlines ways in which this may be built upon eventually to effect a full system. Techniques have been derived to estimate physical parameters such as dip and strike from both DTMs and remotely-sensed imagery. These incorporate various rules to reduce confusion between geological and non-geological features. Although for much of the study area the estimates of dip and strike derived using these techniques lie within $\pm 10°$ of the corresponding values taken from the geological map, there are some notable errors. It is hoped to improve upon these results in future work by including improved image processing techniques (e.g. line following techniques), knowledge-based rules and structural modelling. These, in turn, should bring the goal of a complete image understanding system decidedly closer. In this respect, Wilkinson and Fisher[41] believe that such a system could "eventually out-perform a human at making diagnoses from given inputs, due to rules being soundly based on accumulated statistics".

AKNOWLEDGEMENTS
I am grateful to the Natural Environment Research Council for provision of the Daedalus ATM imagery through grant number GR3/7020. Thanks to Dr. Mike Barnsley for his supervision and support and for polishing my sentence construction, rusted by years of computer dialogue. I am indebted to Andrea Morris for informing me of what constitutes a road and for eventually realizing that rocks are much more interesting at the *top* of a mountain. Thanks to Dr. Allison Reid for discovering and bypassing the peculiar whims of an archaic I^{2}S sysytem. Thanks also to my colleagues at UCL, Lewis, Paul Schooling and Bill Campbell for technical support and welcome diversions.

REFERENCES

[1] Parsons, A.J. and Yearley, R.J., 1986, An analysis of geologic lineaments as seen on LANDSAT MSS imagery. *International Journal of Remote Sensing*, 7, 1773–1782.

[2] Butler, D.M., Gurney, R.J., and Miller, T.L., 1987, EOS: The Earth observing system and polar platforms. *Eos Transactions of the American Geophysical Union*, 68(45), 1589.

[3] Truss, P., September 12–16th 1988, European remote sensing satellite platforms for the 1990s. In *Proceedings of 1988 International Geoscience and Remote Sensing Symposium*, pp. 171–174, Edinburgh, U.K.

[4] Hara, N., Homma, M., Igarashi, T., Tsuiki, A., and Ohta, K., September 12–16th 1988, Results of ADEOS conceptual study. In *Proceedings of 1988 International Geoscience and Remote Sensing Symposium*, pp. 179–182, Edinburgh, U.K.

[5] Goetz, A.F.H., July 10–14th 1989, The high resolution imaging spectrometer (HIRIS) facilty instrument for the first polar orbiting platform. In *Proceedings of 1989 International Geoscience and Remote Sensing Symposium*, pp. 2922–2924, Vancouver, Canada.

[6] EosDIS, , 1988, Baseline report. Technical Report NAS5-30079, Computer Technology Associates, Inc.

[7] Muller, J-P., 1988, Key issues in image understanding in remote sensing. *Philosophical Transactions of the Royal Society London A*, **324**, 381–395.

[8] Tailor, A., Cross, A., Hogg, D.C., and Mason, D.C., 1986, Knowledge-based interpretation of remotely-sensed imagery. *Image and Vision Computing*, **14**(2), 67–84.

[9] Matsuyama, T., 1987, Knowledge-based aerial image understanding systems and expert systems for image processing. *IEEE Transactions on Geoscience and Remote Sensing*, **GE–25**(3), 305–315.

[10] McKeown, D.M.Jr., Harvey, W.A.Jr., and McDermott, J., 1985, Rule-based interpretation of aerial imagery. *IEEE Transactions on Pattern Analysis and Machine Intelligence*, **PAMI–7**(5), 570–585.

[11] Thiessen, R.L., Eliason, J.R., and Rieken, E.R., July 11–14th 1989, Three-dimensional computer analysis and modelling of remote sensing–structural geologic problems. In *Proceedings of 1989 International Geoscience and Remote Sensing Symposium*, pp. 89–92, Vancouver, Canada.

[12] Sauter, D., Fraipont, P.de , and Ruhland, M., 1989, Image processing applied to digital elevation models: A useful tool for structural studies. In *Proceedings of the Seventh Thematic Conference on Remote Sensing for Exploration Geology*, pp. 1073–1080, Calgary, Canada.

[13] Burrough, P.A., 1986, *Principles of Geographic Information Systems for Land Resources Assessment*. Clarendon Press, Oxford, UK.

[14] Howells, M.F., Francis, E.H., Leveridge, B.E., and Evans, C.D.R., 1978, *Capel Curig and Betws-y-Coed: Description of 1:25,000 sheet SH 75*. Classical areas of British Geology. Institute of Geological Sciences, London.

[15] Barnsley, M.J. and Kay, S.A.W., 1990, The relationship between sensor geometry, vegetation canopy geometry and image variance. *International Journal of Remote Sensing*, **11**, 1075–1083.

[16] Wilson, A.K., November 26th 1985, Calibration of ATM data. In *Proceedings of the NERC 1985 Airborne Campaign Workshop*, pp. E25–E40.

[17] Schowengerdt, R.A., 1983, *Techniques for Image Processing and Classification in Remote Sensing*. Academic Press, London.

[18] MAPICS Limited, University College London, 1986, *MAPICS Reference Manual*.

[19] Davis, J.C., 1973, *Statistics and Data Analysis in Geology*. John Wiley & Sons, Inc., New York, 2nd edition.

[20] Morris, K.P., 1990, Evaluating digital elevation models for the identification of geological features in remotely-sensed imagery. In *Proceedings of the 16th Annual Conference of the Remote Sensing Society*, Nottingham. Remote Sensing Society. In press.

[21] Day, T. and Muller, J-P., 1988, Quality assessment of digital elevation models produced by automatic stereo matchers from SPOT image pairs. *IAPRS*, **27-A3**.

[22] Muller, J-P., Dalton, M., Day, T., Kolbusz, J., Pearson, J.C., and Richards, S., 1988, Visualisation of topographic data using video animation. In *Proceedings of the XVIth International Congress of ISPRS, Kyoto, Japan, IAPRS 27-A3*.

[23] Harris, R., 1987, *Satellite Remote Sensing*. Routledge & Kegan Paul, London.

[24] Stefouli, M. and Osmaston, H.A., 1984, The remote sensing of geological linear features using Landsat: Matching analytical approaches to practical applications. In *Satellite Remote Sensing. Review and Preview, Remote Sensing Society*, pp. 227–236, Reading.

[25] Podwysocki, M.H., Moik, J.G., and Shoup, W.C., June 1975, Quantification of geologic lineaments by manual and machine processing techniques. In *Proceedings of NASA Earth Resource Survey Symposium, NASA TM-X-58168*, pp. 855–905, Houston, Texas.

[26] Smithurst, L.J.M., Vaughan, R.A., and Stone, P., 1987, An evaluation of remote sensing techniques for lithological discrimination and lineament analysis in the Ballantrae Complex, SW Scotland. In *Proc. Annual Conf. of the Remote Sensing Society*, pp. 580–591, September 7–11th.

[27] Blicher, A.P., 1985, Edge detection and geometric methods in computer vision. Technical Report No. STAN-CS-85-1041, Department of Computer Science, Stanford University, California, CA 94305, USA.

[28] Majumdar, T.J. and Bhattacharya, B.B., 1988, Application of the Haar transform for extraction of linear and anomalous patterns over part of Cambay Basin, India. *International Journal of Remote Sensing*, **9**, 1937–1942.

[29] Skingley, J. and Rye, A.J., 1986, The Hough transform applied to SAR images for thin line detection. Technical Report Y/251/3077, GEC Research Ltd., Great Baddow, Essex, UK.

[30] Wadge, G. and Cross, A.M., 1989, Identification and analysis of the alignments of point-like features in remotely-sensed imagery. Volcanic cones in the Pinacate volcanic field, Mexico. *International Journal of Remote Sensing*, **10**, 455–474.

[31] McKeown, D.M.Jr. and Pane, J.F., 1985, Alignment and connection of fragmented linear features in aerial imagery. Technical Report CMU-CS-85-122, Carnegie-Mellon University.

[32] Riazanoff, S., Cervelle, B., and Chorowicz, J., 1988, Ridge and valley line extraction from digital terrain models. *International Journal of Remote Sensing*, **9**, 1175–1183.

[33] Parvis, M., 1950, Drainage pattern significance in air-photo identification of soils and bedrocks. *Photogrammetric Engineering*, **16**, 387–409.

[34] Howard, A., 1967, Drainage analysis in geologic interpretation: A summation. *American Association of Petroleum Geologists*, **51**, 2246–2259.

[35] Argialas, D.P., Lyon, J.G., and Mintzer, O.W., 1988, Quantitative description and classification of drainage patterns. *Photogrammetric Engineering and Remote Sensing*, **54**(4), 505–509.

[36] Skidmore, A.K., 1990, Terrain position as mapped from gridded digital elevation data. *International Journal of Geographic Information Systems*, **4**, 33–49.

[37] Survey, British Geological, 1985. Bangor – Solid Geology. 1:50,000 Series: England and Wales sheet 106.

[38] Devereaux, B., 1990. Personal communication.

[39] Drury, S.A., 1987, *Image Interpretation in Geology*. Allen and Unwin (Publishers) Ltd., London.

[40] Nicolin, B. and Gabler, R., 1987, A knowledge-based system for the analysis of aerial images. *IEEE Transactions on Geoscience and Remote Sensing*, **GE-25**(3), 317–329.

[41] Wilkinson, G.G. and Fisher, P.F., April 8–11th 1984, The role of expert systems in remote sensing. In *Proceedings: Integrated Approaches in Remote Sensing, Remote Sensing Society*, pp. 353–360, Guildford, U.K.

Airborne geophysics and remote sensing: some common ground in presentation techniques and interpretation

Colin V. Reeves M.A., M.Sc., Ph.D.
Peter W. Zeil Dipl.geophys.
Department of Earth Resources Surveys, International Institute for Aerospace Survey and Earth Sciences (ITC), Delft, The Netherlands

SYNOPSIS

Airborne geophysical techniques - aeromagnetic, electromagnetic and gamma-ray spectrometer surveys - are long established methods in mineral exploration whose main debt to the advent of satellite imagery has been its use as topographic base maps of last resort. Historically, there has been little common ground between the treatment of geophysical data sets which reflect rock properties at depth, and the surficial overview provided by remote sensing.

The microcomputer technologies of the eighties have started to break down this barrier. In the first place, the techniques of presenting remote sensing data in colour are gradually being adopted and adapted for the presentation of (increasingly large) geophysical data sets with impact and meaning for the non-geophysicist. In the second place, inexpensive systems affordable by small exploration offices now provide the possibility of implementing Geographic Information Systems adequate for overlaying data sets of different origins - remote sensing, geophysical surveys, geochemical surveys - and conveniently integrating their interpretation.

Arguments concerning the cost-effectiveness of the various methods have to be considered, particularly now that the cost-to-the-user of acquiring remote sensing data is increasing towards its true cost.

In the practical situation, it is most often the case for a particular study area that only partial data sets are available, whether of remote sensing data, geophysical surveys, geochemical surveys or whatever, and that data quality (e.g. survey specifications) varies from one part of a study area to another. Logically, decison-making and concept-building

for exploration strategies combines the interpretation of ALL relevant data - with due regard to data quality - to produce an earth-model that can be tested (by investment) in the next exploration phase. This parallels the approach of theory-building in pure science. But, as in pure science, theories (models) can only be proved wrong, not right.

GEOPHYSICS IN THE CONTEXT OF REMOTE SENSING.

The geophysical literature to date does not offer much evidence of an effective symbiosis between geophysical exploration and remote sensing applications. In 1985, for example, a 400 page review section to mark the 50th Anniversary of the journal 'Geophysics' published by the Society of Exploration Geophysicists contained only one 16-page article on remote sensing which noted that 'energy and mining companies *are recognizing* the important role of remote sensing data in developing their exploration strategies' [our italics] [15]. Even this paper suggested that remote sensing developments were largely confined to the technology of sensors and effective interpretation methods lay in the future. A large part of the blame for this corporate shortsightedness can probably be directed at the pre-occupation of the geophysics community at large with the seismic reflection technique which employs much of its talent and consumes over 95% of the annual expenditure on geophysical exploration. Seismic reflection surveys - highly refined and sophisticated as they are - are usually only employed for detailed studies of carefully selected areas, leaving the 'Cinderella' methods of geophysics with the task of area selection in the more regional phases of exploration for hydrocarbons, and all the geophysical aspects of locating mineral deposits and groundwater resources.

While remote sensing (sensu stricto) is confined to the collection of radiation in defined parts of the electromagnetic spectrum (such as visible, infra-red and radar) we wish first to stress some similarities in techniques and objectives between some of the methods of geophysical reconnaissance and remote sensing. These geophysical methods work because variations in the physical properties of rocks in the earth give rise to variations in certain physical quantities that may be measured at a distance from the rocks themselves (i.e. remotely), usually at or above the earth's surface. The exploration geophysicist carries out a survey to produce a map of these variations and interprets it in terms of the hidden geology.

The geophysical sensing is 'remote' because reconnaissance geophysical methods (gravity, magnetic, radiometric and electromagnetic surveys) do not require any direct contact between the geophysical 'sensor' and the sources of (the often complex patterns of) anomalies. Definition and resolution of geophysical anomaly patterns improves with decreasing source-sensor distance, so there is little to be gained - and much to be lost - by putting geophysical sensors at high altitude above the earth's surface. The practical difficulties of working on the earth's surface itself dictate that low-flying aircraft are the most cost-effective means of capturing geophysical data for most types of survey at a reconnaissance scale. The survey results are thus built up by a scanning process - as in true remote sensing - but the sensor has to be physically moved over the survey area in order to build up the picture; only one input value is possible for a given sensor at any given location. The captured data for a single geophysical flight-line can, however, be compared with that of a single scan-line in a remotely-sensed image.

Unlike true remote sensing, however, the 'remoteness' of sources in geophysical surveys is usually due to their being buried well below the ground surface. This is a major strength of geophysical methods - namely that they can detect rock bodies having particular distinguishing physical characteristics even when they are concealed from direct observation at the earth's surface. Cover of some sort is nearly always present upon what may be regarded in any area as the bedrock, ranging from thin soil cover, through thick vegetation, deep weathering, desert sand cover, glacial till, to swamps and water. In fact, the more disabled the surface geologist seems to be, the more the likelihood that the geophysicist will be called in to carry out a survey.

Geophysical data, then, provide perhaps the only type of 'remote sensing' capable of directly detecting geological configurations in three dimensions. The depth of investigation is certainly not confined to the surface layers which reflect the energy captured by conventional remote sensing detectors, except in the case of gamma-ray spectrometer surveys where detectability of sources falls off quickly with depth of burial. On the other hand, the complex anomaly patterns of a geophysical anomaly map result from the summation of the effects of sources at a wide range of depths and there can be no unique interpretation of a two-dimensional anomaly map in terms of a three dimensional array of sources. Nevertheless, if the objective is simply to produce a two-dimensional geological map of the bedrock surface, data processing techniques may be employed to help eliminate the effects of the deeper sources.

As a result of these abilities, airborne geophysical surveys - and pre-eminently airborne magnetometer surveys - have been brought into action routinely over the past forty years, particularly when data-gathering in a new area was in its earliest phases. Statistics show that some 2½ million line-kilometres of data have been acquired annually over the past 20 years [12]. There are undoubtedly modern refinements, but the principles are well established.

REMOTE SENSING AS A NAVIGATION AID

Two of the most tedious problems of carrying out airborne geophysics are (a) navigating the aircraft as closely as possible to the desired flight-path and (b) recovering the path actually flown in order to position the captured data on a map. So-called frontier areas of exploration are often lacking anything by way of a published map which is useful in this respect. Controlled or even uncontrolled air photo-mosaics have been used for many years as a base-map for geophysical survey compilation. The advantages of using satellite imagery in this role - namely the relative lack of geometrical distortion and the availability of a large area in one frame - were soon recognized, even where the lack of definition on early imagery required use of aerial photographs in a supplemental role.

The uses of satellite imagery as a map-substitute appear to be under-valued, but from personal experiences of the authors it can be an invaluable aid to the most basic geophysical procedures where topographic maps are unavailable. Satellite images may even be preferable to topographic maps for airborne navigation since variations in soil colour, vegetation and other subtle physiographic features or even patterns of grass-burning will not appear on even the most up-to-date and detailed cartographic product but are clearly to be seen from the air and on satellite images.

Examples of such applications from geophysical surveying include the use of the earliest Landsat (ERTS as it was then) images for positioning helicopter-borne gravity observation points in the Central Kalahari, Botswana, in 1973, and navigating chartered aircraft to remote base-camps (the exact coordinates of which were known from radio reports of surveyors based there) in northwest Sudan in 1986 and 1987.

It is worth remarking that it is still often only the oldest satellite imagery that is available for remote frontier areas; more recent and sophisticated satellite images have - to the outsider - never gained the currency of those obtained by the early Landsat missions. Superior resolution and larger scale are no compensation when coverage has been selectively concentrated on other areas or is unaffordable, the latter being a point we will return to later.

RASTER METHODS FOR PRESENTING GEOPHYSICAL DATA.

Further common ground between geophysical surveys and remote sensing has been rather hard to find until recently. The advent of inexpensive microcomputers in the early 1980s, however, forced the geophysical community to rethink the answer to many of its data processing needs and provided an opportunity for imaginative software developmnent to serve a potentially large new user community. Particularly in mineral exploration geophysics, the presence of - or the need for - a state-of-the-art computing centre had never been a priority in the way that petroleum exploration has demanded. Many exploration companies - mining houses, contractors, consultants - only entered the computer era with the availability of powerful personal computers (most notably the IBM-PC in 1981), and the strategy was often to put the portable computing power right in the field camp [6].

This resulted in a large number of potential users for geophysical software, in distinct contrast to the previous situation where the processing and presentation of regional

geophysical data was effectively confined to a small number of specialized processing centres around the world, each with its own software package. This naturally stimulated the development of geophysical software for microcomputers since a mass-market made it easier to recover development costs while keeping powerful products attractively inexpensive. For similar reasons, hardware devices of impressive peformance became available within the budget of even a small exploration office. One of the effects of this new operating environment was the re-thinking of presentation of geophysical data.Initially the move was to replace the skills of the draftsman in drawing contour maps and profiles, but progressively the ability to use other types of presentation of map data led to experimentation with techniques more akin to those established for remote sensing data.

The most important of these alternative methods rely on using the grid or raster of geophysical values traditionally created for the contouring process and displaying each grid value as a pixel on either a screen or paper copy display. An immediate consequence of this is usually that a smaller grid cell size is necessary than for traditional contouring where a pixel dimension of 2.5 mm is adequate. Even grid cell sizes of of less than 1 mm on paper (100m at 1:100 000) produce quantities of data which are manageable for normal map sizes with modern PC equipment. Some of the more popular types of pixel or raster maps for aero-magnetic data presentation are described below and illustrated in Figure 1.

Grey-scale raster maps (Figure 1(c)).

Each cell of the raster is ascribed a grey pixel the brightness of which depends on the magnetic value at that grid point. The highest values might be white, the lowest black (or vice versa) with an appropriate use of contrast stretching to maximise the use of the available grey scale. Subtle variations in grey tone over large distances are not detected by the human eye, but local contrasts are recognized. As a result, short wavelength anomalies tend to stand out. At certain magnetic inclinations, typical magnetic anomalies with both positive and negative parts give a three-dimensional effect, as though the map was an undulating surface illuminated from one direction.

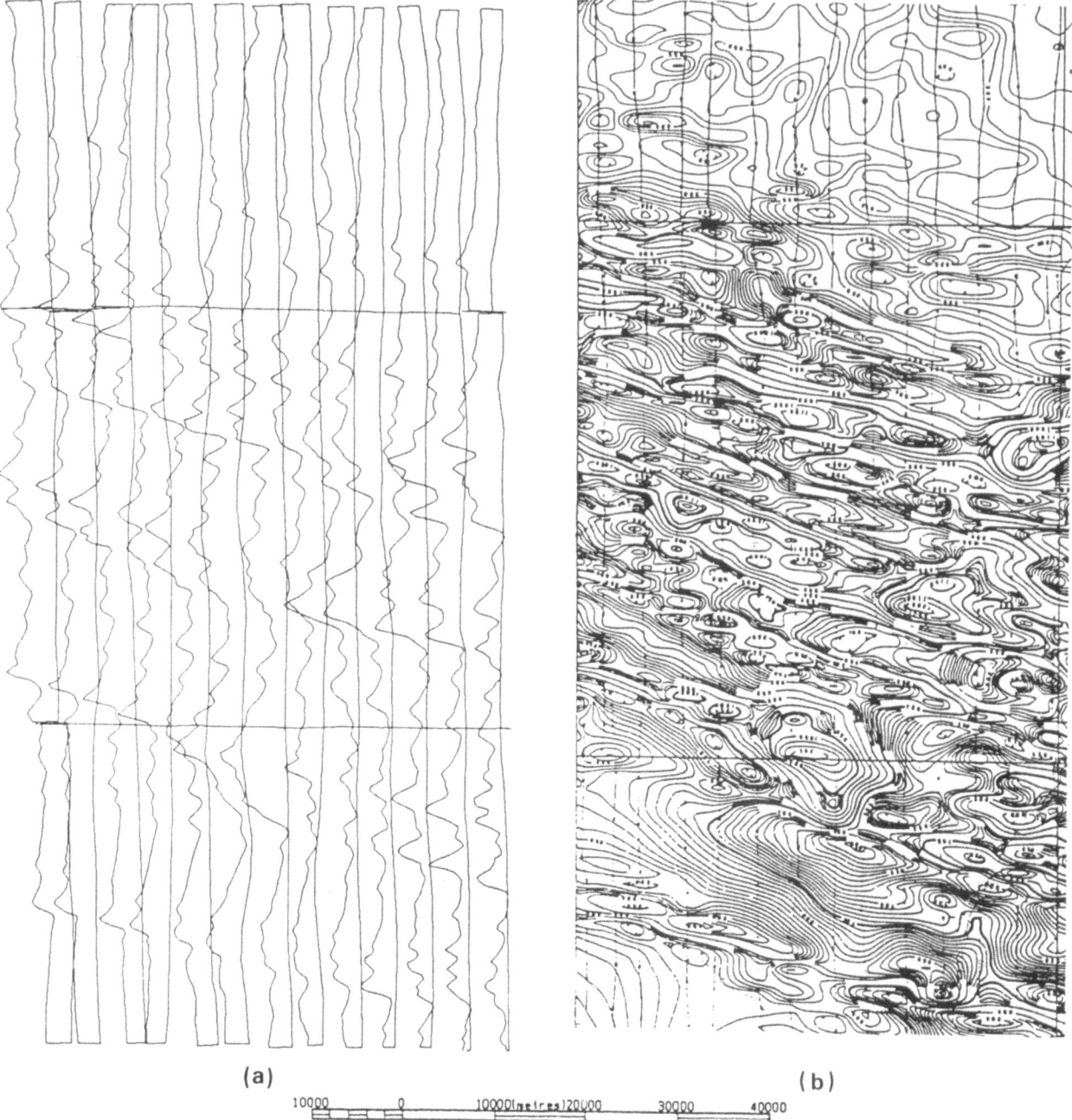

(a)   (b)

10000   0   10000(metres)20000   30000   40000

Figure 1. A selection of presentation styles for aeromagnetic data on a map. The area is in northern Botswana (bounded by latitudes 19° 48'S and 20° 48'S and longitudes 24° 00'E and 24° 30'E) where a swarm of dolerite dykes of Jurassic age cuts the country rocks on a WNW-ESE strike direction. Dykes and country rock are in turn overlain by over 100 m of Kalahari Sand cover. The original data were collected at a ground clearance of 300 m along north- *south flight-lines spaced 4000 metres apart.*

(a). Profile map of the original data and flight-path. For the magnetic scale, a distance equivalent to the nominal flight-line spacing equals 320 nT.

(b). Conventional contour map. Contour interval 10 nT with considerable deletion of contours in areas of high gradient.

Shaded relief raster maps.

The three-dimensional effect can be enhanced by processing so that (for example) even a simple positive anomaly appears three-dimensional. Calculating the first horizontal derivative in the direction of the supposed illumination and presenting the output as a grey-scale raster deceives the eye into seeing the magnetic variations as though they were physical topography. Choice of azimuth and elevation of the illumination source can be made to highlight the desired features and suppress others.

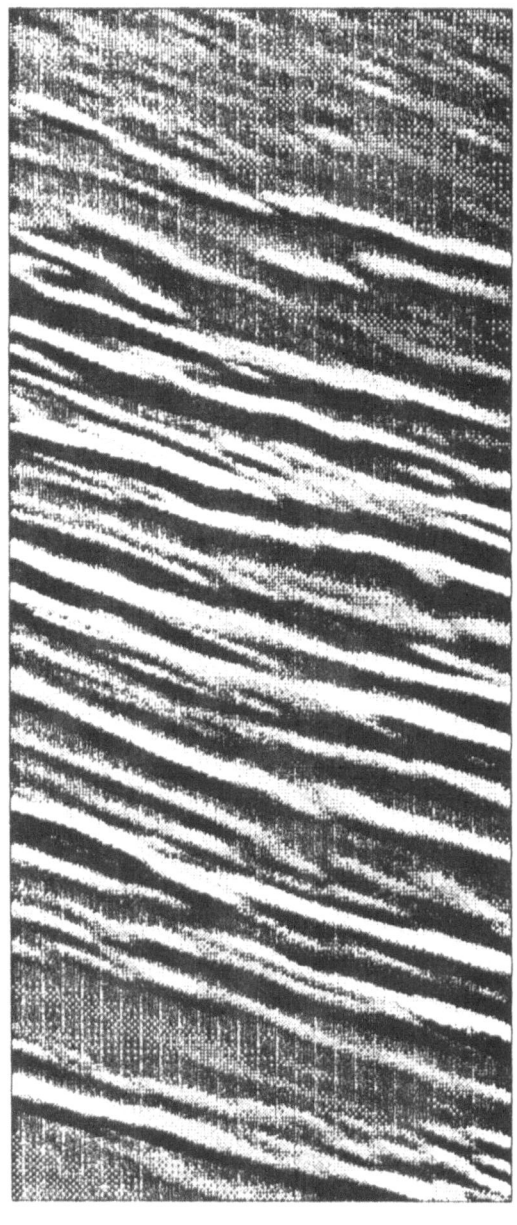

(c)   (d)

10000   0   10000(metres)20000   30000   40000

(c). Grey-scale raster map with areas of high magnetic values white and low values black. Note the broken nature of otherwise linear anomalies in some places due to undersampling in the direction perpendicular to the flight lines.

(After Wu [18]; courtesy of the Director of Geological Survey, Botswana).

Colour raster maps.

More popular in geophysical circles in recent years has been the colour raster map which, in effect, emulates the hand-colouring of intervals between contour lines with a range of colours from the natural spectrum, usually with red

(d). Resolution of adjacent anomalies has here been improved by calculating the first vertical derivative of the magnetic field along each profile and then gridding with respect to the strike direction and presentating the grid as a shaded relief map with an illumination from the NNE. Note the suppression of some of the broader anomalies - probably not related to near-surface sources - which has also been achieved.

as the high values through orange, yellow, green to blue for the low values. Each pixel is coloured appropriately for its magnetic anomaly value. These maps are attractive to the eye and reveal at once areas of 'highs' and 'lows' which are not immediately obvious on a contour map. However they tend psychologically to emphasize long wavelength, high amplitude features which are often not easily interpretable because they arise from deep-seated sources rather than from surface geology. Purists object to the use of different colours to record the variation in a single quantity.

Combined colour raster and shaded relief maps.

Since the shaded relief style of presentation described above tends to emphasize the short wavelength features, it complements the colour raster style which tends to emphasize the long wavelength features. In practice, the two styles are simply combined by overprinting the grey of the shaded relief image onto the colour map. One specific advantage is to reveal low amplitude magnetic features which would otherwise be lost within a single colour level.

Dot-matrix (colour) and laser printers with resolutions of 100, 200 even 400 dots per inch are now commonplace in geophysical data laboratories and have become very popular for map-making applications, more than making up in low cost, convenience and accuracy of scale for the more continuous colour and grey-scale capabilities of photographic products which require expensive, specialized equipment to produce. The principle has been extended to the production of inexpensive geophysical maps   (e.g. 1:1 000 000 scale for the whole Canadian Shield) on conventional four-colour printing presses.

Beyond simple presentation techniques of this type, geophysical rasters can also be processed to improve their interpretability. As has been common in the past, the specific Laplacian properties of potential fields may be exploited to produce some physically meaningful derived maps from the geophysical raster (e.g. an upward or downward continued version of the observed field), the output of which can employ any of the above methods of presentation. In addition, simple space-domain operators of the type pioneered by the image processing community may be applied to geophysical rasters to produce processed maps which are more appealing and useful to the interpreter in the way that edge-enhancement (for example) may aid the satellite image interpreter.

Figure 1(d) shows an example of such a presentation. Here the aeromagnetic profile data have been subjected to a first vertical derivative filter which better resolves the anomalies due to adjacent dykes along each flight-line. This processed data has then been gridded, and the resulting raster is here presented as a shaded relief image illuminated from the north-east to emphasize the expression of the dykes.

Figure 2 shows something of the potential for using image-enhancement techniques directly on the geophysical raster. The area is one of Precambrian rocks in southern Cameroun where dense tropical vegetation and deep weathering of the bedrock obscure the rocks from direct physical examination [16]. In Figure 2(a) the aeromagnetic data is presented as a raster in which the large anomalies due to magnetic ironstones within a greenstone belt are clearly evident, but the precise location of their sources is unclear on account of the low angle of inclination of the earth's magnetic field and the presence of remanent magnetization in the source rocks. Use of a simple operator (Figure 2(b)) to assess the magnetic 'relief' quanitatively - and the use of an appropriate choice of colour scale for raster presentation - focusses the 'pseudo-anomalies' over their sources in a way that can assist even the unskilled interpreter by making an intuitive interpretation approximate to the likely configuration of magnetic rock units in the ground.

COMPARATIVE PERFORMANCE OF GEOPHYCICAL SURVEYS AND SATELLITE IMAGERY.

It should now be clear that not only can airborne geophysics be considered as a specialized type of remote sensing, but also that modern methods of geophysical map presentation share much of the methodology of remote sensing data presentation. It is logical then to compare the effectiveness of the two methods for certain geological objectives. This is done first by way of a number of examples.

Southwest Botswana, southern Africa.

The ubiquitous cover of the Kalahari Sand has frustrated the conventional geological mapping of most of Botswana, while the better-known geology of neighbouring countries promises hidden mineral wealth in the bedrock concealed below the Kalahari (Figure 3(a)). Discovery of diamonds (Orapa) and Copper-Nickel sulfides (Selebi-Pikwe) in the late 1960s encouraged the use of well established geophysical methods to map the hidden geology in the 1970s.

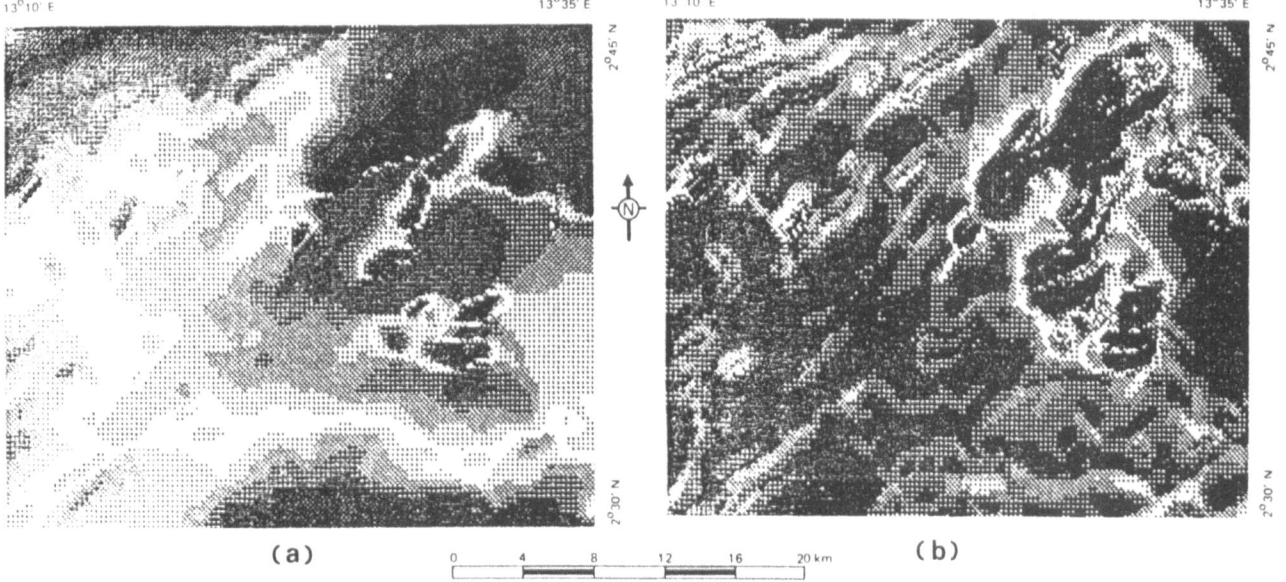

Figure 2. An example of image-enhancement techniques applied to aeromagnetic survey data from Cameroun.

(b). Output of a simple filter operator to assess the magnetic 'relief' quantitatively.

(a). Aeromagnetic data presented as a raster map.

After Way [16]. For full explanation, see text.

Nationwide reconnaissance gravity surveying was carried out 1972 to 1973 and interpreted in 1974-5 [8]. Initial interpretation of the gravity anomalies encountered is shown in Figure 3(b).

Aeromagnetic survey of the Kalahari followed in 1975-6 under Canadian Government funding and an interpretation of the results was published in 1978 [14]. Coverage of the remainder of the country - the relatively well-exposed eastern part - was completed in 1986 under European Community funding (Figure 3(c)).

The Geological Survey of Botswana was an enthusiastic participant in the initial US Landsat programme and complete coverage of imagery was obtained by 1974. A thorough geological interpretation of these images, supported by photogeological studies and field checking [7], was also carried out.

The expectation was that the veneer of Kalahari sand concealed a largely Precambrian basement of ages varying from Archean to late Proterozoic. Continental sediments and lavas of Karoo age (late Carboniferous to mid-Jurassic) could be expected locally above the basement and below the sand. Thickness of the Kalahari sand and of the Karoo sequence was generally unknown, but was clearly a vital factor in determining the prospectivity of the older rocks. In the event, quantitative interpretation of aeromagnetic anom-

alies arising from sources within the basement and also from the late-Karoo lavas and post-Karoo intrusions revealed sand thicknesses no greater than 100 m in many areas.

One of the outstanding features of continental proportion discovered by the geophysical surveys was the so-called 'Kalahari Line' [10]. This feature extends in a north-south direction approximately along meridian 22°E for a distance of over 600 km within Botswana. The appearance of the Kalahri Line on the aeromagnetic survey is shown in Figure 3(c). Interpretation of magnetic anomalies to the east and west of the line indicate that, whereas the Precambrian basement may be expected within a few hundred metres of the ground surface to the east of the line, to the west there are are no magnetic sources (except local Karoo sills and dykes) within about 15 km of the ground surface. The implication is that a hitherto unsuspected sedimentary basin - the Nosop Basin - of dimension about 500 by 200 km exists in southwest Botswana, west of the Kalahari Line, with obvious potential for oil, gas and coal exploration. Seismic surveys and drilling have been carried out here in recent years (to 1989) [17].

Interpretation of Landsat images covering a section of the Kalahari Line is shown in Figure 4(a).Some savannah vegetation patterns, pans, sand dune and semi-arid landforms are evident, but only a few vaguely defined lineaments relate to the underlying geology. A presentation of the aeromagnetic data coverage revealing the Kalahari Line in this area is shown in Figure 4(b). Even with hindsight, no evidence of this major geological structure can be found on the satellite imagery.

In a neighbouring area - the Molopo Farms Complex - a sand-covered intrusive complex of age and structure similar to that of the Bushveld Complex has been the object of investigation for mineral exploration in recent years. Geophysical methods have been used intensively and successfully to map the hidden basement features, but the authors of a summary report state that whereas some NE-SW trending faults can be identified "...remote sensing has been shown to be otherwise ineffective, except where Kalahari Beds cover is absent" [4].

Yilgarn Block, Western Australia.

The Archean rocks of Australia's Yilgarn Block are being explored intensively, particularly for gold mineralization, in recent years. In many areas, remote sensing imagery reveals very clearly the chequer-board pattern of field boundaries and crop types to the exclusion of almost all geological information. There is little or no outcrop information in many areas but the soils are residual and their composition largely reflects that of the underlying bedrock from which they are derived. As a consequence, airborne gamma-ray spectrometer surveys carried out at low altitude (80 m) and close line- spacing (250 m) reveal clear variations in the elemental abudances of Th, U and K. When the gamma-ray spectrometer results are plotted with each of the three primary colours assigned to one of the three radio-elements, a colour image is obtained which shows a clear representation of the sub-outcropping geology.

Lineament analysis for groundwater exploration.

In semi-arid areas with crystalline basement rocks, successful exploration for groundwater supplies for village use is highly dependent on finding either water-bearing fractures in the basement, or areas of deeper weathering where water may be present in bedrock depressions within a medium of reasonable porosity.

In a pilot study for FAO, reported by Astier and Paterson [1], yields of boreholes and wells in an area of Burkino Faso, West Africa, were compared with their distance from lineaments interpreted independently on published maps of aeromagnetic intepretation. The main conclusion of the study was that, within 600 m of a mapped lineament, yields were generally much higher than elsewhere, confirming a close correlation between lineaments of all sorts (faults, fractures, dykes, etc) interpretable from aeromagnetic survey and the likelihood of sucessful water exploration in their vicinity. This concept offers the possibility of a economical groundwater exploration strategy in any area of crystalline basement where an aeromagnetic survey is available.

By contrast, in the same area, relatively few lineaments could be mapped from satellite imagery and their continuity was much less clear.

Mapping dykes and dyke swarms.

Halls [5] has emphasized the importance of mapping dyke swarms as a tool for investigating paleo-stress patterns, particularly in the world's Precambrian shield areas. An impressive swarm of dolerite dykes of mid-Jurassic age was outlined in northern Botswana as a direct result of the national aeromagnetic survey coverage [9]. In an approximately 50 km x 50 km area of exposed Precambrian basement within the dyke swarm, the occurrence of dykes as determined from three independent data sources is compared in Figure 5.

Figure 5(a) shows all dykes and lineaments recognized from the interpretation of satellite imagery [7]. In Figure 5(b), dykes interpreted from the aeromagnetic survey of 1986 are shown, while Figure 5(c) shows the results of geological mapping supported by conventional air-photograph interpretation published in 1968 [2].

The ability of the aeromagnetic survey to recognize the continuity of the dykes is clearly seen, and this ability extends beyond the central Kalahari inland drainage watershed (near the western margin of the figure) where the continuous sand cover commences. Few of the dykes are recognized as such in the satellite imagery, while the photogeological work is very precise in exposed river valleys, but lacks continuity even across the areas between river courses. Hence, only the national aeromagnetic coverage could reveal the peculiar geometrical pattern of this swarm, which was the clue to its tectonic significance [9].

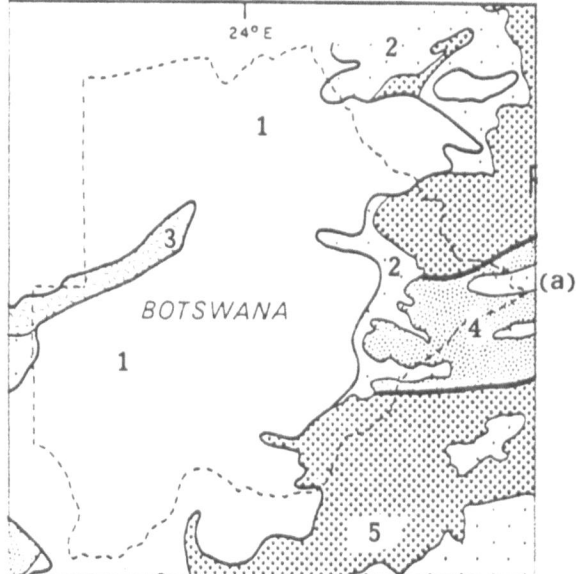

(a)

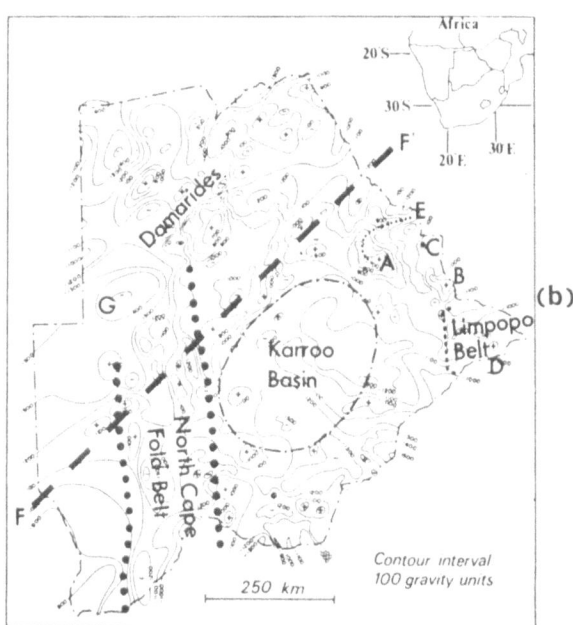

(b)

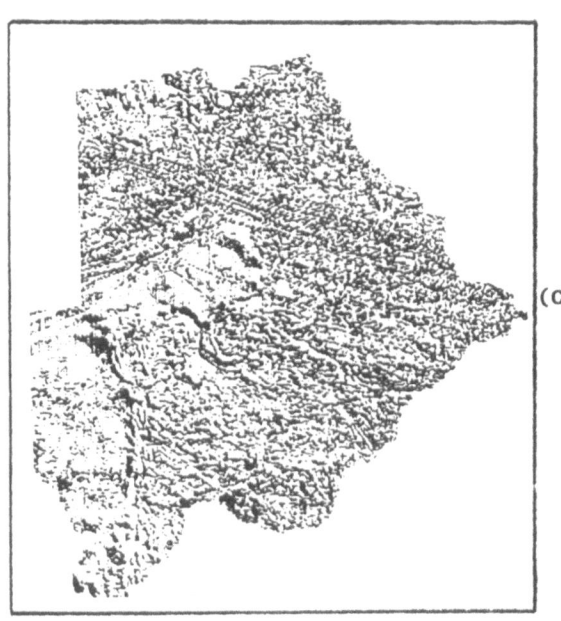

(c)

COMPARISON OF RESOLUTION AND COSTS.

While the similarities between remote sensing data and airborne geophysics have been emphasized, one important difference should be stressed. That is that whereas the intensity of reflected light can vary in a step-wise fashion - a boundary between a 'dark' area and a 'light' one need not be in any way gradational - geophysical potential fields are Laplacian - they vary smoothly and without any abrupt discontinuities and, since the sum of their second derivatives in three orthogonal directions must be zero, information gathered in two dimensions can be used to infer variations that would have been measured in the third dimension.

The wavelengths of smooth variations in potential field values are closely related to the distance between the sources of these anomalies and the 'detector' measuring them. When surveying in areas of near-surface or outcropping basement, it therefore follows that the resolution of adjacent sources contributing to the anomaly pattern is improved by reducing the source-sensor separation or flying as low as possible. On the other hand, simple sampling theory dictates that to record anomalies adequately, samples must be captured at intervals no greater than about half the source-sensor distance. Little is to be gained by decreasing this sampling interval, while increasing it leads quickly to the dangers of aliasing. Airborne survey costs, however, are primarily proportional to distance flown, so the spacing between flight lines tends to be kept as wide as possible, usually compromising the quality of the data in the interests of reduced cost. The result is that sampling *along* flight lines is usually adequate while sampling *between* flight lines leads to some aliasing. Surveys are usually designed so that flight lines are at right-angles to the geological strike so as to minimise this problem.

While the foregoing argument suggests that geophysical rasters generated from flight line data should be of a cell dimension comparable to about half the source-

- -

Figure 3.

(a). Simplified geology of Botswana. 1 = Kalahari Sand cover; 2 = Karoo cover; 3 = late Precambrian basement; 4 and 5 = Archean basement. (After [10]).

(b). Preliminary interpretation of the national gravity survey coverage of Botswana undertaken 1972-3 (from [8]).

(c). National aeromagnetic survey coverage of Botswana presented as a shaded relief image illuminated from the northeast.

sensor distance, in practice a compromise has usually been made between adequate sampling along the lines and less-than-adequate sampling across them to give a typical raster having three to five points per flight-line spacing. With airborne surveys of line-spacings 500 to 1000 metres, the resulting raster element size then falls in the range 100 to 300 m. This compares with along-line sampling at time intervals of one second, equivalent to 50 to 70 m at normal survey aircraft ground speeds, indicating undersampling of the original profile data.

In recent years, the quest for improved resolution has led to a tightening of specifications for airborne surveys in many countries (e.g. Finland, Sweden and Australia). Reducing the flying height to 80 m and the line spacing to 250 or 200 m significantly improves the resolution of anomalous sources and leads to rasters of effective cell dimension 50 to 80 m, comparable with the cell-size of early satellite imagery. These tightened survey specifications are achieved at no significant cost increase per line-kilometre; prices of US$10 to US$15 have been virtually constant for some time. But acquiring data on more closely-spaced lines of course significantly increases the cost per unit area.

Nevertheless, the advantages of increased resolution seem to outweigh the disadvantages of increased cost and users claim that they are now able to go directly from high-quality aeromagnetic surveys to ground follow-up of targets whereas in earlier years an intermediate phase of detailed airborne geophysics was necessary. In Australia particularly there appears to be a strong demand from the mining companies for surveys with these improved specifications [13], particularly now that the national reconnaissance aeromagnetic coverage of the entire continent nears completion. Clearly investments of millions of dollars on airborne survey acquisition are only made where this is seen as directly helpful to the exploration objective, and the 'hard data' acquired are worthwhile.

In this context, modern satellite imagery is often perceived as 'too expensive' in a hard-nosed commercial world. To a community weaned on early Landsat data which was virtually free to the interested user, the trend towards higher resolution and even stereo imagery appears to be accompanied by a trend towards costs-to-the-user more in line with the real costs of acquisition. These costs, in turn, are far removed from any proportionality to the quantity of data acquired; mostly they relate to the high initial launch costs which have to be recovered over the life-time of the

mission, or else subsidized by governmental investments on behalf of national prestige or security. In this setting there is no certain way of telling what is true cost, but there is no doubt that the mineral exploration market will only pay an amount equal to what it considers the value of the data to any given project.

From an educational standpoint, the cash barriers are effectively much higher and the value of imagery cannot be demonstrated to students when the images cannot be afforded, as is often the case for a specific project area.

DATA INTEGRATION IN INTERPRETATION

While it is easy to argue the relative merits of various types of data set, scientifically it must be argued that the only valid interpretation is one which takes into account *all* the available data for a given area, with due regard to the quality of the data and the completeness of the data sets. To produce an interpretation is seen as equivalent to producing a scientific theory in which a large number of seemingly disparate observations are explained by one relatively simple concept with a minimum number of arbitrary assumptions. In pure science, such a theory is only valuable if it makes predictions about the results of new experiments. These experiments may then be carried out to test the validity of the theory.

Resources exploration can be seen in this same perspective. All available data, such as satellite imagery, outcrop observations, geophysical and geochemical surveys, are interpreted to produce a conceptual model of the unseen subsurface which is consistent with all the available data. In the ideal case it may predict the presence of an ore body which gives rise to some of the surface geophysical, geochemical and other anomalies. This model may then be tested in the next phase of the exploration programme, for example by drilling a hole.

An essential pre-requisite for this type of combined interpretation is good communication between workers in different disciplines of the earth sciences. It is not useful for the geophysicist to produce mathematically accurate models which are clearly at variance to the known geology, nor it is helpful for geologists to ignore geophysical indications that the subsurface is different from what is seen in outcrop. Individual data sets may be interpreted independently *in the first instance*, but the interpretation is certainly not complete

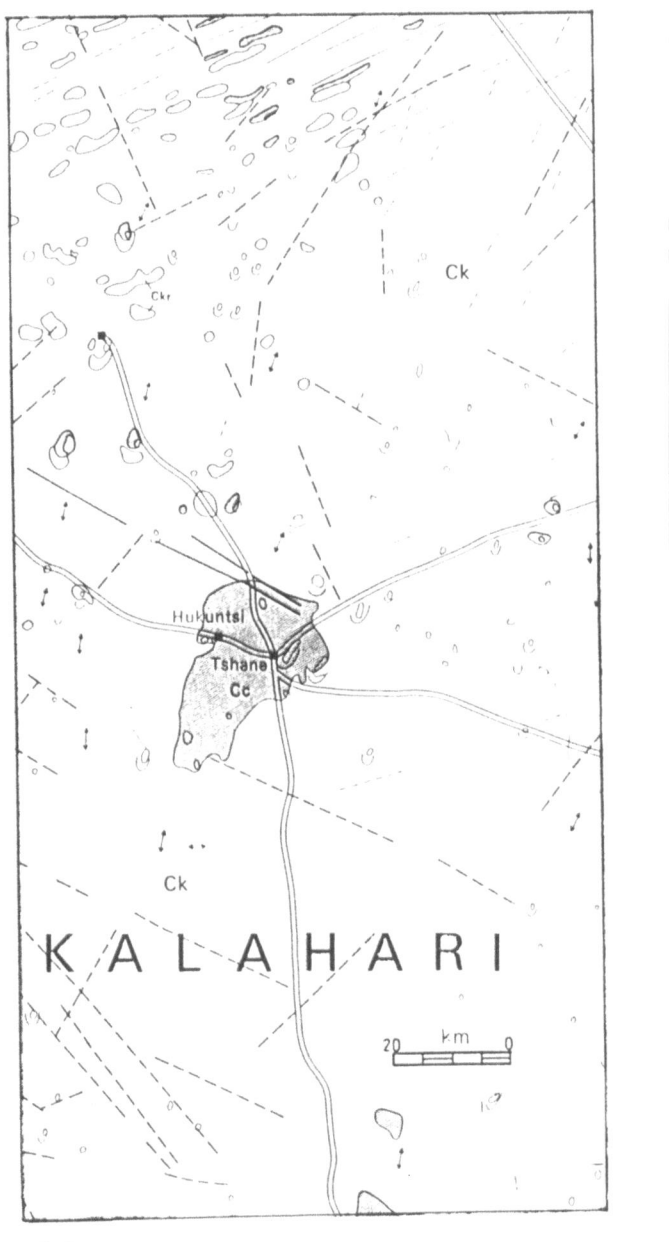

(a)

(b)

Figure 4. Two aspects of the Kalahari Line, southwest Botswana:

(a). Geological interpretation of Landsat imagery, supported by aerial photograph interpretation and field checking (from [7]).

until the various interpretations have been reconciled and combined and special attention brought to bear on those areas where they are at variance. The variance indicates incomplete understanding which it may be very profitable to resolve. Not only may this reveal new, unforeseen elements of the geology, but it may also reveal misconceptions in one or more of the interpretation methods which could then be eliminated more generally.

The integration of many data sets is, however, not so easy. Paper maps are very often at various scales and projections and not on transparent media that allow them to be

(b). Aeromagnetic survey data presented as a shaded relief image illuminated from the northwest after migration to the magnetic pole.

overlain with convenience. While good drafting and repro-duction facilities can minimize these difficulties, it also signals a potentially fruitful application for Geographic Information Systems (GIS) which have emerged as powerful tools from the digital computing world in recent years.

The backbone of GIS is a relational data-base where all incoming map information - either in raster or vector (line) format - is archived with reference to one geographic coordinate system. This enables the execution of two essential operations: One is the ability to overlay two or more selected data planes with ease and to view the results on the

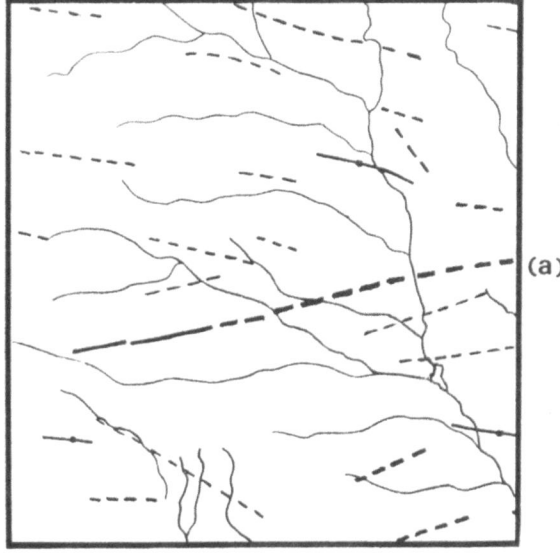

(a)

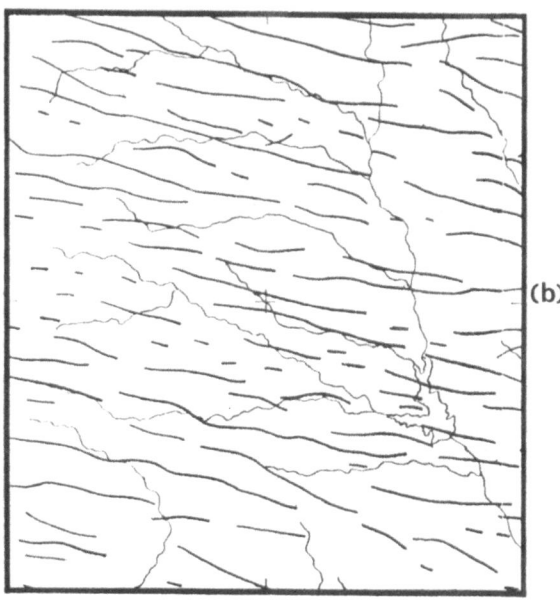

(b)

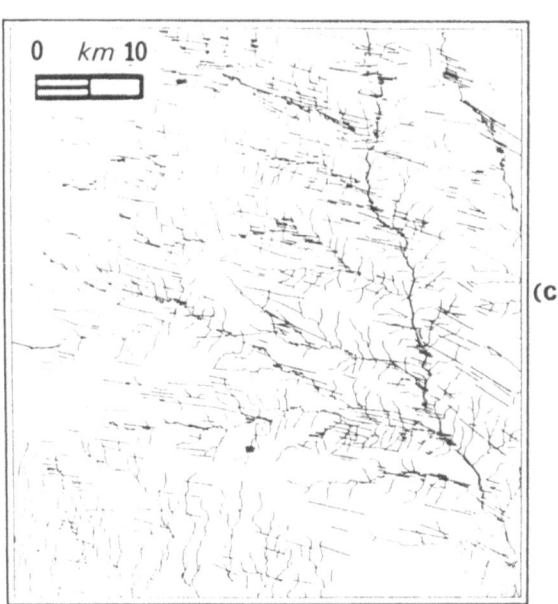

0   km 10

(c)

computer screen or in hard copy. The second is the ability to ask questions which require a systematic search of the data planes and perhaps to seek a relationship between many sets of information.

We demonstrate the advantages of the interpretation of combined data sets with two examples which are presented in full in the special issue of the ITC Journal prepared for the present meeting [11].

Letlhakeng

In a groundwater exploration campaign in Botswana, LANDSAT TM and aeromagnetic data were used to trace lineaments as potential targets for fractured rock aquifers. The prevalent cover of Kalahari sand hampers considerably the analysis of lineaments from satellite imagery. Often the recent drainage systems follow tracks of linear features and in those cases they can be marked, but mostly for short distances only. Since sedimentary cover does not contain any magnetic sources, faults or fracture zones intersecting the magnetic basement are commonly displayed indirectly in magnetic surveys [1]. They can manifest themselves for example as:

- sharp gradients forming a linear boundary between areas of different magnetic level, relief or texture;

- disruptions and/or deflections of magnetic trends. These are commonly wrench faults or shears, often with distinguishable lateral movement.

The overlay of the two data sets by the use of a geographic information system helps to define linear features more accurately than in one data set alone. The combined presentation is achieved by using a structural processing of TM channel 4 (shaded relief) as the intensity component, along with the hue image derived from magnetic amplitudes ranging from 400 nT (violet) to 750 nT (red). In this way, both data planes can be inspected separately, but simultaneously. A logical relation between brightness (TM)

Figure 5. Three expressions of a dolerite dyke swarm in the catchment of the Shashe River, eastern Botswana.

(a). Geological intepretation of Landsat imagery, supported by aerial photograph interpretation and field checking (from [7]).

(b). Interpretation of dykes from aeromagnetic survey of the same area (present authors).

(c). Photogeological and field mapping of dykes by Crockett [2].

Note the progressive filling of the reservoir behind the Shashe dam ((c) = 1968, (a) = 1976, (b) = 1986.

86

and magnetic relief is absent, but both sets of data contain information about lineaments.

Mt Isa area, Queensland.

The physical properties of geophysical potential fields admit the possibility that, if a simple model of the ground is assumed such that the geology is composed of rectangular vertical prisms of uniform physical property (e.g. density) and dimension equal to the geophysical raster, then the potential field survey data may be inverted to produce a value for each prism of the variable physical property which may be presented as (e.g.) a pseudo-density map. Where a second set of survey observations is available, such as a magnetic survey, then a pseudo- susceptibility map may be prepared similarly.

The power of extending this type of rock property mapping to more than one rock property was demonstrated by Everaerts [3, 11]. He took both magnetic susceptibility and pseudo-density maps of a Precambrian basement area in Queensland, Australia and combined them using different colour scales for each property. It can be shown that Precambrian rocks tend to group into classes which may be termed ferrimagnetic and paramagnetic. The latter are so weak in their magnetism that they may be considered effectively non-magnetic for survey purposes. The relatively non-magnetic areas of the susceptibility map were left white, while increasingly dark shades of magenta were used for increasingly magnetic units. Somewhat similarly, areas showing average rock density were left white, while less-than-average-density areas were shaded increasingly cyan and above-average-density areas were shaded increasingly yellow. Combining the two maps now results in the combination of the two colour sets such that, for example, areas of non-magnetic rocks of average density remain white, while magnetic rocks of low density are purple (magenta + cyan) and magnetic rocks of high density are orange (yellow + magenta).

CONCLUDING REMARKS

The case for considering regional geophysical surveys, particularly airborne surveys, as a type of remote sensing has been stated, and some of the advantages of these data over satellite imagery (which is more conventionally considered as 'remote sensing') have been demonstrated for a numbe · of

special cases. In other cases, no doubt, advantages can be shown for satellite imagery; for example, where two rock types of similar physical property come into contact, we would not expect a geophysical anomaly, but there may well be a visual expression of the contact.

We would not wish to suggest that either data type should be used to the exclusion of the other, however. From a scientific standpoint, any valid model we may build of the geology must be consistent with *all* data presently to hand. We have emphasized the common ground between presentation techniques for geophysical and remote sensing data afforded by modern computer technology which may make this integration more appealing.

In practice, many data sets may be available, but most of them will be incomplete: only certain areas of a concession will be covered, or the quality of the data may vary from one survey or image to another. Implementation of geographic information systems offers new possibilities for convenient display and combined interpretation of many such data sets which have yet to be fully exploited.

In the medium term, we anticipate that the result of this approach will be an increased enthusiasm for acquiring airborne geophysical data to complete survey coverage of all land areas of the world at a reconnaissance scale, and to acquire data of higher quality over selected provinces having significant mineral potential.

REFERENCES

[1].   Astier, J.L., and Paterson, N.R., 1989. 'Hydrogeological interest of aeromagnetic maps in crystalline and metamorphic areas'. Paper 59 in Proceedings of Exploration '87: Third Decennial International Conference on Geophysical and Geochemical Exploration for Minerals and Groundwater, edited by G.D.Garland, Ontario Geological Survey, Special Volume 3, 960 pp.

[2].   Crockett, R.N., 1968: Quarter degree sheet 2127A, Shashi. Geological Map, scale 1:125 000, Geological Survey of Botswana.

[3]. Everaerts, M., 1989. 'Interpretation of geophysical data from the Mt Isa area, N.W. Queensland, using a new form of rock-property map', poster presentation at the Geological Society/Royal Astronomical Society Joint Association for Geophysics meeting on 'Image analysis and computer graphics in the display and interpretation of geophysical data' at the British Geological Survey, Nottingham, September 27-28.

[4]. Gould, D., Rathbone, P.A. and Kimbell, G.S., 1987: 'The geology of the Molopo Farms complex, southern Botswana'. Bulletin 23, Geological Survey of Botswana.

[5]. Halls, H.C., 1982: The importance and potential of mafic dyke swarms in studies of geodynamic processes. Geoscience Canada, Volume 9 number 3, p.145-154.

[6]. MacLeod, I.N., 1989: 'Computing in Canadian Exploration'. Mining Magazine, November 1989, pp 460-463.

[7]. Mallick, D.I.J., Habgood, F., and Skinner, A.C., 1981: A geological interpretation of Landsat imagery and air photography of Botswana. Overseas Geology and Mineral Resources, No.56, 35 pp.

[8]. Reeves, C.V., and Hutchins, D.G., 1976: 'Crustal structures in central southern Africa', Nature, London, vol 254, pp 408-410.

[9]. Reeves, C.V., 1978: 'A failed Gondwana spreading axis in southern Africa', Nature, London, vol 273, pp 222-223.

[10]. Reeves, C.V., 1985: 'The Kalahari Desert, central southern Africa - a case history of regional gravity and magnetic exploration' in W.J.Hinze (ed), The utility of gravity and magnetic surveys, Society of Exploration Geophysicists special volume, pp 144-156.

[11]. Reeves, C.V., Zeil, P.W., and Zhou Yunxuan, 1990: 'Interpretation of airborne geophysical surveys: some applications of image processing and geographic information systems in systematic exploration strategy', ITC Journal 1990-2 (in the press).

[12]. Reford, M.S., 1980: The magnetic method. Geophysics, Volume 45, pp 1640-1658.

[13]. Smith, R.J., and Pridmore, D.F., 1989: 'Exploration in weathered terrains - 1989 perspective'. Exploration Geophysics, Volume 20, pp 411-434.

[14]. Terra Surveys, 1978: 'Reconnaissance Aeromagnetic Survey of Botswana, 1975-7, Final Interpretation Report', Special Publication of the Botswana Geological Survey and the Canadian International Development Agency, 199 pp + appendices.

[15]. Watson, K., 1985: 'Remote sensing: A geophysical perspective'. Geophysics Volume 50, pp 2595-2610.

[16]. Way, Khin Maung, 1988. Digital image processing of airborne magnetic data for regional geological survey, MSc thesis abstract, ITC Journal 1988-2, p 207.

[17]. Wright, J.A., and Hall, J., 1990: 'Deep seismic profiling in the Nosop Basin, Botswana: cratons, mobile belts and sedimentary basins'. Tectonophysics (in press).

[18]. Wu Chaojun, 1989. Unpublished MSc thesis, International Institute for Aerospace Surveys and Earth Sciences, Delft, The Netherlands.

* * *

Interpretation of satellite image data integrated within a Geographic Information System Toolkit (GIST) for regional geological mapping and mineral exploration in northern Xinjiang, China

Y. Shao B.Sc., M.Sc.
K. Xiao B.Sc., M.Sc.
H. Guo B.Sc., M.Sc.
Institute of Remote Sensing Applications, Academia Sinica, Beijing, China
J. Davie B.Sc., M.I.C.E., C.Eng.
T. E. Beaumont B.Sc., M.Sc., M.Phil., Ph.D., C.Eng., M.I.H.T., M.I.M.M., F.R.S.Soc.
Scott Wilson Kirkpatrick & Partners, Basingstoke, Hampshire, England
P. S. Griffiths B.Sc., Ph.D., F.G.S., M.I.M.M., C.Eng.
Griffiths Remote Sensing, London, England

SYNOPSIS

As a response to the practical needs of geologists and other specialists concerned with resource development, a Geographic Information System Toolkit (GIST) has recently been designed which, by combining functions of digital image processing, image interpretation and a GIS, enables remotely sensed data to be fully integrated with other sources of survey mapping and associated information which may be held in a relational database. The facility has proved viable in both supporting and integrating all stages of a project requiring both quantitative and qualitative analysis of remotely sensed imagery and other mappable information. It can be used for the manipulation, registration and comparison of multiple data sets, for the analysis of imagery, including a capacity to undertake sophisticated visual interpretation on-screen, and for the editing and production of derivative GIS, topographical and thematic maps. This paper examines the development and use of GIST in a technical cooperation programme between scientists and engineers from the People's Republic of China and the United Kingdom for regional geological mapping, gold mineral exploration and resource evaluation in a study area centred on the Altai Mountains piedmont of northern Xinjiang, China. The paper concludes with a review of the general value of GIST-type systems as a means of combining well-established and new techniques of applied environmental and geological remote sensing.

INTRODUCTION

Purpose and Organisation of the Paper

This is an account of some achievements of a programme of technical cooperation between the State Science and Technology Commission of China with the Overseas Development Administration of the United Kingdom. This programme provided for geographic information system development and satellite geological remote sensing by a team from Scott Wilson Kirkpatrick & Partners (UK) and the Institute of Remote Sensing Applications (Beijing). The work was based on study of a Test Area in Xinjiang Uygur Autonomous Region (Figure 1), northwestern China.

In essence, GIST (Geographic Information System Toolkit) is a geographic information system combined with a digital image processing system designed to handle remotely sensed data, other forms of map data and associated non-spatial data. The design process is now well advanced and all main functions are operational. GIST has many potential uses in natural resource and

environmental investigations, and is particularly well suited for use in projects involving geological remote sensing.

This paper describes the use of GIST for geological remote sensing with reference to system development by a team comprising specialists in both geological remote sensing and computing science. Part of the point of this paper is to show that such different specialists must work together to make the best use of new developments in technology. Geological results of the work in Xinjiang are described only where needed to support present purposes. A preliminary account of the geology has already been written (Griffiths et al, in press), and a final account is in preparation.

The Xinjiang project is summarised in the remainder of this introductory section. In the second section, the GIST system is described, from its underlying design philosophy through to details of system design. The following two sections describe different aspects of the geological work carried out with GIST, using different parts of the whole study region to make particular points. Most attention is given to the Fuyun area, which occupies the central part of the Test Area (Figure 1), and which serves to discuss the use of GIST for integrating point and line information obtained from a variety of data sources. More briefly, a small area around the Ulungarr River, in the southeastern part of the Test Area, and a rapid study of a much larger area of northern Xinjiang both serve to illustrate the use of GIST for the computationally more difficult task of mapping a combination of lines and areas (polygons). The final section is a brief discussion of the general ability of systems based on the design principles of GIST for geological mapping and resource exploration.

The Xinjiang Project

Geological work in a Test Area in northern Xinjiang (Figure 1) was based mainly on digital processing and visual interpretation of satellite imagery (Landsat MSS and TM, SPOT HRV and NOAA AVHRR data), supported by a brief field campaign in 1988, and use of available geological, topographical and geophysical maps and literature. For the most part, the work was treated as an exercise in reconnaissance geological mapping and mineral exploration, building on principles discussed by Griffiths et al (1987). The importance of using fieldwork to support such an exercise has also been referred to recently by Leith and Alvarez (1985). The main results of the geological investigation are summarised below.

The geological information content of remotely sensed imagery is partly expressed by spatial variations in physiographical (terrain) conditions. Accordingly, the results of this study are best summarised within a physiographical framework. Relevant works in English include Coleman (1989), Feng et al (1989), Taner et al (1988), Tapponier & Molnar (1979), Yang et al (1986) and Zhao (1986).

The Test Area, which covers about 15,000km², is divisible into three main physiographical zones within an altitudinal range of 600m to 2400m above sea level. These zones are shown on Figure 1.

Zone I   The Altai Mountains
Zone II   The Altai Piedmont
Zone III The Jungar Depression

An account of the terrain conditions and the geology of each zones follows.

Within the Test Area, the Altai Mountains (Zone I) form a deeply dissected, high-relief, rather densely vegetated alpine region. Available maps and literature indicate that this area comprises Cambrian to Devonian gneisses, schists, amphibolites and related metamorphic rocks, intruded by extensive granitic to granodioritic intrusions of Carboniferous and Triassic age. The intrusions are mainly associated with late Palaeozoic orogenesis and are often strongly deformed, passing locally into granitic gneisses. Terrain conditions make it difficult to recognise individual rock units on the satellite imagery. However, aeromagnetic contour maps can be used together with the imagery to map some features of rock type variation.

Most of the Test Area falls within Zone II, the Altai piedmont, which represents the best local conditions for geological remote sensing. The zone chiefly comprises a dissected plateau marked by steppe to semi-arid conditions. The geology is similar to that of the Altai Mountains, but (a) stratified rocks are rather younger, being Devonian and Carboniferous, (b) intrusive bodies are less extensive, and (c) weakly metamorphosed rocks are relatively more common. The latter comprises psammitic to pelitic metasediments and andesitic to basaltic metavolcanites. Although such rock units are already well defined on existing geological survey maps, study of satellite imagery does suggest where some revisions could be introduced and, more significantly, indicates that many of the rock units correspond with the surface expression of individual thrust sheets.

Combined study of the imagery and other available information also suggests that the associated thrusts are one important control on the distribution of metalliferous mineralization, including gold, copper and iron (haematite).

The Jungar Depression (Zone III) is a semi-arid to arid region mostly occupied by continental sedimentary deposits of Tertiary to Recent age. Within and close to the Test Area, this intermontane basin has no obvious great resource potential. However, study of geomorphological and structural relationships between this basin and adjacent areas, including the Altai piedmont, may be a useful method of reconnaissance for new sources of petroleum and groundwater within deeper parts of the basin.

GIST APPLICATIONS DEVELOPMENT

General Principles

GIST was originally conceived to overcome the many problems faced by land resource scientists, engineers, planners and project managers unable to use existing facilities to fully integrate and analyse survey and associated information acquired from different sources and formats and especially that information becoming increasingly available from improved resolution Earth observation remote sensing satellite imagery. In meeting these practical requirements, a system has been developed consisting of a software environment linking image processing and interpretation, a relational database and digital mapping operations to provide a GIS Toolkit capable of storing non-spatial information and working simultaneously with raster, vector, attribute and terrain model data. Adaptation of a modular design concept also allows for the straightforward addition of new applications modules as appropriate.

Previous experience has shown that the key to successful implementation of software projects is insistence on development being driven by applications. (Beaumont et al 1988). Furthermore, in establishing a lifetime long enough to justify development costs, advances must be predicted in the computing field and incorporated into the design so full advantage may be taken of significant innovations. These fundamental software design considerations involved initially the preparation by specialists of applications reports detailing the way in which remote sensing data is currently used and secondly how a GIS facility could improve or indeed revolutionise such activities. Subsequently, advantage has been taken of technological advances by developing and adapting new parallel processing transputer facilities to completely emulate mini-computer and specialist frame store hardware functionality in a low cost personal computer environment. This has resulted in the GIST software being made available in a low cost and easily maintained configuration

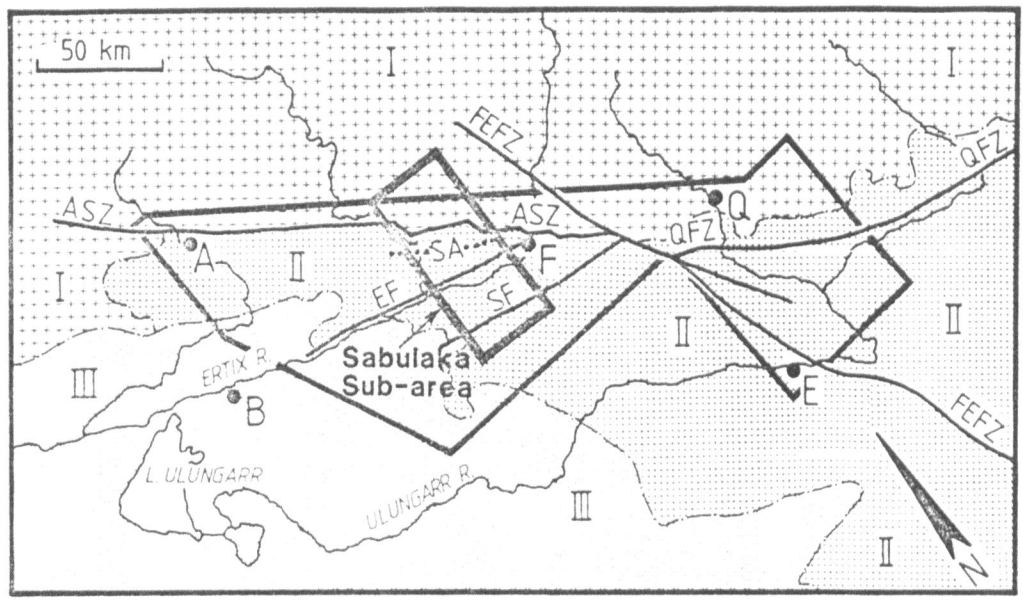

boundary of Test Area ⌒∿ rivers and lakes

boundaries of physiographical zones

● towns (A-Altai, B-Beitwun, E-Ertai, F-Fuyun, Q-Qinghe)

selected structures (ASZ-Abagum shear zone, EF-Ertix fault, FEFZ-Fuyun-
  Ertai fault zone, QFZ-Qinghe fault zone, SA-Suburt antiform, SF-Selego fault)

PHYSIOGRAPHICAL ZONES

I Altai Mountains   II Altai-Ertai Piedmont   III Jungarr Depression

Figure 1. Physiographical setting of the Test Area

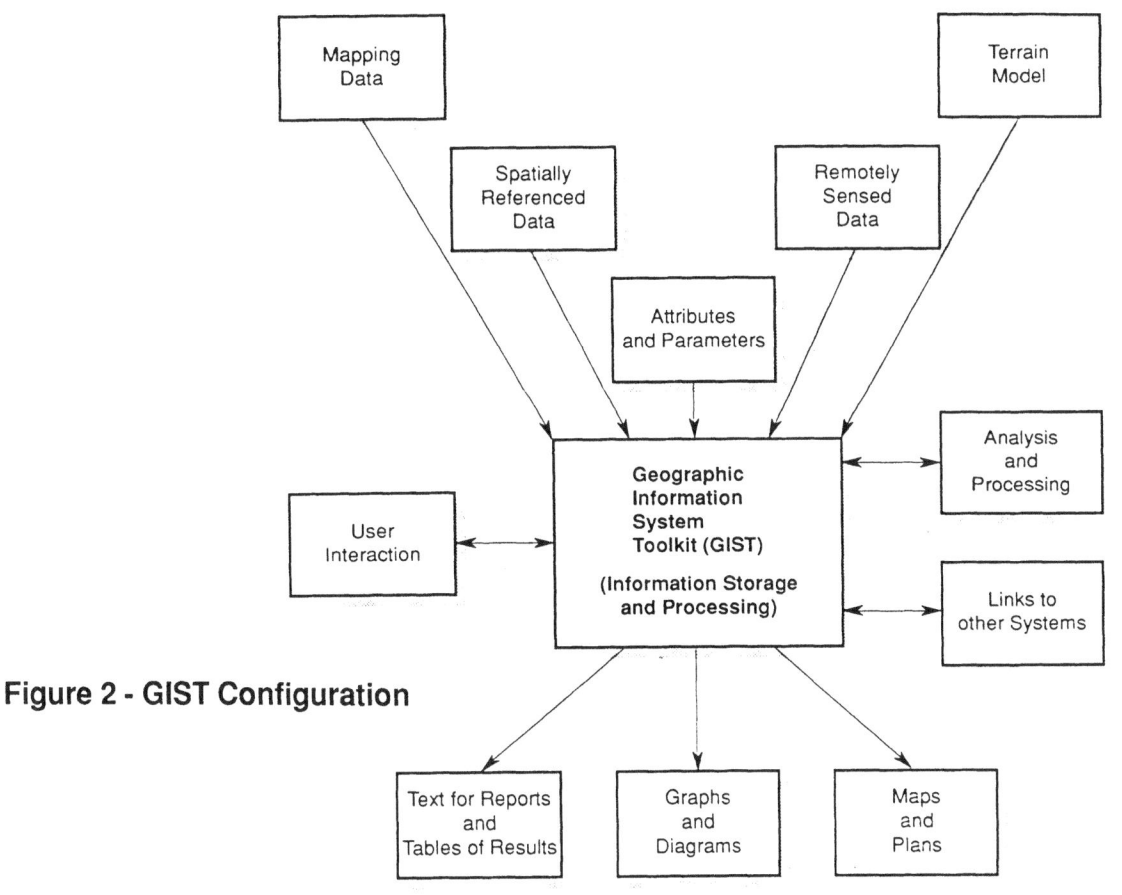

Figure 2 - GIST Configuration

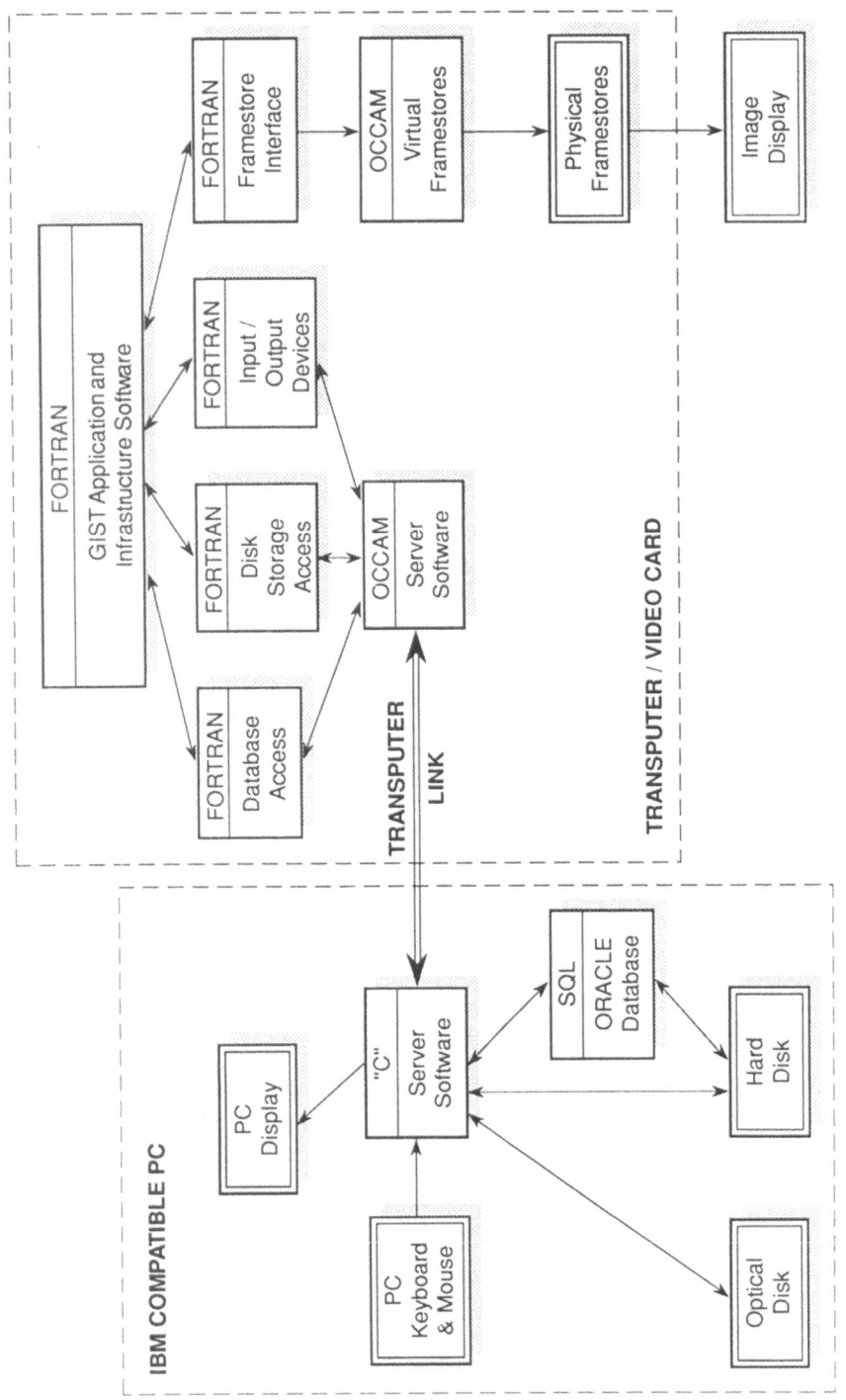

Figure 3 - PC / Transputer GIST System Software and Devices Block Diagram

which has had a significant impact for natural resource and environmental monitoring applications in developing countries.

System Configuration

Analysis of the applications identified the need to interpret features on the image directly from the screen monitor, using a cursor and digitising tablet, which would be automatically entered into the GIS database for future display, analysis and integration with other sources of information. Other priority requirements developed into the configuration, as shown in Figure 2, may be briefly summarised as follows:

Data Input - ability to accept input from a variety of data sources available in or transferable to a digital format including typically, satellite imagery, digitised aerial photography and video recordings, thermal infra-red and airborne scanner imagery, digital terrain and elevation models, radar imagery, digital maps and survey data, manually digitised and other non-spatial data such as geological borehole records and time series information from meteorological or gauging stations and the like.

Raster and Vector Data - ability to work concurrently with raster and vector data including simultaneous display either in plan view or, if a digital terrain model is available, in true perspective which provides a useful aid to visualisation.

Coordinate Transformations - In view of the number of potential information sources and the desire to maintain accuracy, all spatial data is stored in its original map projection. This requires the facility to transform the data into any common and desired projection for display or output. The technique chosen for data transformation is one which is more usually found in finite element analysis work. Each transformation is decomposed into a set of quadratic Lagrangian iso-parametric shape functions which approximate the transformation over a rectangular area. The areas are formed into a linear quadtree so that the accuracy of the transformation can be increased to a value appropriate to either the display device or the pen plotter. The conversions are made in real time.

Information Storage - The unique philosophy behind the database storage development has been to separate the graphical representation of a feature from the details of the characteristics of the feature stored in the relational database. This enables a single line to be both a geological fault in its own right and also to be part of say the boundary of a geological unit. The geographical representation of the line is stored only once, but with multiple attributes, ie all information is stored uniquely. This has the advantage that should the fault be moved within the course of final image interpretations and mapping, only the single graphical representation need be modified.

Communication - ability to communicate with other software packages for special analysis involving not just a transfer of data via disk files but an integration within GIST.

Data output - ability for data within GIST, or the results from analysis of that data, to be output in the form of high quality reports, tables, graphs or maps. Availability of a wide range of line types and stipples for ornamentation enables near cartographic quality of mapping to be achieved from standard plotters.

The software design has been implemented in modular form so that each module could be written in the standard machine independent language most appropriate to its function. Modular design also allows software which is machine dependent to be isolated, making it possible to "port" the software onto different machines. Furthermore, control of the software has been amalgamated into a single module to permit the introduction of artificial intelligence at a later data. Artificial intelligence techniques have the potential to simplify the user interface of the software thereby enabling the applications specialist to manipulate the system without extensive computing knowledge. GIST is an acronymn for Geographic Information System Toolkit, a name reflecting the underlying approach which has been to supply the user with a set of tools to solve their particular problem. Such an approach moreover, is further assisted by the system talking to each specialist in their own "language".

The majority of the software which has been developed is independent of any particular application, so that although the system has been designed to solve specific problems, a comprehensive GIS has none-the-less been produced. This means that the effort required to incorporate additional applications should be minimal and likely to decrease with each new application.

Data management

In accepting data from any source, various data-sets have to be registered, including those derived from interpretations of different imagery covering the same area, but which are in different projections. GIST stores all input data in its original projection (ie coordinate system) to preserve accuracy with each data-set being transformed into a more desirable common output projection by first relating the data to standard latitude/longitude and then to the desired output projection. This is advantageous in that fewer transformations need setting up. Using the GIST method, only one new transformation is required. GIST also has a number of standard projections incorporated in addition to a facility to build up transformations utilising ground control points to provide much greater flexibility and cope with all non-standard input projections. The system derives the transformation formulae and once all the required information has been extracted from the imagery, the image may then be corrected and original data transformed for presentation.

The ORACLE relational database management system is used at the core of GIST to provide comprehensive flexibility for storage of non-spatial information. Using this facility to create tables without specialist knowledge of the software itself, enables data to be easily stored from various sources. In the Xinjiang Province study the following information tables were constructed:

Table No.
1   Station Detail : including position (in lat/long) and corrected altitude.

2   Field Data : including, inter alia, traverse date, geologist, field image, vehicle odometer reading, altimeter reading, other measurements, lithology, rock units and comments.

3   Traverse Data

4   Dips : including dip angle and direction and plunge angle and direction.

5   Metal Minerals : including element name, tenor (ppm) assessment and locations.

6   Soil Geochemistry : including location of samples and chemical element contents.

Data Analysis

Data collected by the field geologists was manually entered into the system, the unlimited text facility enabling full descriptions to be stored, in addition to available contoured aeromagnetic survey mapping which was input by manual digitising. Once entered into the relational database, linkages may be made across the data-sets allowing the user to undertake topological analyses including simple enquiries, complex spatial queries, and to use the stored information for evaluation purposes ranging for example, from traverse lengths, difference in traverse station elevations to producing three dimensional models of the aeromagnetic contour and associated information.

Visual interpretation and annotation of imagery is easily accomplished using the full image processing facilities and the screen cursor to express points, lines, networks or polygons and through graphical representation for the defined class of interpreted feature chosen from a series of user defined colours and symbols. Interpretation is assisted by simultaneous reference to available vector mapping information, algorithm functions to check for closure of polygons and full editing facilities to enable modification of the graphical representations. All on-screen interpretation is automatically entered into the relational database as are image classification boundaries when use is made of the auto-vectorise facility. The advantage of being able to use text information in database tables to support visual interpretation is illustrated in Tables 1 - 3 for the Xinjiang project where three fieldwork tables were constructed for each observation station. In addition to storing relevant information, appropriate calculations and statistical analyses were able to be performed as for example in the

distribution, orientation and length of lineaments interpreted from imagery and other data-sets. Similarly, stereonets, rose diagrams and histograms were produced for the orientation and distribution of dips and plunges measured at the fieldwork observation stations.

Map Production

Comprehensive post processing mapping software incorporated into the system allows near cartographic quality output to be obtained from the interpretation graphics and other digitised data. Currently displayed classes are incorporated into a file containing the graphical data from a user defined area of interest. Since this information is no longer related back to the database retaining the original data structure, modifications are possible to allow for "artistic licence" and for the production of maps of more conventional appearance. Full editing features are provided to enhance the map, examples for the Xinjiang project being illustrated in Figures 4-6 and 9-10. Furthermore, the system also contains a full drawing register facility enabling, if required, several output files to be combined into one map.

Transputer Applications Development

GIST is designed for field-use in developing countries. An IBM PC or compatible can now be installed and maintained almost anywhere in the world, and in most cases repair facilities are available locally.

The relational database central to the operation of GIST is likely to make heavy use of the PC processor, so that an independent processor was required for running the remainder of the software. A transputer processor expansion card is available and the GIST software, written in standard Fortran 77, can be recompiled to run on a transputer.

Transputer GIST uses a screen display of 1280 x 1024 pixels with 256 colours per pixel. When displaying satellite imagery, a block of four pixels of slightly different colours can be 'dithered' together to produce the visual effect of a single larger pixel with one of a range of many thousands of colours. Timing tests indicated that this dithering effect could be created using transputer software in times comparable with the software zoom times of the MicroVAX GIST system. The results of this hardware choice is that the video hardware is simpler, and the same workstation can be used far more effectively to work with vector linework.

Studies of current hardware developments by Scott Wilson Kirkpatrick revealed that a transputer and its associated memory can be installed on a single PC AT expansion card, and that there was also sufficient space on the same circuit board to install all of the video memory for the display.

With the transputer and video expansion card, all of the processing power required for both GIST data analysis and the generation of screen displays are supplied by a single transputer.

The power of the transputer processor should enable all functions to be carried out at speeds comparable with those of a MicroVAX system. If greater performance is needed in the future, the modular nature of the software will mean that specific elements of the software, for instance the generation of 3D perspective views, could later be transferred to an additional multi-transputer expansion board.

The problems associated with mass storage and data transfer requirements for hundreds of megabytes of data were overcome by adapting the latest optical disk technology initially utilising the IBM 'write once read many times' (WORM) optical disk drive into the system and, subsequently, a further advance in this technology which allows both multiple reads and writes.

The technologically advanced PC transputer system developed to operate the full functions of the GIST software which is now being installed for operational use in China is summarised in Figure 3.

STUDY OF THE FUYUN AREA

Introduction

The Test Area is covered by three 1:200,000 scale Chinese survey map sheets (Altai, Fuyun, Ertai). The Fuyun sheet area (Figures 4 to 8) is used to discuss use of GIST for integrating point and line information obtained from several data sources. This area merits particular attention because of the variety of data available for use and because it includes a major copper (Cu-Ni-Co-Au) deposit at Halatongke and a gold prospect near Sabulaka. The following subsections relate stages of the investigation to the use of GIST.

Data Availability

The following satellite imagery covering the Fuyun area was used:

Landsat 3 MSS scene   153/027:20JUN77
SPOT HRV multispectral scene   219/255:22MAY87
SPOT HRV panchromatic scene   219/255:06JUL88

Reconnaissance fieldwork was based on preliminary interpretation of EOSAT Landsat MSS image hardcopy. Fieldwork tested this interpretation and provided complementary lithological and structural data which cannot be obtained from the imagery. The field data, including measurements of dips of foliation and stratification, were entered into the ORACLE Database within GIST. The Landsat MSS and SPOT HRV imagery were also obtained in digital form for processing within the GEMS 35 image processing facility and for detailed interpretation on-screen within GIST.

Chinese surveys of the Fuyun area have produced geological, topographical and aeromagnetic contour maps at 1:200,000 scale. The standard projection used for systematic mapping at this scale is the Transverse Mercator (Gauss Conformal). These maps were digitised and incorporated within GIST.

Geochemical and (minor) mineral occurrence data obtained by Chinese surveys were available as locations specified in latitude and longitude. These data were entered into the GIST system.

Image Processing

In the present context, the most important aspects of image processing are rectification and enhancement.

An important practical point is that, within GIST, it is not necessary to rectify an image to the required output projection in advance of its use for interpretation. In this case, the images were interpreted unrectified within GIST, while interpretation work performed on-screen within GIST was subsequently rectified to fit the Transverse Mercator projection.

Image enhancement procedures provide images which are as suitable as possible for visual interpretation as both the image data and terrain conditions allow. The menu-driven functions of the GEMS 35 were used to produce standard false colour composites of two subscenes of the Landsat MSS image, both processed by application of linear contrast sketches and an edge enhancement to each band used (4, 5 and 7). A similar procedure was adopted from the single band of the SPOT HRV panchromatic image.

Image Interpretation

As part of the exercise in GIST development, image interpretation work in the Fuyun area concentrated on the use of GIST's line plotting functions for interpretation of linear geological structures, including a variety of fracture types, foliation and stratification traces, and boundaries between geological units. The main results of the work are shown on Figure 4, where they are plotted together with (a) the drainage network derived from the digitized topographical map and (b) the locations of minor mineral occurrences and geochemical anomalies stored within GIST.

Data Integration and Structural Analysis

Figures 5 and 6 illustrate two other possible combinations of data stored within GIST. Figure 5 combines structural features from the Chinese survey geological map with the contoured aeromagnetic map. Figure 6 adds contours and major drainage channels from the topographical map to the locations of geological fieldwork stations.

These presentations of combined data are aids to understanding the study areas's geology and resources in the context of all available data. In the present case, for example, the relationship between geological structure and metalliferous mineralization is worth investigating.

All the maps show that the structure of the Fuyun area is characterized by a complicated fault system. Even the topographical contours (Figure 5) reveal major breaks of slope which correspond with major faults parallel to the Altai Mountains front (see also Figure 1). Field observations

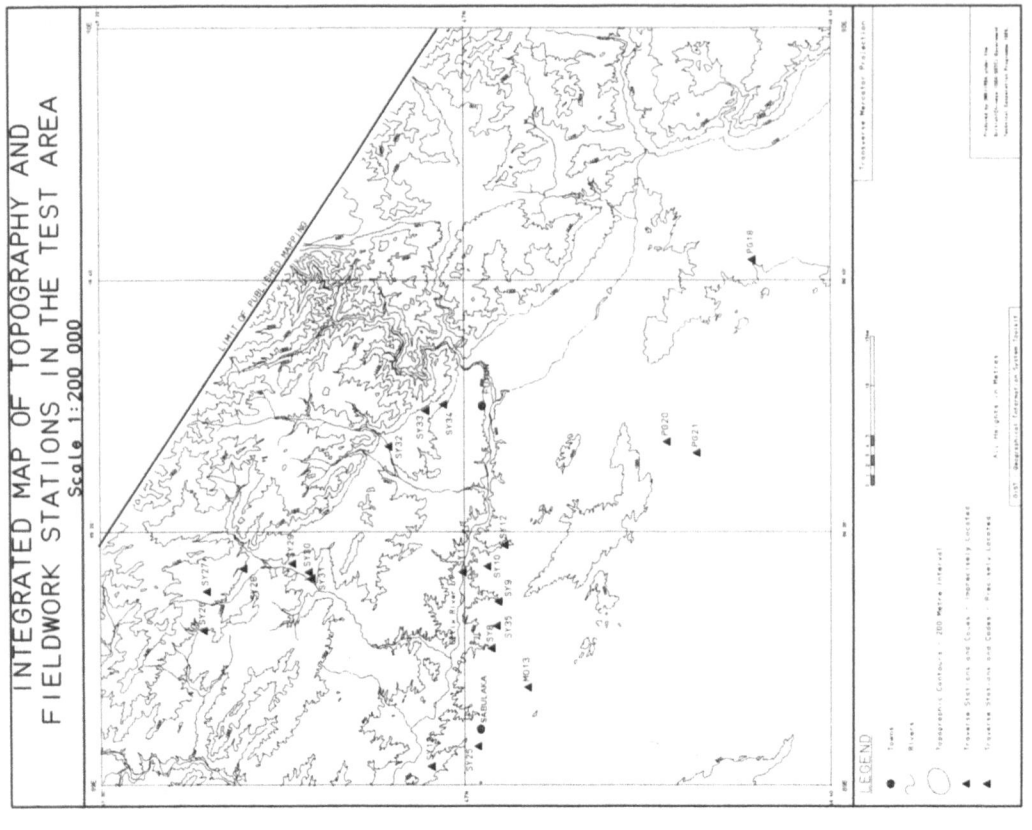

INTEGRATED MAP OF TOPOGRAPHY AND
FIELDWORK STATIONS IN THE TEST AREA
Scale 1:200 000

Figure 5

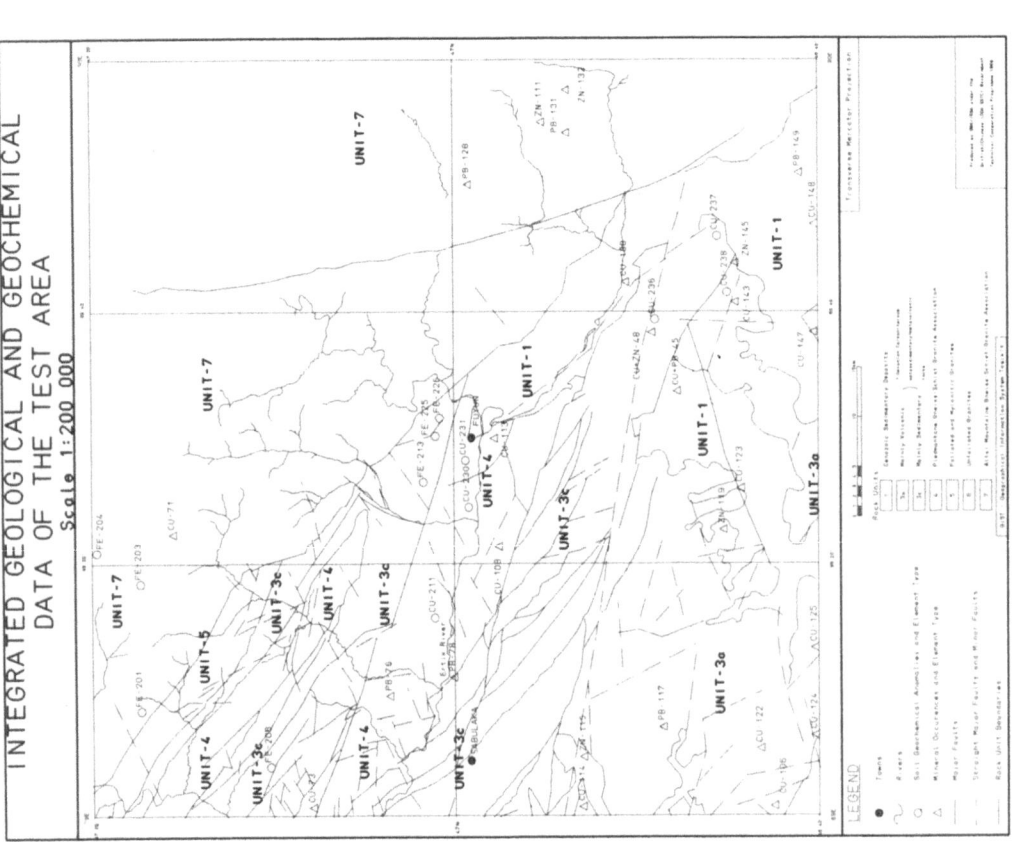

INTEGRATED GEOLOGICAL AND GEOCHEMICAL
DATA OF THE TEST AREA
Scale 1:200 000

Figure 4

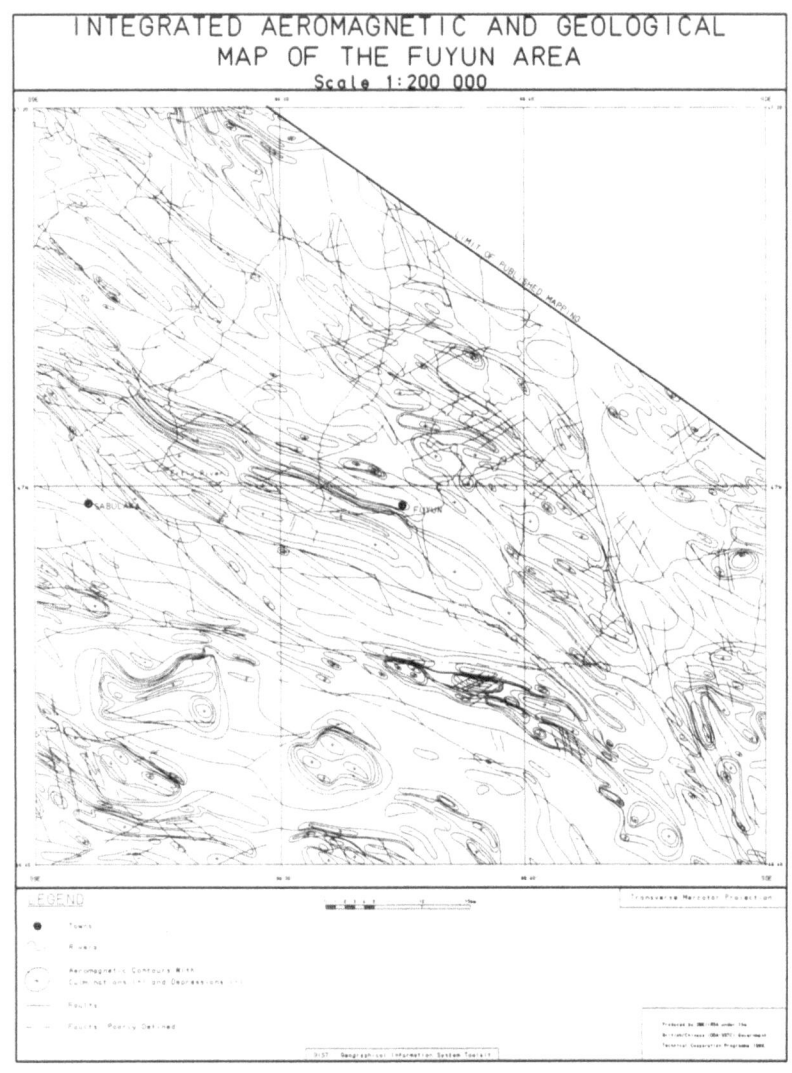

Figure 6

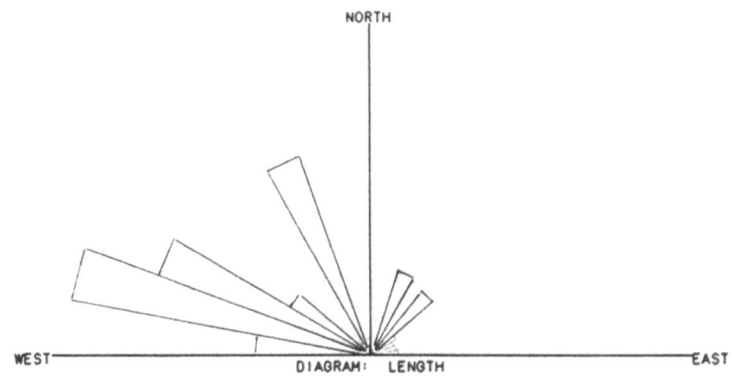

Figure 7   Fuyun Area

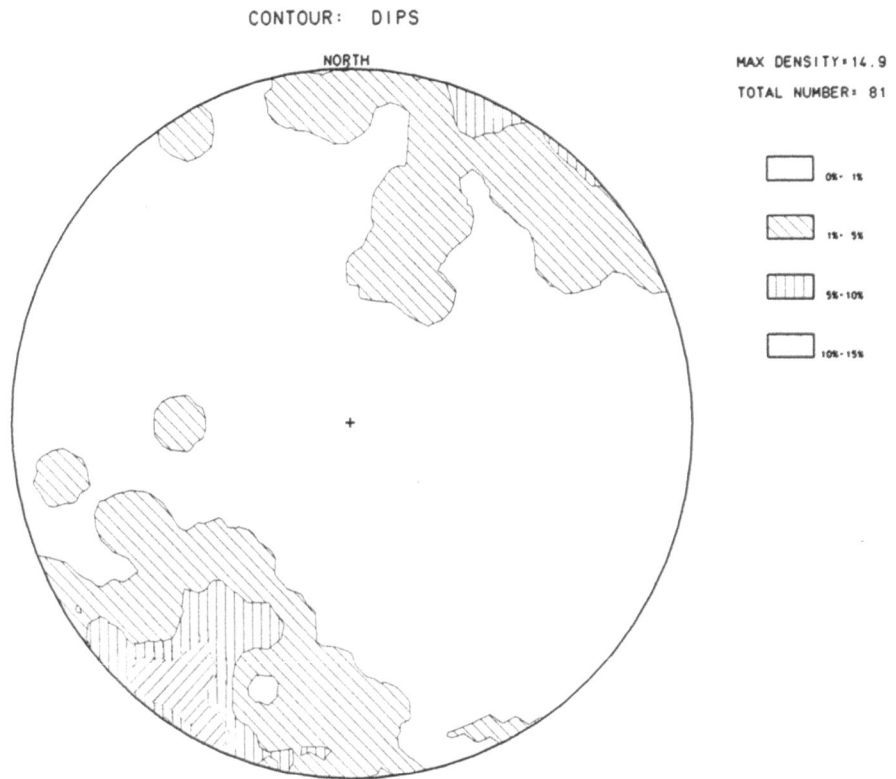

Figure 8   Altai / Fuyun

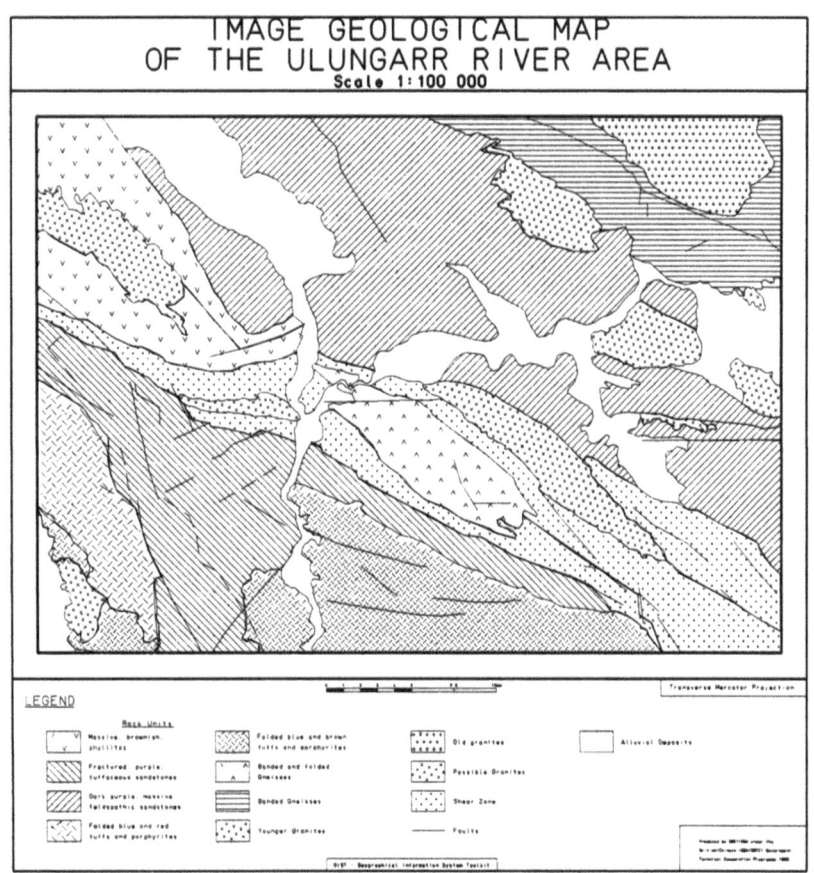

Figure 9

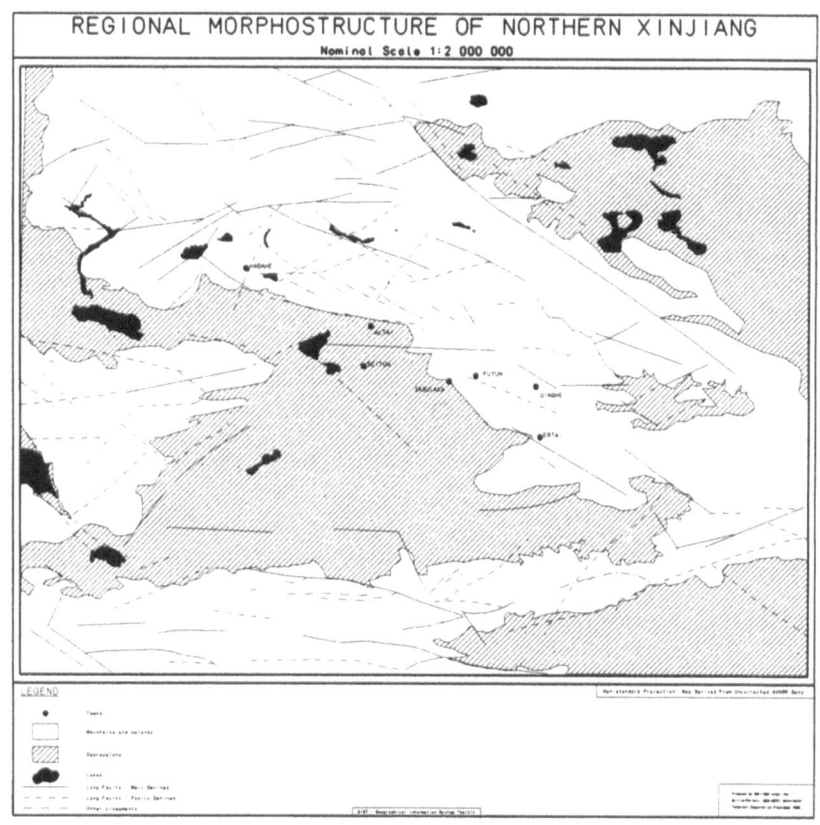

Figure 10

99

and published information suggest that these major, NW-SE aligned faults are thrusts displaced to the SW, at least in the late Palaeozoic and possibly also at later times.

GIST has been used to compute and display structural data in various ways which help clarify the structural development of the area. Maps such as Figure 4 and 5 give an impression of the spatial variation in the pattern of faults. This information can be complemented through use of diagrams, such as Figures 7 and 8, which depict the statistical or overall angular relationships between structures. Figure 7 is a rose diagram of fault trace orientations, derived by applying GIST's computational facilities to vector line data stored in the database. In this case, the vectors used are fault lines digitized from the published geological map. Figure 8, also produced using the computational and data storage facilities of GIST, is a contoured equal area net plot of poles to the dips of schistosity, slaty cleavage and bedding measured in the field.

Figures 4 to 8 show that the general structure of the Fuyun area exhibits a marked geometrical uniformity. Field observations and comparison with the known structure of adjacent areas, suggest that late Palaeozoic deformation was marked locally by the development of a series of steeply inclined thrust and associated fold nappes that, overall, strike WNW-ESE (120°-140°) and were displaced to the SSW. Many minor faults represent conjugate shears associated with the thrusts, while the pattern of linkage of the traces of some long faults suggests that a complex thrust system is linked in three dimensions through a series of splays and branch lines (Boyer & Elliott 1982). With the exception of intrusive bodies, the extent of every important mappable rock unit is defined chiefly by these interconnected faults. In many cases individual rock units apparently comprise individual thrust sheets. It is not possible to define structural geometry with much more precision from this reconnaissance study, except to say that thrust fault development was apparently accomplished by both brittle and brittle-ductile shear, the former being best associated with the weakly metamorphosed rock units of the Altai piedmont and the latter with the more gneissose and mylonitic rocks of the Altai Mountains.

It is oversimplistic to relate mineralization in this area only to the structure. Controls on metalliferous mineralization, notably including gold, in northwestern China are likely to be determined by a tectonic history embracing island arc accretion, subduction and granite emplacement (Allen 1988, Wang 1989). Field observations of the prospect at Sabulaka, SW of Fuyun, do suggest that gold is associated with limonitic alteration of fractured, clastic metasediments (including volcaniclastic rocks). The present study shows that mineralization occurs close to the trace of a long thrust fault and that several other minor mineral occurrences, as well as the Halatongke copper mine, are associated with thrusting and brittle shear of metasedimentary and metavolcanic rocks. A more detailed understanding of the relationship

between gold deposits and shear could be based on a recent review by Hodgson (1989). The important point here is that geological remote sensing, supported by use of GIST, has produced a fair case for guidance of future mineral exploration in the Altai piedmont by the structural model outlined above.

OTHER STUDIES

Introduction

The Fuyun study introduces the basic principles of GIST for geological mapping and resource exploration based on a variety of data sources. The products which illustrate that study are based on the manipulation of line and point data. The use of GIST for the computationally more difficult task of mapping a combination of lines and areas (polygons) can be demonstrated by the following two examples of the work carried out in Xinjiang.

The Ulungarr River Area

Figure 9 covers a small area (c. 1500km²) of the Ulungarr River basin in the southeastern part of the Test Area. The mineral potential of this area is unknown, but is it interesting because it is crossed by a broad, complex shear zone which incorporates, among other things, andesitic to basaltic metavolcanic rocks and granitic-granodioritic intrusions.

Figure 9 is based mainly on interpretation within GIST of Landsat 5 TM scene 141/028:06JLY87. In 1988, fieldwork was carried out along the western margin of this map area. Other supporting information was limited to the enhancements performed within GIST including false colour composites and IHS transforms, mainly based on use of TM bands 2, 4 and 7.

The objective of image interpretation was to produce a geological map from TM imagery which would complement the existing survey mapping by providing a clearer picture of the relationship between the distribution of of rock units and important elements of the structure. When working with hardcopy imagery, this may be achieved by the annotation of various line types, such as 'faults' and 'rock unit boundaries'. Definition of the rock unit is completed by use of codes and/or symbols. Figure 9 illustrates how GIST can be used to perform all these functions, beginning with line annotation and area (rock unit) definition and resulting in the production of a usable map.

Regional Morphostructural Mapping

Figure 10 represents the use of GIST for comprehensive mapping of both lines and areas applied to the analysis of NOAA AVHRR imagery in a region embracing 1,400,000km² of the Altai Mountains and the surrounding intermontane basins. The principle underlying the study is that, in a neotectonically active region such as this (Tapponnier & Molnar 1979), there are likely to be some simple and direct relationships between geological structure and terrain. Interpretation of satellite imagery and other data can be used to study these relationships and their possible tectonic significance (Morisawa & Hack 1985) and, perhaps, also serve as a reconnaissance method of hydrocarbon and groundwater exploration in the intermontane basins.

Figure 10 represents one element of a morphostructural study that is still in progress. GIST is being used for a rapid interpretation of unrectified NOAA AHVRR imagery and for production of a simple map which shows the relationships between major fracture alignments and the region's main terrain units.

DISCUSSION

Geologists, depend upon maps. As it is necessary to obtain a broad understanding of both the surface and subsurface of the Earth, there has been a requirement to become familiar not only with various geological maps but also, for example, topographical, geomorphological, geophysical and geochemical mapping, as well as in more recent times different types of remotely sensed imagery. Ideally, it should be possible to make an integrated study of all the wide ranging forms of map data that may be available for a project area and relevant to the investigation in hand. Conventional, paper-based maps can be very difficult to use in an integrated way, partly because of the inconvenience of handling and studying several, possibly large sheets of paper, and because of the likelihood that each of the map types being used is likely to have its own format, scale and map projection.

The principal capabilities of a GIS are its digital storage of spatial data the manipulation of the data-sets to common formats, scales and projections, and their superimposition in forms suited to both numerical and visual comparison. Furthermore, the ability of the computer database to store and integrate both spatial and associated non-spatial attribute information provides a means of analysis and modelling which hitherto has been unattainable. This has been well illustrated in the utilisation of GIST on this and a number of other projects, each of which has had the same underlying requirement of making use of diverse data sets including a variety of maps and imagery at diverse scales. Two such projects involved the amalgamation of existing geological mapping and new image interpretations for water resource studies of the two Caribbean islands of St Vincent and Grenada, and a similar study for the Li Shui catchment area of Hunan Province, in China. Another, more subtle aspect of this point is that rectification of the imagery need not be made until after the image interpretation stage. The regional morphostructure study described earlier was based on unrectified AVHRR data and the resulting interpretation could then be transformed to any desired output projection. The reason behind this is that attendant degradation of raster data and consequent loss of detail which occurs on geometric correction of the data is avoided.

In overcoming the conventional problems of data integration, GIST can support and integrate the main stages of a project requiring both quantitative and qualitative analysis of remotely sensed imagery and other sources of mappable information. Moreover, the comprehensive ability to handle the same spatial data in both raster and vector format is of particular value in remote sensing geological investigations through allowing visual interpretation and annotation of imagery within the system, while the results of annotation (geological linework and mapped rock units) are stored in the database in a form suitable for subsequent further manipulation and analysis. As the value of human analysis cannot be overstated for most geological mapping investigations (Estes et al 1983), in practical terms the availability of such a sophisticated capacity in GIST for recording the results of visual interpretation is also generally regarded as an important development.

Currently it is possible to make a comparison between GIST and other GIS facilities developed for geological and mineral exploration investigations. In Canada, for example, recent use of a system has been described for successful studies in support of gold mineral exploration (Bonham-Carter et al 1988). Whilst the system deployed was typically able to take information from a variety of sources in order to characterise gold mineralization within a given area of interest, there appears to have been limited ability to make use of remotely sensed imagery. GIST is more comprehensive in that it incorporates not only an image processing system, but also a capacity for sophisticated image interpretation/annotation. Furthermore the design of a GIS system reflects the perceptions of how data should be or is likely to be analysed in 'real' operational conditions. Part of the present perception is that geological projects may benefit from the use of visual interpretation of remotely sensed imagery to achieve these goals. The result is that the system is well-adapted for use at all stages of a mineral exploration project and can make especially good use of satellite imagery in its most appropriate role, which is for reconnaissance mapping, target definition and editing of existing geological maps.

Acknowledgements

All authors wish to thank the ODA, who sponsored the programme of collaboration especially Ian Brooks, and colleagues in SSTC and IRSA, especially Mrs Zheng Li Zhong, for permission to publish this work.

REFERENCES

ALLEN, MB 1988. Palaeozoic terrains of West Jungarr, Xinjiang Province, China. Abstract. Trans. Inst. Min. Metall., 97.

BEAUMONT, TE, BURKE, JJ, DAVIE, J, LEESE MJ and VINCENT, SPR, 1988. GIST, a Geographic Information System Toolkit for Water Resource and Engineering Applications. Proc. 16th International Congress of Photogrammetry and Remote Sensing, Kyoto, Japan, ISPRS, Commission IV (Intercommission III/IV), 725-738.

BONHAM-CARTER, GF, AGTERBERG, FP & WRIGHT, DF 1988. Integration of geological datasets for gold exploration in Nova Scotia. Photogramme. Eng. & Rem. Sensing, 54, 1585-1592.

BOYER, SE & ELLIOTT, D, 1982. Thrust systems. Bull. Am. Assoc. Petrol. Geol., 66, 1196-1230.

COLEMAN, RG 1989. Continental growth of northwest China. Tectonics, 8, 621-635.

ESTES, JE, HAJIC, EJ & TINNEY, LR, 1983. Fundamentals of image analysis: analysis of visible and thermal infrared data. Chapter 24 in: Colwell, RN (ed.) Manual of Remote Sensing, 2nd Edition, American Society of Photogrammetry, Falls Church, Virginia, USA.

FENG, Y, COLEMAN, RG, TILTON, G & XIAO, X 1989. Tectonic evolution of the West Jungarr region, Xinjiang, China, Tectonics, 8, 729-752.

GRIFFITHS, PS, CURTIS, PAS, FADUL, SEA & SCHOLES, PD 1987. Reconnaissance geological mapping and mineral exploration in northern Sudan using satellite remote sensing. Geological Journal, 22, 225-249.

GRIFFITHS, PS, ODINGA, M & SHAO Y in press. Geological remote sensing from satellite imagery in northern Xinjiang, China: some preliminary results. Institute of Remote Sensing Applications, Beijing.

HODGSON, CJ, 1989. The structure of shear-related, vein-type gold deposits: a review. Ore Geol. Reviews, 4, 231-273.

LEITH, W & ALVAREZ, W 1985. Structure of the Vaksh fold-and-thrust belt, Tadjik SSR: geological mapping on a Landsat image base. Bull. geol. Soc. Am., 96, 875-885.

MORISAWA, M & HACK, JT (eds) 1985. Tectonic Geomorphology. Allen & Unwin, London.

SMIRNOV, VI, 1977. Ore Deposits of the USSR. Pittman, London.

TANER, IK, KAMEN-KAYE, M & MEYERHOFF, AA 1988. Petroleum in the Jungarr basin, northwestern China. J. Southeast Asian Earth Sci., 2, 163-174.

TAPPONIER, P & MOLNAR, P 1979. Active faulting and Cenozoic tectonics of the Tienshan, Mongolia and Baykal regions. J Geophys. Res., 84, 3425-3459.

WANG, QM 1989. Tectonic features and aspects of mineralization in the region north of Tibet (part of Qinghai, Gansu and Xinjiang Provinces), China. Trans. Inst. Metall., 98, B83-B90.

YANG ZUNYI, CHENG YUQI & WANG HONZHEN 1986. The geology of China. Clarendon Press, Oxford.

ZHAO, S 1986. Physical Geography of China. Science Press, Beijing; John Wiley, New York.

Table 1 : PRECISE STATION DETAILS

Column Name	Description
Station code	S78
Latitude	47° 44
Longitude	89° 21
Corrected – Altitude	108/m

Table 2 : PRECISE STATION MEASUREMENTS

Column Name	Description
Station Code	S78
Traverse Date	16/7/88
Geologist	SY
Measurement – No	So L1
Dip. Angle Degree	86
Dip. Direction	235
Plunge Angle Degree	57
Plunge Direction	064

Table 3 : PRECISE STATION FIELD DATA

Column Name	Description
Station Code	S78
Traverse Data	26/7/88
Geologist	SY
Field Image	MSS & SPOT
Odometer Reading	10 298 km
Altimeter Reading	1080m
Lithology	Phyllitic, siliceous mudrocks, mainly light olive grey, some very thin (<5cm) interbeds of "pelites-semipelites".
Structure	Strike of So is fairly constant, while dip varies quickly across low amplitude folds. There is crenulation cleavage. All rocks exhibit micro-jointing, which locally exhibit conjugate shear on So.
Surface Condition	20% Lichen
Image Appearance	Dark Blue
Comments	
Photo samples	PH11
Field Sketch No.	No 1 Relationship of Crenuln. cleavage to So.

Semi-automatic structural mapping in arid terrain from remotely sensed images

G. Wadge B.Sc., Ph.D.
A. M. Cross B.Sc., M.Sc.
C. Angelikaki B.Sc., Ph.D.
N.E.R.C. Unit for Thematic Information Systems, Department of Geography, University of Reading, Reading, England

SYNOPSIS

Remotely sensed digital imagery of many arid terrains contains abundant structural geological information. This information is typically photointerpreted from topographic shading clues about bedding plane attitudes, rock competence based on local relief and fault offsets. Geological structural mapping can be performed from pairs of images using stereophotometric techniques. However, this is still relatively expensive and difficult to attempt from spaceborne sensor data. Systems such as JPL's IGIS system assume the prior availability of digital elevation models for their operation. We report here the development of an approach to structural mapping of geological information using single images with no ancillary DEM. Using techniques more commonly used in computer vision we perform the following tasks : extraction of the image subscene that contains most of the structural information using textural segmentation, mapping topographic "edges" that correspond to lithological boundaries, searching for valley/bedding plane "V" criteria and mapping dip directions there, quantifying bedding plane orientations using photoclinometry We have tested our ideas on an image of LANDSAT Thematic Mapper data from a part of the alpine Atlas mountains of south central Tunisia.

The mapping of geology from aerial photographs at regional and large scales is long-established[1]. In particular, the interpretation of geological structure is enhanced by the ability to view the surface in 3D using stereoscopic properties of aerial photographs and extract simple geometrical relationships such as the dip and strike of a bedding plane from three points on the plane. This concept has been developed by the assistance of computers such that mathematical structural models can be created and used interactively with photogrammetric equipment for deriving quantitative measurements from air photo pairs[2]. The advent of digital remotely sensed images should have increased the ability of geologists to extract structural information using computer techniques. However, progress has been slow mainly because the photointerpretation techniques of the human are difficult to simulate by computer and the ancillary requirement for digital topographic data (DEM) is an obstacle to ease of implementation.

Stereo pairs of images from the SPOT satellite can be used to create DEMs[3] but the need for two images and intensive computer processing is inhibitory. DEMs can be created from photogrammetric data or from topographic maps but there are still large areas of the world for which the quality or availability of DEMs or the data to produce them is poor or lacking. There have been a number of attempts to utilise co-registered remotely sensed data and a DEM for the purpose of extracting structural information or recognising geomorphological elements of the scene. The IGIS system[4] developed at JPL uses image processing and computer graphics techniques to derive strike/dip measurements, cross section construction and stratigraphic section measurements from co-registered Thematic Mapper and DEM data. The user of this system must map lithological boundaries using a trackball cursor, there is no automatic element to this basic photointerpretation task. It is only when this is done that a relationship between the geometrical information of the DEM can be related to the geology of the scene. Chorowicz *et al.*[5] used DEM data in a more fundamental way to identify geomorphological features. Firstly, slope classes are identified for each pixel and then pattern recognition rules are applied to strings of these classes to derive geomorphological features such as strike-ridges. Interaction with remotely sensed spectral data is by superimposition after this classification process using the DEM.

The work we report attempts a similar task to that of the IGIS system but without the use of an ancillary DEM. We wish to extract useful quantitative information on geological structure from a single digital remotely sensed image using shading information from the structurally-controlled topography. We use the term semi-automatic for our approach to indicate that it is a combination of human interactive photointerpretational skills and automatic computer vision/image processing operations. We do not deny the value of the combined DEM/remotely sensed approach which because of its greater information content must be able to achieve superior results. Rather we ask the question - what can be achieved using the remotely sensed data alone? Given the current relative difficulty in obtaining co-registered remotely sensed and DEM data for exploration areas, is this approach a useful, practical alternative? We do not provide complete answers to these questions here, work at the time of writing is continuing and a full discussion of these ideas lies in the future.

STRATEGY

We have chosen a single LANDSAT Thematic Mapper (TM) subscene (512 x 512 pixels) for a region in south central Tunisia as a testbed for our ideas (Fig.1). This scene is of a plunging, anticlinal hill, Jebel Morra, exposing Upper Cretaceous sedimentary rocks mainly limestones, marls and dolomites[6]. The area is arid with minimal vegetation cover and sharp gulley profiles typical of intermittent stream erosion. Within this scene, as will generally be the case, only parts display useful structural information. This is principally around the margins of the anticline, the interior being extremely dissected (Fig. 1a). In order to extract this region of useful information we perform a segmentation of the scene based on textural variations. Within the subarea of interest we use algorithms borrowed from computer vision studies to follow topographic boundaries such as ridge crests and valley bottoms and label them accordingly. Structural clues such as "V-ing" criteria are mapped automatically. However, to quantify these into dip and strike values of bedding planes additional information is needed. This comes from photoclinometric methods of inverting the reflectance signal of the remotely sensed data to give the orientation of the surface.

SEGMENTATION

The human photointerpreter is selective in terms of the major physiographic domains to be mapped and the parts of a scene that are chosen for detailed interpretation. Textural pattern is the principal form of evidence used for this task. We have performed a segmentation of the scene based on texture as the first step of our mapping. The method used is Law's "texture energy measures" based on convolution of the image with a set of texture masks[7]. Local statistics of the results are then used in a per pixel classification. In practice there are two stages of segmentation. The first successfully segments the high variance texture of the mountain chain from the smoother textures of the pediment surfaces (PED) to the north and south (Fig. 1b). The

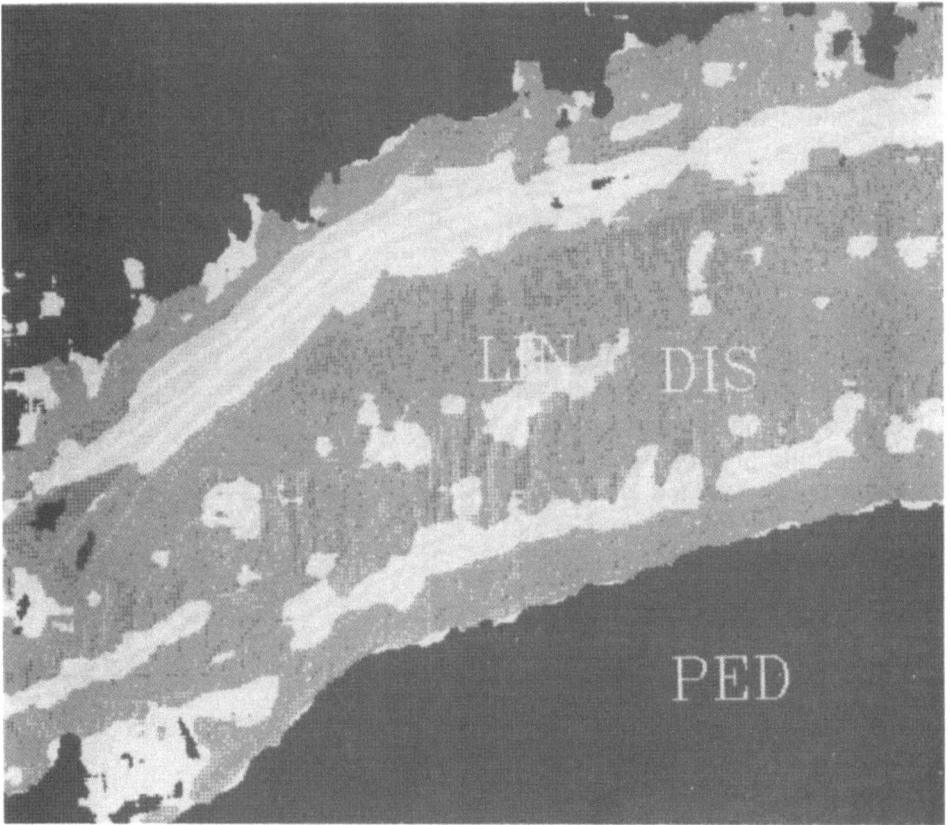

Figure 1. (a) LANDSAT Thematic Mapper (band 3) image of Jebel Morra in south central Tunisia. (b) Segmentation of this scene using a texture energy measure easily distinguishes the low energy pediment surface (PED) surrounding the higher energy mountain terrain. Most of the structural information in the mountain occurs in the marginal region (LIN). This can also be segmented using a N-S directional texture mask from the more dissected area (DIS).

second pass uses a mask with a N-S directional bias to discriminate, less successfully, between the very finely dissected core of the mountain (DES) from the more linearly-eroded envelope of rocks around its flanks (LIN). It is this latter region that contains most of the useful structural information (Fig.1b).

MAPPING OF TOPOGRAPHIC EDGES

Having extracted the part of the scene with the most useful topographic information we then need to determine discrete topographic boundaries or edges. These will usually take the form of major breaks of slope associated with ridges and valleys. The topographic boundaries can be treated as edges from an image processing point of view. A Sobel edge operator is applied to the single-band Thematic Mapper image to produce an edge gradient image. The local orientations or aspects of this image are then computed to produce an edge orientation image. These three images : TM band3, edge gradient and edge orientation are then used in a semi-automatic routine that we have written to map topographic edges.

This works in the following manner. The user interactively generates a seed point lying on an edge displayed in the TM image. The pixel corresponding to the orientation octant (e.g. N,NE,E etc.) of this point is searched together with its two neighbours. For example, if the orientation is NE, then its neighbours are N and E. The pixel with the greatest edge gradient of these three is selected. If this gradient is greater than a threshold and is not a previously encountered pixel then a move to that pixel is made and the process is repeated. In this way the edge is followed, and the resultant line is written to the screen. When the edge can be followed no further in one direction, the edge in the opposite direction from the seed point is followed. Upon completion this line is written to a formatted file. This can be labelled as either high (ridge) or low(valley) from the edge orientation information if the solar azimuth is known and entered by the user. A new seed point can then be chosen. This semi-automatic approach is quick and accurate, probably more so than the

equivalent task performed by a human interpreter. However, the routine can make errors, particularly at transverse breaks in the topography, though this could be corrected with further high-level rules being applied.

STRUCTURAL MAPPING

A resistant rock unit whose strike has been "mapped" by the previous procedure will present a "V"-shape outcrop plan when it encounters a cross-cutting river valley. The exact shape of the "V" will depend on the dip of the unit relative to the topography of the valley. We have written a routine that detects such Vs from the edges mapped previously. This procedure works on the basis that an idealised V consists of three changes of direction or veers (Fig. 2a): the first and third being of similar magnitude and the same sense, the second being of greater magnitude and the opposite sense. Veers are measured by proceeding along an edge and for every pixel recording the difference in edge orientation between its two neighbours. A second pass along the edge flags local peaks of veer. A third pass then looks for triple peaks in close proximity that conform to the model above.

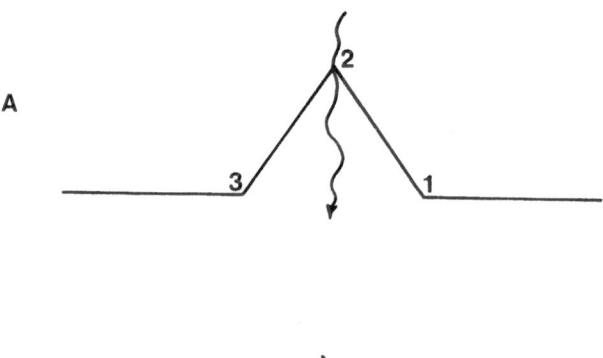

Figure 2 (a) Model of veers in direction along a topographic edge at the intersection of a valley to form a "V". (b) More typical example of valley-Vs encountered in the image leading to errors of omission.

This scheme was only partly successful. Errors of omission occurred because many valley-Vs lacked sharp first and third veers (Fig. 2b). Errors of commission occur at very minor deviations of the edge not associated with valleys. The V-finding procedure could be improved by using the actual orientation (i.e. tangent) of the line defining the edge, rather than the associated edge orientation. Once requested the V-finding is automatic. Similarly, the dip direction from these Vs can be generated. Firstly the apex pixels of the Vs are flagged. An arrow bisects the limbs of the V and is drawn from each apex pixel that lies on a topographic high. For beds dipping at angles less than the local gradient, this gives a rough measure of the dip direction (Fig.3).

PHOTOCLINOMETRY

The term photoclinometry can be used in the specific sense of extracting topographic information from a single photometric image[8]. Shape from shading is the equivalent concept used in computer vision studies to explore the relationship between variations in surface reflected light and the local orientation of the surface[9]. Horn[10] devised a method for systematically calculating orientation from viewing geometry and reflectance values using a "reflectance map". The radiance received by a satellite sensor such as Thematic Mapper is a function of terrain illumination geometry and instantaneously variable incidence angle effects can be ignored[11]. Specifically, it is a

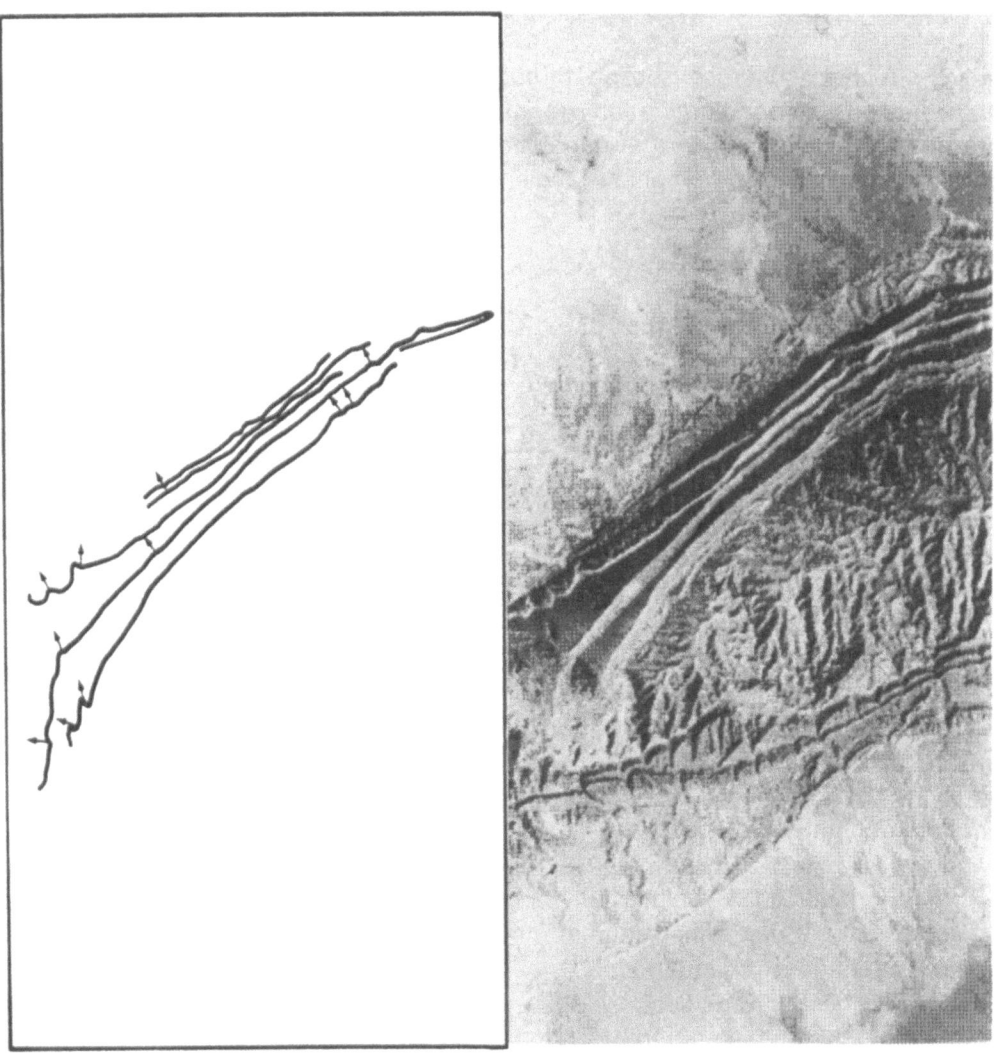

Figure 3. Example of output results from topographic edge and dip direction mapper on left, compared with unsegmented original scene.

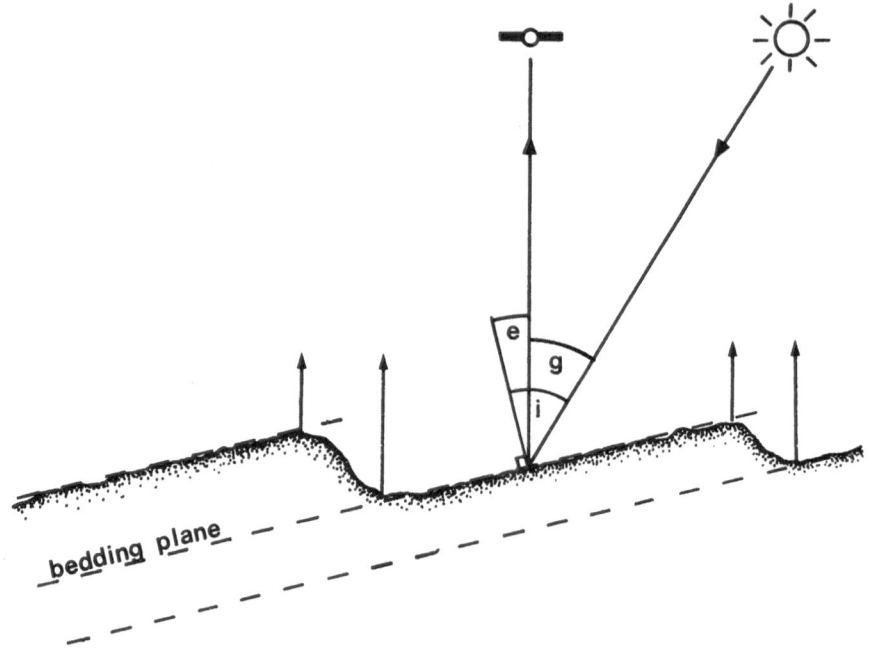

Figure 4. Schematic cross section through northern flank of Jebel Morra to illustrate local surface illumination geometry, location of mapped topographic edges and the relationship to the planar bedding surfaces. i = angle of incidence, e = angle of emittance and g = phase angle. Arrows mark scene locations of mapped topographic edges.

function of the angles of incidence (solar illumination direction relative to the local surface normal), emittance (sensor direction relative to the surface normal) and phase (solar illumination direction relative to sensor direction). Satellite-received radiance is also dependent on intrinsic material properties of the surface, namely albedo and colour. A method of independently determining the spectral albedo information using band ratios after atmospheric correction is given by Eliason et al [12] which allows the image component solely due to topography to be extracted. This can be converted to local pixel slope values if the solar illumination geometry is known.

Values of surface orientation calculated in this way are not unique. Some other source of information is needed to constrain multiple solutions. The dip directions calculated from the valley-Vs could provide such a constraint.

For example, a dipping bedding plane that has two dip directions associated with it along strike could have unambiguous orientation

values assigned to it from the photoclinometry approach. A problem with this is that in the system developed so far potential topographic edges to individual lithological units have been mapped and not the units themselves. Figure 4 shows that in the simplest cases bedding units could be defined from neighbouring topographic edges alone. Also the user would need to make the decision that the surface being mapped did represent a meaningful planar surface (e.g. bedding).

DISCUSSION

We have yet to fully implement some of the ideas explored in this paper, and have yet to prove the usefulness of photoclinometry in quantifying structural orientation. The ability to segment and label features in images such as that of Jebel Morra that are relevant to structural interpretation has been demonstrated. Validation of the results of this approach can be achieved by comparison with field measurements of dip and strike. The creation of a DEM will allow the three-point method to be applied and compared.

One of the principal arguments against the utility of photoclinometry is mutual illumination[13]. This is the reflection of light not only back to the sensor but back to other patches of surface which means that reflectance may not be a function of local orientation but of global scene geometry. The magnitude of this effect depends on the reflection function of the surface and the relief of the surface. The only way to assess the effect in scenes such as Jebel Morra will be empirical tests against validatory data as above. Assuming that photoclinometry does give acceptable results by these criteria we see the principal challenge of further work will be the identification of structural planes. In a relatively simple case such as Jebel Morra the photointerpretation of tilted planar surface as bedding planes is fairly straightforward. The task of automatic identification is more difficult. Multispectral information could be used to define lithological units, though in the case of Jebel Morra Thematic Mapper provides very little discriminatory information between rather similar lithological units. However it is the identification of the surface as part of a plane of known significance that is crucial. Human interpretational skills are probably best employed at this step in any semi-automatic system, although further work will explore the utility of an expert system for encapsulating this high-level knowledge.

References

1.   Allum J.A.E. Photogeology and regional mapping. London: Pergammon, 1966.
2.   Dueholm K.S. and Pillmore C.L. Computer-assisted geologic photogrammetry. Photogrammetric Engineering and Remote Sensing, vol.55, 1989, p.1191-1196.
3.   Day T. and Muller, J.P. Digital elevation model produced by stereomatching SPOT image pairs: a comparison of algorithms. Image and Vision Computing, vol.7,1989, p.95-102.

4.   McGuffie B.A., Johnson L.F., Alley R.E. and Lang H.R. IGIS computer-aided photogeologic mapping with image processing, graphics and CAD/CAM capabilities. Geobyte, October 1989 p.9-14.
5.   Chorowicz J., Kim J., Manoussis S., Rudant, J.-P., Foin P. and Veillet I. A new technique for recognition of geological and geomorphological patterns in digital terrain models. Remote Sensing Environment, vol.29, 1989, p.229-239.
6.   Domergue C., Dumon E., De Lapparent A.F. and Lossel P. Sud et Extreme-Sud Tunisiens. 19th International Geological Congress Regional Monographs 2nd series: Tunisia, no. 7.
7.   Pietikainen M., Rosenfeld A. and Davis L.S. Experiments with texture classification using averages of local pattern matches. IEEE Transactions on Systems, Man and Cybernetics, vol. smc-13, 1983, p.421-426.
8.   Watson K. Photoclinometry from spacecraft images. U.S. Geological Survey Professional Paper 599-B, 1968, p.1-10.
9.   Horn B.K.P. and Brooks M.J. (Eds.) Shape from shading. Cambridge, Mass.: MIT Press, 1989, 577pp.
10.   Horn B.K.P. Understanding image intensities. Artificial Intelligence, vol.8, 1977, p.201-231.
11.   Hugli H. and Frei W. Understanding anisotropic reflectance in mountainous terrain. Photogrammetric Engineering and Remote Sensing, vol.49, 1983, p.671-683.
12.   Eliason P.T., Soderblom L.A. and Chavez P.S. Jr. Extraction of topographic and spectral albedo information from multispectral images. Photogrammetric Engineering and Remote Sensing, vol. 48, 1981, p.1571-1579.
13.   Forsyth D. and Zisserman A. Shape from shading in the light of mutual illumination. Image and Vision Computing, vol.8, 1990, p.42-49.

Acknowledgements

This work is supported under N.E.R.C. contract no. F60/G6/12.

Geobotanical remote sensing in China

Xu Ruisong
Guangzhou New Technical Institute of Geology, Chinese Academy of Sciences, Guangzhou, China
Xu Houshen
Guangzhou New Technical Institute of Geology, Chinese Academy of Sciences, Guangzhou, China
Ye Suqun
Guangzhou New Technical Institute of Geology, Chinese Academy of Sciences, Guangzhou, China
Lu Huiping
Guangzhou New Technical Institute of Geology, Chinese Academy of Sciences, Guangzhou, China
Zhang Lixia
Institute of National Meteorological Bureaux, China

SYNOPSIS

In China, hundreds of people are working on the theory, techniques and applications of geobotanical remote sensing. These researches began in the 1970s, and have now made considerable advances. Theoretical studies have concentrated on characteristics of geobotanical physiology and ecology, vegetation spectra and optimum imaging wavelengths, and on factors which interfere with geobotanical effects and how to compensate for them. The results have shown that when certain elements are either very abundant or absent this can lead to positive or negative biogeochemical effects. Anomalies in moisture contents and pigment of plants can arise and plant cells can be broken or deformed. When the geochemical balance is disturbed unusual plant communities appear, the overall leaf reflectance increases and blue or red shifts of reflectance occur, producing anomalous colours in many kinds of remote sensing imagery. Research in techniques has concentrated on acquisition and processing of remote sensing information, including information systems and expert systems. One result of these researches has been the development of the Fine Split Multispectral Scanner operating in the 0.4–13 µm region. Geobotanical remote sensing has been widely applied in geology, resources exploration and environmental surveying of diseases in plants and crops, and great benefits to society, the economy and the environment have resulted.

INTRODUCTION

Geobotany has been used for thousands of years for finding mines, but geobotanical remote sensing dates from the sixties of this century, with the development of space science. In China, geobotanical remote sensing began in the 1970s, and there are now hundreds of scientists and technicians from tens of institutes and universities working on the theory, techniques and applications of this science. The researches are financed mainly by Natural Science Funds at local or national levels. The studies combine theory with practice, techniques with their application, and research targets are determined according to the needs of society, the economy and the environment.

BASIC RESEARCH

Geobotanical physiology and ecology

In this research domain, we mainly study the physiology and ecology of plants poisoned by harmful elements. The results show that when harmful elements are higher than a critical threshold, the pigment and moisture contents of leaf blades are reduced, the cells break or deform, chemical balance between a range of elements is destroyed, and the molecular structure is changed. Changes are induced in both leaf blades and leaf veins, as well as in the assemblages of plants. If the content of Mo in rock or in soil is higher than the threshold, then the contents of Cu, Zn, Fe, Mo, P, N and especially Mg are reduced in plants, and photosynthesis is affected. In molybdenum mining areas, the content of "grey pigment-deficient Mg" pigment in plants is far higher than usual. At the same time, the moisture and chlorophyll contents are reduced, cells break, and the linkage type and spin direction of molecular C–O, C–L and C–H bonds in the pigment suffer some changes, resulting in red spots on leaves, which also become yellow and rough. Phenological changes such as early blossoming, lengthened flowering and late fruiting may also be induced. The traditional belief is that Mo can assist nitrogen fixation and improve growth of legumes, but recent researches have shown that high Mo contents correlate with low N in soil and plants, and with reduced legume growth. When Mo exceeds 500 ppm, few legumes die, but when Mo is greater than 750 ppm, almost all legumes vanish. Tian Gonglian and other Chinese scientists discovered that plants are most sensitive to pollutants when they are growing luxuriantly. During this period, in comparison with other seasons, variations in the green vegetation index are most clear. They also discovered that when Cd exceeded 50 ppm and SO exceeded 0.5 ppm paddy rice was poisoned and clear changes in spectral characteristics occurred.

Spectral characteristics

In China, geobotanical spectral characteristics have been systematically studied during comprehensive remote sensing projects in Tengchong, Etan, Xingjian and elsewhere. As a result, "*Data*

Compilation of Chinese Earth Resources Spectral Information" and other works on applied remote sensing have been published, and a data bank of Chinese earth resources spectral information has been established. The spectra which have been measured include airborne spectra (Fig. 1) and ground spectra measured in the field and laboratory (Fig. 2). After a great deal of study, we discovered that plants which have been poisoned

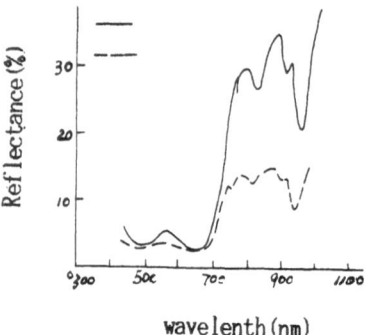

Fig. 1 Airborne reflectance spectra of broadleaf forest (*continuous line*) and bamboo forest (*dashed line*) that have been poisoned. Location, Nankun Mountain, Guangdong Province, China; instrument, HG-1 airborne spectrometer; resolution, 1.5 nm; height of aircraft, 1500 m; date, 21 December, 1988

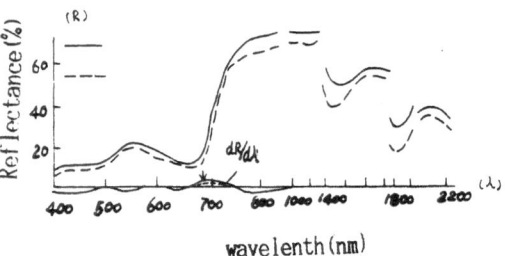

Fig. 2 Laboratory reflectance spectra of *Dalbergia hancei* in area of Dinghu molybdenum mine, Guangdong Province, China. Leaf reflectance in mine area (*continuous line*, 1200 ppm Mo) is higher than in background areas (*dashed line*, 10 ppm Mo). Instrument, UV-340 spectrometer; resolution, 0.1 nm; date, November 1987

by ions or compounds of harmful elements in the atmosphere, water, soil or rock, when compared with spectra of healthy leaves show the following characteristics:-

1) The fluorescent absorption peaks of normal leaves appear at 405 and 455 nm, while with poisoned leaves the peak occurs at 420 nm.

2) In visible and near-infrared bands, the reflectance of most poisoned plants is from 5 to 30% higher than healthy plants.

3) In the thermal infrared, the temperature of poisoned plants is 1–5° lower than normal.

4) Poisoned plants have a higher microwave back-scatter than normal plants.

5) Compared with healthy plants, the visible spectra of poisoned plants have a characteristic blue shift of 1–5 nm. A reverse shift occurs in the near-infrared.

Image characteristics

Chinese scientists have been making systematic studies in order to provide geobotanical remote sensing products suited to their application. Remote sensing data were acquired from a range of sources, and quantitative studies carried out on the reflectance, thermal radiance and geometry of the images, including geobotanical anomalies. Table 1 summarizes the main results. At the same time, these images were processed in analogue and digital forms, and colour composite images were generated. Poisoned plants show different hues and patterns from healthy plants, as shown in Fig. 3.

Research in optimum band selection

To assist in the design of Chinese natural resources satellites, many scientists and technicians are involved in research on the selection of the optimum spectral bands for geobotanical

Fig. 3 TM false-colour image of area around Dinghu Mo mine, Guangdong Province, China, acquired on 6 December, 1988. False-colour composite produced by combining TM bands 5 (red), 4 (green) and 2 and 3 (blue). *A*, vegetation poisoned by Mo and S; *B*, abnormal hues of *Pinus massoniana* on a circular granite body; *C*, normal vegetation. Approximate scale, 1:200 000

remote sensing. The results show that the optimal wavelengths are 0.43, 0.55, 0.68, 0.72, 0.80, 1.25, 1.65, 2.20, 3.50, 11.80 μm and the centimetre band in the microwave region. Moreover, the narrower the bandwidth, the better the definition of important spectral features. The optimum bandwidth appears to be from 2 to 40 nm. Table 2 summarizes the main results.

Table 1 Image features of geobotanical remote sensing

Plant	Species	Reflectivity	Moisture	Vigour	Temperature	Geometry	Hue
Coniferous woodland	*Pinus massoniana*, etc.	H	L	L	L	Rough	H
Broadleaved woodland	*Itea chinensis*, *Rhodomytus*, *Tomentosa*, etc.	H	L	L	L	Smooth	H
Mixed woodland	Mixture of above	H	L	L	L	Fairly smooth	H
Lower plants	Dilinears, var. *dichotona*, *miscanthus*, *floridulus*, etc	H_v L_{in}	L	L	L	Smooth	H

H, value of poisoned plant higher than for healthy plant; L, value of poisoned plant lower than for healthy plant; H_v, higher value for poisoned plant at visible wavelengths; L_{in}, lower value for poisoned plant at infrared wavelengths.

Table 2 Optimum bands for geobotanical remote sensing

	Centric wavelength, μm										Centimetre band 1–30 cm
	0.43	0.55	0.68	0.72	0.80	1.25	1.65	2.20	3.50	11.80	
	Optimum wavelength, nm										
	10–25	10	2–5	20	20	20	50	50	100	500–1000	
Physical meaning	Feature of reflection, geometry, vegetation			Feature of reflection, humidity, geometry and blade structure					Thermal feature		Electrical and structural features
Meaning in plant	Type of pigment and content, element content, degree of finish, structure and shape			Water content, cell structure, blade structure, molecular structure and vibration			Water content, molecular structure and vibration		Temperature, molecular structure and vibration		Temperature, water content, molecular structure and vibration
Penetration of atmospheric window, %	>95			<80	60–95			>80	60–70	>80	100

Research into interfering factors

Geobotanical remote sensing is concerned with the biosphere, and this is disturbed by many dynamic variables whose effects cannot be measured directly or even calculated theoretically. The boundary conditions are difficult to control, so research is difficult and the results relative, subject to random error, and often of merely local significance. Based on a very large number of studies, however, the statistical laws governing the factors that interfere with geobotanical remote sensing can be discovered. The authors and other Chinese scientists have systematically studied some of the disturbing factors, for example atmospheric and climatic effects, water, soil rock, temperature, humidity, topography, time, plant species and age, in terms of spectral and image features. Some results are illustrated in Table 3 and Fig. 4.

Table 3 Main interfering factors in geobotanical remote sensing

	Interfering factors								
	Atmosphere	Climate	Water	Soil	Rock	Topography	Time	Plant species	Plant age
Interference effect	Difference of transparency of atmosphere, window effect, blade spectra	Clear sky, overcast sky, rain, etc., lead to variations in temperature, humidity, soil organic matter, pH, zH, covering the poisoning function	Water content and quality disturb plant growth and cover the poisoning function	Type, quality, organic matter, zH, pH disturb poisoning function	Disturb poisoning function	Different types lead to different water, temperature, nutrients, disturb poisoning function	At different periods poisoning function is different	Different absorbed elements disturb poisoning function	Different physiologies lead to different poisoning function
Optimum condition	Clear sky transparency is same	Clear sky and same climate	Select dry season	Same soil type and quality	Same rock type	Select same type of topography as comparison area	Select late autumn and dry season	Select indicator plant	Select adult and immature forest

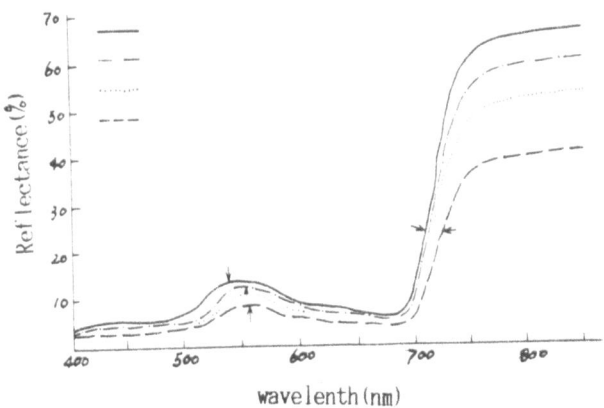

Fig. 4 Laboratory reflectance spectra for *Pinus massoniana* of different ages and in different seasons, Dinghu Mo mine area, Guangdong Province, China. Reflectance of young growth is higher than that of mature forest and is shifted towards blue (*arrow*). Reflectance in October is higher than in January and April, and a blue shift is also apparent. *Continuous line*, young *Pinus m.*, October 1988; *dashed line*, mature *Pinus m.*, October 1988; *dotted line*, mature *Pinus m.*, January 1988; *dashes and dots*, mature *Pinus m.*, April 1988

TECHNICAL APPROACH

Information acquisition

Based on studies of spectral characteristics, image features and optimum band selection, new techniques of information acquisition for geobotanical remote sensing have been successfully developed by Chinese scientists and technicians. These include a satellite-borne microwave system, the 71-channel Airborne Imaging Spectrometer (Table 4), Multi-Polarized Airborne Side-Looking Radar, a laser height indicator, and a series of full-channel spectrometers. These instruments, especially the 71-channel Airborne Imaging Spectrometer, place Chinese geobotanical remote sensing among the most advanced in the world. In addition, the Chinese Academy of Sciences has a resources satellite ground receiving station, and has launched a series of satellites including a scientific laboratory satellite, a Territory satellite and a meteorological satellite during the 1980s. The launch of a resources satellite is planned for the early 1990s. There are also other satellite ground receiving stations, such as that in Guangzhou, and China has a range of high-quality airborne remote sensing aircraft.

Information processing techniques

Techniques in geobotanical remote sensing information processing can be divided into qualitative and quantitative analysis, the former using optical processing and visual interpretation, the latter including micro-density slicing, quantitative retrieval of spectral signatures by computer, and the quantitative analysis and image processing of remote sensing imagery with computer image processing techniques (CIPT) such as vegetation indices, character charts of temperature, character charts of humidity and character charts of reflectance. Techniques of differential and integral analysis, Chebyshev linear fitting, factor analysis and the extraction of principal components are used to extract spectral information. Image analysis techniques include production of ratio composites, principal factor analysis, biological indices and multivariate information combination. The Chinese Academy of Sciences, Energy Resources Administration, Geology and Mine Administration, National Meteorological Bureau, the National Remote Sensing Centre and some universities have large and medium sized image processing systems as well as PC-based systems.

Table 4   71-channel airborne imaging spectrometer

Band, μm	0.4–0.7	0.7–1.1	1.4–2.5	8–13
Channel numbers	12	20	32	7
Wave width, nm	20	20	20	400–800
Meaning in geobotanical remote sensing	Pigment, structure and absorbing function	Water content, cell structure, stretching of O–H and C–O–C bonds and vibration function		Temperature, stretching and vibration of O–H and C–O–C bonds, assymmetrical vibration and net absorption

Information systems and expert systems

Suitable hardware to host comprehensive national information systems are set up at a number of sites, including the Geographic Information System Laboratory led by Professor Chen Supen, the Oil Institute of Remote Sensing of the Ministry of Energy Resources, the Institute of Remote Sensing at Beijing University, and the Institute of Remote Sensing Applications of the Chinese Academy of Sciences. Experimental information systems, developed internally or from abroad, are supported, and there has been rapid development of microcomputer-based information systems. Research into expert systems for use in geobotanical remote sensing was started during the 1980s. The present authors carried out a systematic study of spectral characteristics and image features related to the biogeochemical effects of gold deposits in Guangdong Province and Hainan Island. This work was supported by the Geographic Information Systems Laboratory of the Institute of Remote Sensing Applications, Chinese Academy of Sciences, and was financed by National Natural Scientific Research funds within the context of China's Seventh Five-Year Plan. During this study, a microcomputer-based information system of data relating to the biogeochemistry of the gold deposits and their related elements was established, including an expert system based on a prediction model for gold resources. Using this system and model, geobotanically anomalous regions related to gold and polymetallic mineralization have been located in mountainous and densely vegetated terrain in Guangdong Province and Hainan Island.

APPLICATIONS RESEARCH

Geological applications

The assemblages and growth patterns of plants are often distinctive of the rocks on which they grow. Plants grow particularly luxuriantly over faults. One conclusion from our studies is that basalts normally support broad-leaved trees, while granites support coniferous masson pines. Based on simple geobotanical features like this, remote sensing has been used for rapid map-making in some research areas. During these studies, a wide range of problems in rock properties and geological structure were solved efficiently. For example, in Fig. 3, the circular anomaly with a yellow–red colour is a circular granite body which has an identical hue.

Resources detection

Geobotanical remote sensing can be used in exploration for a wide range of precious, base and energy materials. Using geobotanical remote sensing in 1988, the present authors discovered new gold-deposit anomalies (Fig. 5) on the periphery of known gold mines in an area of tropical rain forest with more than 80% vegetation cover on Hainan Island, China. Field

Fig. 5   TM false-colour composite of area of southwest Hainan Island, China (red, bands 5 and 7; green, band 4; blue, bands 1, 2 and 3). Whole area is covered by same type of tropical rain forest; (a) gold mine; (b) , (c) and (d) gold mine anomalies; (e) normal vegetation. Image acquired 26 November, 1987

spectral measurements in the Talimu Basin oil and gas region, carried out in April 1990, indicated that plant spectra for the oil/gas area were different from those for the same plants elsewhere, as shown in Fig. 6. In the eastern Zhurger Basin, researchers from the Institute of Remote Sensing, Academia Sinica and from the Guangzhou Institute of Geology and New Technology found that *Equisetum ramosissimum* in oil/gas regions has a reddish colour, while it appears green in regions without oil or gas. These different tones are seen clearly on Fine Infrared Multispectral Scanner imagery, and

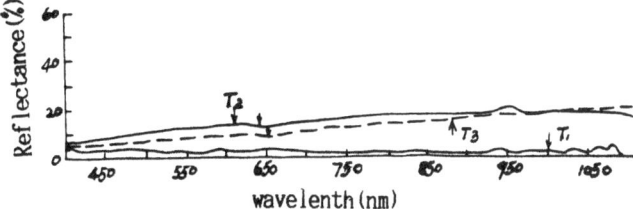

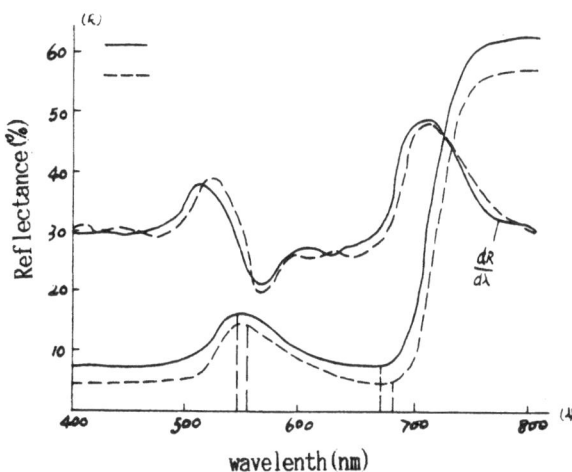

Fig. 6 Field reflectance spectra measured in Suosuocai area of northern oil and gas region of Talimu basin, China. Plants in oil/gas region have higher reflectance at short wavelengths and lower reflectance at long wavelengths than plants elsewhere, and blueshift is evident (*arrows* in region of 650 nm). T_1, crude oil spectrum; T_2, vegetation near oil well; T_3, vegetation outside oil/gas region. Instrument, H-10 field spectra radiometer; resolution, 50 Å; date, 28 March, 1990.

Fig. 7 Laboratory reflectance spectra and spectral ratios of polluted and healthy leaves of *Ficus microphylla* in Guangzhou city, China, measured with UV-340 spectrometer in January, 1985. Polluted plants have higher reflectance, exhibiting blue shift. *Continuous line*, polluted plant; *dashed line*, normal plant

extensions to known oil and gas basins have been located by biogeochemical methods.

Environmental surveillance

Plant growth, the leaf surface structure, contents of moisture and pigment, and molecular structure and vibration directions are all affected when plants are contaminated by pernicious elements in rocks, soil, water and atmosphere. Most of these changes result in differences in reflectance spectra and image character. The more serious the contamination, the larger the differences. The spectra and image characteristics of healthy and contaminated plants from Guangzhou city were studied in 1985 (Fig. 7) and a subdivision of degrees of atmospheric pollution in Guangzhou was achieved based on the direct relationship between degrees of atmospheric pollution and the spectral character of leaf surfaces. In the early 1980s, Tong Qinxi, Tian Guoliang and other scientists made a systematic study of the spectral characters of leaf surfaces of a range of plants poisoned by a range of contaminants in Tianjing city. They discovered that contaminated plants appear pale red, purplish red or grey–green in colour–infrared aerial photographs, while normal plants are bright red. This allowed rapid efficient measurement and surveillance of the extent and degree of

contamination around Tianjing City. Many cities in China are now using remote sensing for real-time surveillance of environmental contamination.

Surveillance of crop and forest diseases

Apart from the possibility of detecting plant damage due to anomalous elements, the effects of certain diseases can also be monitored using remote sensing. For example, researchers at Beijing Agricultural University studied the infrared spectral characteristics of wheat affected by leaf rust.

ACKNOWLEDGEMENTS

This paper covers only a part of the work on geobotanical remote sensing in China. The authors are very grateful to Professor Chen Supeng and Professor Tong Qinxi for their constant guidance and encouragement throughout our studies. The authors are also grateful to other colleagues for providing valuable information on geobotanical remote sensing.

Mineral exploration by use of infrared multispectral remote sensing in China

Zheng Lanfen
Tong Qingxi
Institute of Remote Sensing Applications, Chinese Academy of Sciences, Beijing, China
Li Yan
Xinjiang Institute of Geography, Chinese Academy of Sciences, China
Chi Guobin
Ding Xuan
Institute of Geochemistry Science, Chinese Academy of Sciences, China
Xue Yongqi
Shanghai Institute of Technical Physics, Chinese Academy of Sciences, China

ABSTRACT

A new airborne infrared multispectral technology has been used for study of the gold and other mineral exploration in the North–Western Part of China during past few years.In this study the following steps were involved:the measurements and analysis of spectral characteristics,airborne remote sensing data acquisition and processing , extracting information on mineralized features and the assessment of techniques.

For datathe acquisition a newly developed 12 channels Fine–split Infrared Multispectral Scanner(FIMS) was used.Following radiometric and geometric correction,the remotely sensed FIMS data were Processed and analysed by ratioing ,principal component and other methods.Due to the absorption features of OH,CO$_3$and ferric–ferrous contents in rocks and minerals, alteration or mineralization information are clearly displayed by the distinguishable colour anomalies and other image features.These results were verified by field investigation,rocks and minerals analysis.It was found that most existing gold and other mineral deposits showed good coincidence with the information extracted by analysis of remotely sensed data.

INTRODUCTION

Recent developments in remote sensing technology have made great progress in applications to geological and mineral exploration.The general procedure of geological study for an area is always from synoptical to detailed. Landsat MSS,TM and SPOT images are very useful for the general investigation of regional geology,while the airborne remote sensing techniques such as aerial colour and colour infrared photography,thermal infrared and multispectral imagery and side–looking airborne radar are most useful for detailed studies of limited areas and mining districts.

As a result of remote sensing spectral properties studies during the 1980s, a series of sensors with high spectral resolution have been developed . The new multispectral super–multispectral scanners and imaging spectrometers represent the latest stage of development of remote sensing technology and greatly increase the capabilities of remote sensing application. Geological and mineral exploration benefit wealthily from the use of such new remote sensing technology and many new studies in this field have been done in the last few years.

In the second half of the 1980s a new type of multispectral scanner has been developed by the Chinese Academy of Sciences(CAS) .As well as the development of technologies,a series of applicational studies,such as geological and mineral exploration and other new applications have been carried out .The purpose of using remote sensing techniques is to discover some exploitable mineral deposits in an area.Most of the surface or near–surface deposits in areas with dense population have been found or well investigated .It is reasonable the analysis of new mineral expolration will move from accessible regions to remote areas.Xinjiang Uigur Autonomous Region, the largest Province of China, is just such remote area and is rich in mineral resources and relatively unexplored.

In this study our task is to evaluate the

effectiveness of this newly developed infrared multispectral remote sensing technique for geological applications, especially for detecting alteration zones and mineralization zones.

DISCRIPTION OF STUDY AREAS

Two study areas, the western Zhunger area, and the Altai Tiemuerte area were selected for test and assessment of the newly developed infrared multispectral techniques. Both areas are located in the North of Xinjiang Uigur Autonomous Region of China and geologically belong to the North Xinjinag minerogenetic zone.

The western zhunger study area is a triangular zone controlled by three major geological structures ,the Dalbut,Anqi and Mayiler fault systems and three large granite bodies ,the Haftu,the Acbastao and the Miao−erogu rock bodies . This area has very sparse vegetation and good surface exposure of rocks .The predominant rock types areneutral−basic volcanic,basic and ultrabasic rocks,especially the neutra−acidic rock.

These are most suitable for polymetallic and gold mineralization.Gold mineralization in this area is always associated with intense alteration.The following types of alteration were observed in the field of study areas; serpentinization in ultrabasic rocks; pyritization,chloritization,carbonatization and silicification in basalt and neutralbasic volcanic rocks ; clay alteration in neutral−acidic rock and sandstone.Also,this is the most important gold production area in Xinjiang.

The Altai Tiemuerte area is located in the Kelan synclinorium of the Altai fold system.According to the geotectonic units ,the area lies in the center of the inverted flank of the northeast Altai synclinorium,which belongs to the north zone of polymetallie mineralization of the southern fringe of Altai,and its exposed strata are composed of middle and lower Devonian series.It is classified into four lithological units which streche from northeast to southwest.Mineralization is controlled by a major fault system,which runs through the contact zone of III and IV lithological units in the area.Polymetallic mineralization is generated in the third lithological unit,which mainly consists of tuff, marble, quartz biotite schist, chlo−rite quartz schist, skarn,calcsiltstone.Research work has proved that polymetallic mineralization is directly related to skarns and impure magnesian,limestones and that outcropping gossans may reach a width of 25−50m,even hundreds meters on the surface.

The geographical situation of two study areas is shown in Fig.1

TECHNICAL APPROACH

It was found by studying of the spectral signatures that most rocks and minerals, especially the alteration zones are characterized their distinguishable absorbed and reflection features in the short−wavelength infrared spectral region from 1.6 to 2.5μm. This spectral signature of rocks and minerals results from both electronic transitions and molecular vibrations mainly due to water,hydroxyl and carbonate absorption.The Fine−Split Infrared Multispectral Scanner(FIMS) developed by the Shanghai Institute of Technical Physics of the Chinese Academy of Sciences was used for tests of airborne remote sensing exploration of mineral deposits. For assessment of the new technology a comprehensive research programme has been worked out.The block diagram of the procedure of this study projectis shown in Fig.2.

For airborne data acquisition the newly developed Fine−Split Multispectral scanner(FIMS)was used .The FIMS was installed on board a specially modified remote sensing aircraft (the Cessna Citation S / II aircraft). Besides the image data of the FIMS, auxiliary data of aircraft,such as the time of flight ,the attitude of the airborne platform from the INS of aircraft are also simultaneously recorded on the on−board data tape recorder.The above mentioned auxiliary data are very useful for pre−processing of the remote sensing imagery .

The FIMS was designed and developed specially for geological use ,especially for rock type, alteration and mineralization zone recognition.The spectral region for the scanner was selected from 1.6 to 2.5μm.This spectral region was split into 12 narrow bands according to the spectral characteristic and different absorption and reflection peaks of various rocks and minerals . The major parameters of the FIMS are given in the following table (Table I).

All data acquired by the airborne scanner were processed on a digital image processing system.For image pre−processing and processing the following categories have been carried on:

Fig.1   Location of study areas:
a, western Zhunger area; b, Altaı
Tiemuerte area

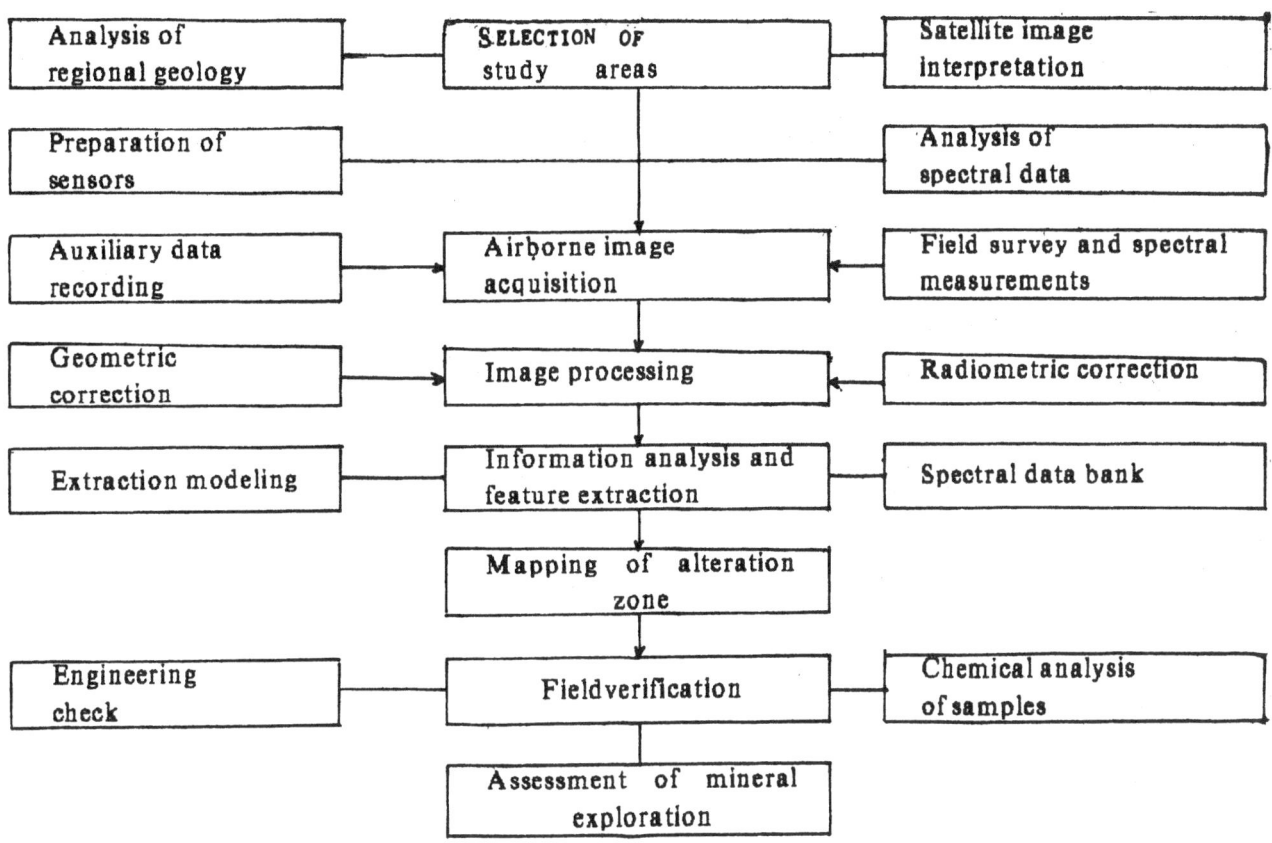

Fig.2   Block diagram of working procedure for mineral exploration in Xinjiang
area, China

Table I. The major technical parameters of CAS FIMS scanner

spectral region	1.6–2.5μm
Numbers of channels	12
Bandwidth of channels	50–100nm
FOV	90^0
IFOV	3–6mrad
Quantization	8 bits
Number of pixels per scan line	256–512

Table II. The high correlation coefficients of FIMS data (band to band).

No. of band	1	2	~3	4	5	6
1	1.000					
2	0.897	1.000				
3	0.910	0.996	1.000			
4	0.921	0.994	0.996	1.000		
5	0.935	0.919	0.929	0.946	1.000	
6	0.928	0.901	0.913	0.925	0.978	1.000

Table III. The typical absorption bands for some rocks and minerals in Xinjing study area.

Rocks and Minerals	Refletance range (%)	Strong Absorption peak (μm)	Weak Absorption peak (μm)
Auriferous altered rock, Auriferous quartz	78–84	2.22	2.35,2.44,2.47
Altered basalt	55–78	2.35	2.20,2.25,2.48
Altered tuff	53–72	2.34	2.20
Tuff	–50	2.21	2.35
Quartz magnesite rock	70–75	2.31	2.385
Talc magnesite rock	–60	2.31	2.34,2.46
Basalt	–55		2.22
Calcite	75–85	2.34	1.97
Chlorite	60–70	2.25	2.34
Epidote		2.33	2.20,2.25
Serpentine	38–46	2.32	
Muscovite	45–55	2.20	2.24,2.35
Limonite	30–38	2.20	
Kaolinite		2.209	2.16,2.32
Illite		2.218	2.251,2.342
Montmorillonite		2.205	2.209,2.232
Alunite		2.208	2.152,2.318
Gypsum		2.075	2.215
Jarosite		2.264	
Dolomite		2.335	2.265
Sericite		2.208	

a) Geometric correction for geometric distortions along the scanning and flight direction or the tangent distortion or the "spring" effect .

b) Radiometric correction along the scanning direction and far atmospheric effects.

c) Due to the extremily hight correlation of data between the spectral bands of FIMS (Table II), decorrelation processing is most important for information analysis and extraction in the first stage of processing .

Spectral differences between various rocks are the physical basis of information extraction. According to these differences, the information will be reconfigured following certain principale (algorithms). Classification is one of the information extraction methods. It is obvious that the more spectral differences there are, the simpler would be the method for the extraction of information, and a perfect result would also be obtained more easily. Otherwise, a suitable method has to be selected when the differences are not clear enough. According to the data , acquired in certain control areas, the ratioing method, principal component method, method of mineral absorption indexes (MAI) and the colour space transformation or IHS transformation are most useful and effective for the information extraction in processing the data.

The ratioing method is a simplest and most frequently used method of information extraction. Due to the differences of area, the differences of earth objects and the differences of data, the selection of bands for processing would be quite different. Physically, the ratio image represents a relative reflectance image or the ratio of reflectance of the same objects in two bands.

The principal component method is an essential method for extraction of the information of mineralization and alteration features. Principal component analysis in image processing is a method of transfering the digital data from spectral band space to the selected principal component space for identifying the feature information. It can be seen from a mechanism study of principal components analysis that the principal components are constituted by linear combination of every band. The enhancing and merging information exist in each principal component but the amplitudes of enhancement and merge are different.Each principal component contains different information types of enhancement and merge. Principal component analysis would show the ef-

fect of comprehensive enhancement, after recombining the intrinsic information. Therefore, the principal component colour composite image would greatly improve the information identification or the interpretability.

The method of mineral absorption indexes is a method of processing based on the quatitative evaluation of the absorption by mineral components in rocks, minerals and soils. The mineral absorption index (MAI) is characterized by the absorption depth at a particular absorption band. For the processing and analysis special algorithms have been developed and used.

RESULTS AND DISCUSSION

The analysis of Spectral Properties of Some Rocks and Minerals in Study Areas.

For the determination of the spectral characteristics of rocks and minerals in the study areas, two thousand groups of spectral measurements have been taken and analysed before the airborne data acquisition. By feature analysis of spectral data measured in field and in laboratories, characteristic absorption bands were distinguished and determined for the rocks and minerals of Xinjing study area. Some results are shown in Table III.

It was found that the spectral signature of rocks and minerals depend on the content of chemical elements.In the spectral range of $0.4-1.3\mu m$ the spectral signature is mainly caused by the transition of few positive ions and electrons.

In $1.3-2.5\mu m$ spectral range the characteristic absorption bands are determined by molecular vibration of hydroxyl,water and car bonate oxidation.

The absorption band of water appears in the $1.4-1.9\mu m$ range .For rocks and minerals the absorption band of water becomes stronger as the water content is increased.

If OH is incorporated with the ion of AI the absorption band occurs at $2.2\mu m$ and if the OH is incorporated with the ion of Mg the absorption band occurs near 2.3um. The CO_3^- absorption bands appeared at $1.84-1.90$, $1.98-2.02, 2.14-2.18, 2.32-2.36$ and $2.5\mu m$ range among these bands the $2.32-2.36$ and near $2.5\mu m$ bands are most obvious.

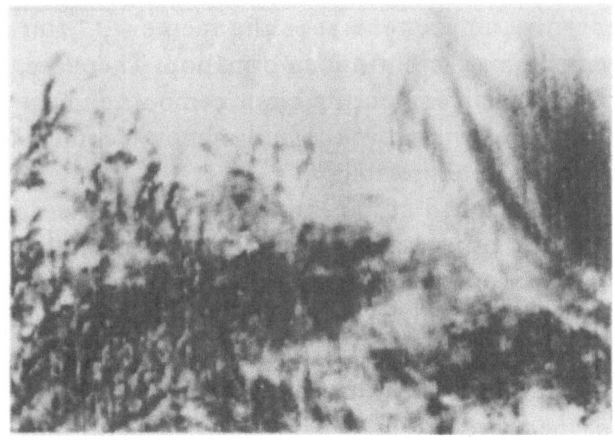

Fig. 3   Alteration and mineralization, gold mining, western Zhunger, Xinjiang

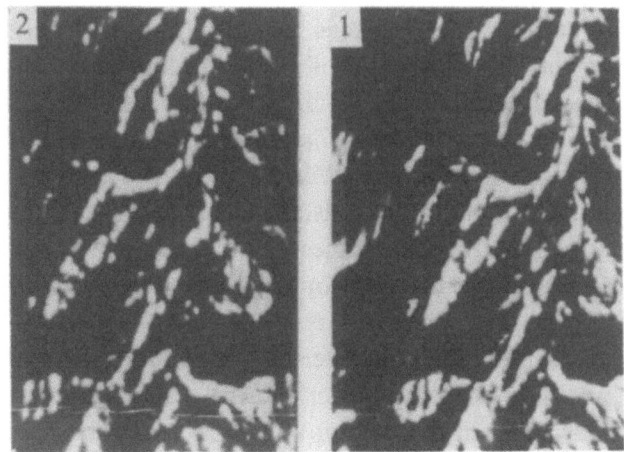

Fig.6   Processed FIMS images of Altai Tiemuerte area: 1, colour composition of principal components: 2 image of information extraction (rendered as black and white)

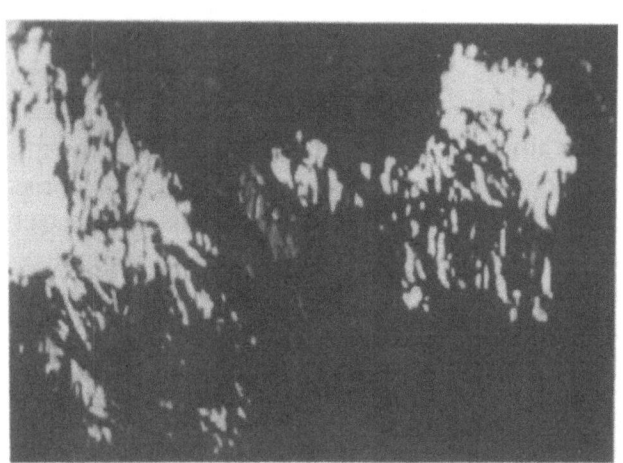

Fig.4   Zone of alteration and mineralization around granite body in western Zhunger, Xinjiang

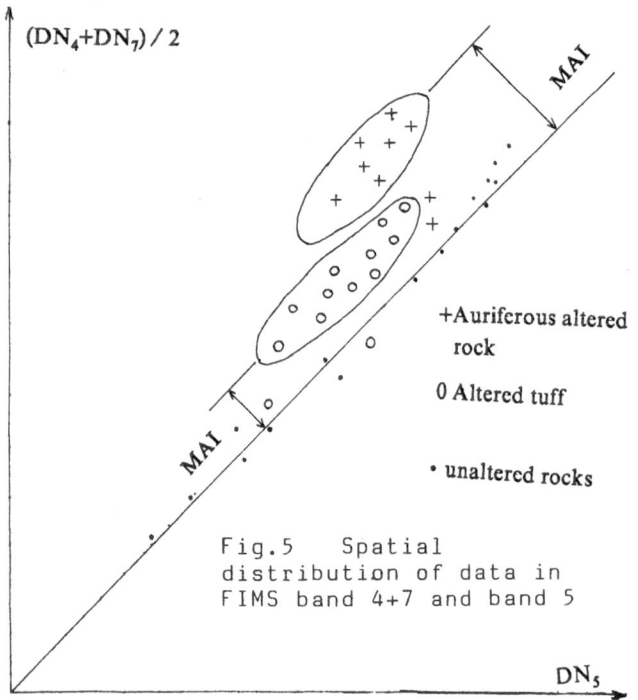

+Auriferous altered rock

0 Altered tuff

· unaltered rocks

Fig.5   Spatial distribution of data in FIMS band 4+7 and band 5

Meanwhile,the spectral reflectance of rocks also depends on the mineral components.For silicate rocks the spectral reflectance increases with increasing content of SiO_2.On the contrary the reflectance decreaseswith decreasing SiO_2content.For carbonate rocks the spectral reflectance increasing with increasing content of MgO and CaO .

Besides the above mentioned features,the reflectance of rocks and minerals also dependson granularity,surface colour,surface texture and degree of weathering.

By analysis of 810 groups of spectral data the wavelength range of 1.6—2.5μm,measured from 92 samples of rocks and minerals in Xinjiang study area ,the 12spectral bands of FIMS were selected for geological and mineral explorationas listed in Table IV.

Mineral Feature Information Extraction

The principal component method is most useful for identification of mineralization of gold and other metals.This true not only for the western Zhunger area but also for the Altai Tiemverte area .Figure 3 and Figure 4 represent some typical examples of FIMS imagery for the Westen Zhunger study area. These principal component images show that alteration and mineralization information are clearly displayed .The intense alteration of rocks such as basalt and tuffaceous sandstone is displayed in distinguishable colors. The purple and greenish—yellow hues in figure 3 are associated closely with gold mineralization.In our case the above represented hues are unique indicators of alteration and mineralization .Also it is very clear in figure 4 that the alteration and the gold mineralization zone ,indicated by dark—red and yellow colours,are located as a ring—shaped belt. around a small—sized granite rock body.

According to the spectral data measured in the field at western Zhunger, Xinjiang,China ,the spatial dis of data in FIMS band are drawn.It is discovered,that the position of spectral data and the degree of deviation off the base—line correspond closely with the types and intensity of alteration. In case of using the bands 4,5,7 or the bands of 2.143, 2.200 and 2.250 μm,the data of unaltered rocks ,such as tuffaceous sandstone silicalite and basalt are distributed closely along the base—line while the altered rocks and minerals such as silicified chloritized altered tuff and auriferus altered rocks are laid off the base—tine. It can be seen from Fig.5 that the distancebetween the long axis of scattered ellipse of the particular rocks or minerals in the space of spectral bands can be presented as the Mineral Absorption Index(MAI).

By using the method of Mineral Absorption Index , 5 types of minerals are easily distinguished and extracted; montmorillonite and kaolinite minerals;talc minerals ;chlorite minerals; pyrophyllite and allunite minerals; and carbonate minerals. There are two steps in the extraction of information from FIMS images of the Altai Tiemuerte area. The purpose of this analysis is to discriminate three lithological units in the area and to extract the mineralized information in the lithological unit III from the FIMS data. Though the two steps by using the same bands (bands 3,4,5,11) and method (principal component analysis) ,the division of three lithological units was mapped in a dis direction and the band combination which reflect several main type of rocks in unit III were determined. The extraction of mineralized feature information was based on the steps above ;its mapping transformation depended on the distribution of gossan in the band combinations.Fig.6 shows one of the typical research results .The right half(Fig.6—1) is a colour composition image of principal components in which three different color belts in the black background are clearly show.The right side of image is a belt which is composed of white,blue and a little pinkish color tone.The middle be lt is composed of yellow color with red and green spots scattered .The left half of image (Fig.6—2) is a processed image of extraction of mineralized feature information.The extracted information is composed of yellow,green and small pinkish spots on a blue background .In order from left to right these three belts represent lithological units II,III and IV . Field survey and examination shows that the boundaries of the three units are coincidente with, the three colour belts distinguished in the image .On the basis of processed results of separation of lithological units the dis of gossan could be more accurately determined and the mineralized feature information would also be easily extracted by principal component analysis.

In the left image, some spots with white and yellow, yellow—green,pinkish colours are displayed on a homogeneous blue background. The

Table IV. The selected spectral bands of FIMS and their geological significance

No.Band	wavelength (central) (μm)	Bandwidth (nm)	Geological significance
1	1.600	100	Statistical important
2	2.035	100	NH^+ absorption
3	2.087	100	Clay absorption
4	2.143	100	Reflection of clay mineral
5	2.200	100	Absorption by clay mineral
6	2.205	50	OH^-,CO_3^{--},absorption(Al—OH)
7	2.250	50	OH,CO_3^{--} absorption
8	2.280	100	OH,CO_3^{--} reflection
9	2.300	50	OH^-,CO_3^{--} absorption (Mg—OH)
10	2.330	50	OH,$CO_3^{--}$$CACO_3$ absorption
11	2.380	100	OH,CO_3^{--} reflection
12	2.450	50	OH,Mg—OH,Al—OH

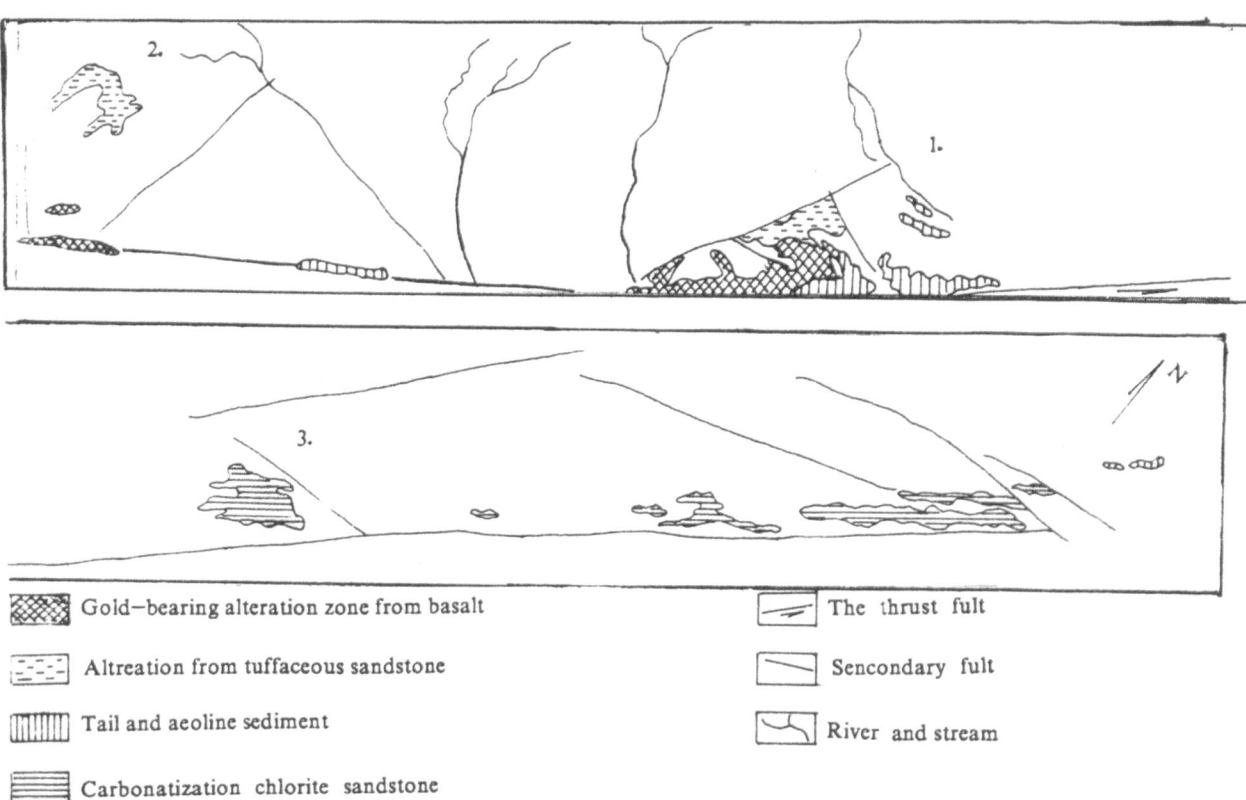

Gold—bearing alteration zone from basalt   The thrust fult

Altreation from tuffaceous sandstone   Sencondary fult

Tail and aeoline sediment   River and stream

Carbonatization chlorite sandstone

Fig.7   Alteration mapping by information extraction techniques of FIMS data

lithological unit II is mainly with white spots. The unit III is in white, yellow and yellow—green, while the unit IV also is blue. The results exactly conform with the stratigraphical zonea in this region.

According to the analysis of large numbers of pixels of principal components images in Altai Tiemuerte area ,it was found that the dis of pixel values in principal components coordinates could be divided into a number of colour regions,which were quantitied against the CIE colour standard. These were the yellow ,yellow—green ,light yellow—green,green ,light green—yellow and orange colours .It may be seen from the analysis of the above results ,that iron ore ,gossan ,skarn and garnet chlorite quartz schist are distributed in coordinate area with yellow colour ,metaquartzite and quartzite are distributed in the yellow —green area,while the tuff and impure crystalline limestone are in the green area .Making a comprehensive analysis of the rock and mineral samples ,it would be considered that the yellow anomaly of colour actually shows the mineralized feature information.Mineralized belt and intensely altered rocks,such as skarn and garnet quartz schist are located in the yellow coloured area of the image ,the less altered rocks ,such as chlorite quartz schist occure in the transition colour area between yellow and green .Rocks without mineralization ,such as tuff,impure crystalline limestone show the green colour mainly.

Mapping of Alteration and Geological Verification

As was known from the geological background and the mineralization situation of study areas,the main gold deposits are auriferrous quartz associated with contact zones of structurally controlled intrusives in Devonian to upper Carbonifereous strata. On the both sides of auriferous quartz veins intense silicification, arsenopyritization,pyritization and sericitization are developed. The gold bearing rocks in this area are mainly basalts and tuffaceous sandstones.Analysis of imagery acquired along the direction of the main ore controlling faults showed that the alteration and mineralization zone are mostly extended along the faults. As a result a number ofalteration zones, distributed along the main structure were mapped.The alteration map of this study area is presented in Fig.7.

After the analysis of imagery, verification by field geological survey has been carried out .This verification included the digging of prospect pits, rock sampling and the chemical analysis of samples .Samples for further chemical analysis have taken from both the main auriferrous quartz vein and the peripheral altered rocks.The results of analysis have shown that in the south—western part the average gold tenor of the verified target area(area 2,Fig.6) is about 5.93g／T. In the North—eastern study area, a target area (area 3, FIg.6) ranged 700m long and 100m wide has been identified with silicification, carlonatization and chloritization.Field verification and chemical analysis indicated that the gold tenor of this target area has in range of 0.5—18.85g／T. This is an area with obvious gold mineralization.

An excellent correspondence is seen between the principal component values and the chemical contents of rocks in the Altai Tiemuerte area. The results of this relationship analysis are shown in Fig.8.

CONCLUSIONS

1.Airborne remote sensing for geological and mineral exploration has been carried out in the West area of China,an area with arid natural conditions and relatively good exposure of rocks.

2.The development of a new 12 channel Multispectral Scanner in shortwavelength infrared region has greatly promoted the progress of direct exploration of minerals by remote sensing techniques.

3.For effective use of the new rsmote sensing data ,image processing and analysis are the first step of procedure.It was found that geometric and radiometric correction was essential for data pre—processing and that decorrelation processing ,ratioing, principal component analysis and the mineral absorption index method are very useful for information extraction techniques.

4.By spectral measurement and analysis,the spectral characteristics of the major rocks and minerals of the study areas in the short—wavelength infrared region have been analysed .The spectral absorption properties of veriousrocks and minerals are the main basis of alteration and mineralization information extraction.

5.Using information extraction techniques,the alteration,mineralization areas and even different important types of rocks are displayed clearly by

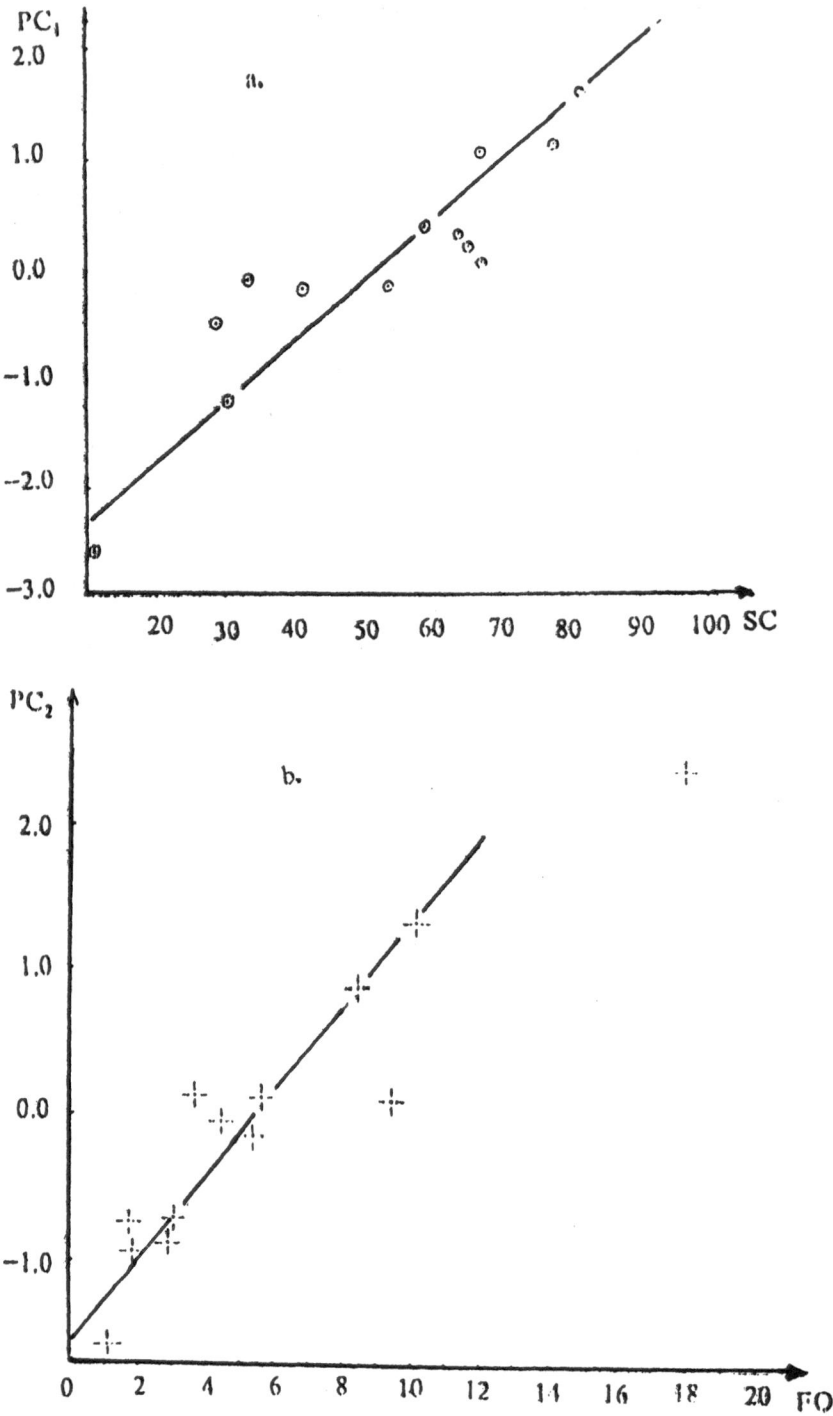

Fig.8   Relationship between chemical
composition content of rocks and principal component value in Tiemuerte area.
PC_1 and PC_2 , first and second principal component.   SC and FO, contents
of silica and ferric oxide in rocks

128

the col our anomalous in the images.Most of the anomalous information has been verifed and supported by field geological work and the chemical analysis of samples.

REFERENCE

1.Abrams, M,J.R. Ashley L,C. Rowan,A.F.H. Goetz and A.B Kahle (1977) Mapping of hyd rothermal alteration in the Cuprite Mining Di strict Nevada. Using aircraft Scanner imagery for the 0.46−2.36µm spectral region: Geology 5:713−718

2.Abrams J,B.M,o.Smith,and P.E.Johnson (1986).Spectral mixture modeling :A new analysis of rock and soil types at the viking Lander 1 site ,J.Geophys.Res 91:8098−8112.

3.Burrows,D.R.: Wood, P.C.: Spooner.E.T.C.1 986.Carbon isotope evidence for amagmatic origin for Archaoan gold−quartz vein ore deposits Nature (Lond−on). Vol.321.No.6073.P.851−854.GBR.(EL)

4.Cole.M.M.1985.Simple remote sensing in prospecting for gold, uranium andbase metals in desert areas in Australia and Africa,some case studies .In Davis,G.R., chairperson: Prospecting in areas of Desert Terrain .Inst.Min.and Metall .,London ,GBR,P.233−248.(EL).

5.Collins, W.,et al.1981.INfrared airborne spectroradiometer survey resuits in Western N evada area. Columbia University Aldrich Laboratory of Applied Geophysics final report to NASA.contract JPL 955832.

6.Craly Covault.1981.Shuttle data reveal new earth features.AWST Dec.21

7.Crovley .J.K.D.W.Brickey and L.C. Rowan (1989) Airborne Imaging SpectrometerData of the Ruby Mountains Montana: Mineral Discrimintaon using RelativeAbsorption Band−Depth Images.Remote sensing Environ 29:120−134

8.Elvidge C,D. and lyer R.J.R (1985) Influence of rock−soil spectralvariatim on the assessment of green Biomass, Remote sens. Environ. 17:265−279.

9.Elvidge C.D.(1987) Using of calibration Targete in the Measurement of 2.22µm Mineral Absorption features In thematic Mapper Data, Proceedings of the FifthThematic conference on Remote sensing for Exploration Goology

10.Gladwell, D.R.; Lawrence.P.;Dancziger,M.19 85.The application of rapid.semi−quantitative clay mineral determination at the Cortez gold mine. Nevada.InCook,j.J.,chairperson;Morris−j ones,D.R,chairperson et al.:Proceedings of the I ternational symposium on remote sensing of environment; Remote sensing for exploration geology.Environ.Res.Inst. Mich., Ann Arbor,M I.USA.P.403.(EL).

11.Goetz A.F.H.et al.1982.Mineral identification from orbit;initial results from the Shuttle Multispectral Infrared Radiometer.Science.Vol. 218.Dec.3.

12.Goetz A.F.H.et al.1983. Airborne imaging spectrometry;A means for directidentification of surface materials.Proceedings of the 1983 IternationalGeoscience and Remote Sensing Symposium.

13.Goetz,A,F,H,G vane,J solomon and B,N Rock(1985) imaging spectrometry forEarth remote sensing .Science 228:1147−1153

14.Goetz,A,F,H et al (1987) High resolution imaging spectromete :Science opportunities for the 1990′s,In Earth observing System,vol.IIc, Instrum,Panel Rep,NASA,cvashington ,D.C.

15.Gillespie AR.et,al (1986) Color enhancement of highly Correlated images ,I, Decorrelation and HSI Contrast stretches Remote sens,Environ,20:209−235

16.Hastings,D.A.1983. Combining Landsat and geophysical data for mineral and groundwater expolration in the Tucson area,Arizona, an d Sonora.Geophysiscs,1984,Vol.49,No.5,P.617.U SA,(EL).

17.Honey ,F.R.;Tapley,I.J.;Wilson.P.1984. Internation of macro−scale structural features in Western Australia from NOAA−AVHRR imagery.Proceedings of theInternational Symposium on Remote sensing of Environment ,Vol .3,P.381−394.III

18.Jackson,R,D(1983) Spectral indices in n−space ,Remote sens.Environ .V.13P.409−421

19.Krohn,M.D.1986.Spectral properties (0.4to 25 microns) of selected) rocksassociated with dissminated gold and silver deposits in Nevadaand Idaho JGR.Journal of Geophysical Researchj.B,Vol.91,No.1,P.767−783.USA,(EL).

20.Krohn,M.D.;Bethke,P.M.1984.Near−infrared spectral features of ammoniumminerals;applications for remote sensing of hot spring deposits.Abstractswith Programs−Geological Society of America,Vol.16,No.6,P.566.USA(EL)

21. Krohn, M.D.1984. Interpretation of Thermal−Infrared Multispectral Scanner images of the Osgood Mountains ,Nevada.Proceedings of

the InternationalSymposium on Remote sensing of Environment,Vol.3,P.735–737.III,(EL).

22.Kruse F.A (1988) Using of airborne imaging spectrometer data to map minerals as sociated with hydrothermally altered rocks in the northern Grapevine Mountains;Nevada and California,Remote sensing Environ.24:31–52

23.Lang ,H.R.et al.1987.Multispectral remote sensing as stratigraphic and structural tool,Wind River Basin and Bing Horn Basin areas, Wyoming .The Ameraican Association Petroleum Geologists bulletin.Vol.71,No.4.

24.Milton,N.M.,Collins,W.et al1983.Remote detection of metal anomalies on Pilot Mountain,Randolph County North Carolina. Econ.Geol.Vol.78

25. Metz, R.A, Blake, D.W.;Wotruba,P.R.1984 .Review of SPOT imagery simulationat the Copper Canyoon Cu–Au mining area, Lander Countr,Nevada.In SPOT simuiation applications handbook.Am.Soc.Photogramm.,Falls Church,VA,USA,P.83–91.(EL).

26.Raines,G.L.;Allen,M.S.1985 .Application of remole sensing and geochemistryto mineral exploration in Saudi Arabia; a case history.Prospecting in areas of desert terrain. Inst.Min.an d Metall.,London,GBR.P.233–248.(EL).

27.Taranik,J.V.;Sabins,F.F.1984.Remote sensing imagery for mineral exploration . In Lintz,j.,Lr.:Western geological excursions; Vol.3.Mackay Sch.MinesDep.Geol.,Reno,NV,U SA,P.205–223.(EL).

28.Taranik,J.V.1981. Advanced aerospace remote sensing system for global resource applications. Proceedings of the 15th International Symposium on Remote Sensing of Environment,Ann Arbor.

29.Tucher,D.H.;Wilson.P.1986.Magnaetic and thermal linears;some economicimplications for gold mineralization in the Yilgarn Block,Western Australia.Abstracts–Geological Society of Australia,Vol.15,P.193–194.AUS.(EL).

30.Vane,G, and Goetz,; A.F.H (1988) Terrestv ial imaging Spectroscopy, Remote sensing Environ.24:1–29

31.Williams,n.1986. Archaean gold;Exploring for magmatic origins.Nature(London),Vol.321, No.6073,P.851–854.GBR,(EL).

Exploration for hydrocarbon and geothermal energy

New understanding of geological structure based on a remote sensing study at a geothermal field in southern California, U.S.A.

J. C. Gutmanis M.Sc., F.G.S.
H. M. H. Lee M.Sc.
GeoScience, Ltd., Ascot, Berkshire, England

SYNOPSIS

During development of a geothermal field at East Mesa in S.California a remote sensing study was carried out, using Landsat Thematic Mapper imagery, SIR-A and Seasat radar imagery, and conventional air photographs. Although bedrock is largely mantled by wind-blown deposits, the satellite imagery displayed lineaments which in many cases could be correlated with subtle linear zones of more dense scrub growth observed on the air photographs. These zones of enhanced vegetation growth were assumed to mark the surface traces of faults striking transverse to the main fault set in the region. The proposed faults were integrated with subsurface data from dipmeter logs and well cuttings. The new surface data also provoked a re-interpretation of existing geophysical data, especially seismic profiles. A revised structural model of the field was developed, in which the form of the reservoir appeared more closely related to the fault pattern than previously. A 40 MW power plant has now been built.

INTRODUCTION

East Mesa "Known Geothermal Resource Area" (KGRA)[1] is one of many geothermal anomalies located south of the Salton Sea in the Imperial Valley of Southern California (Figure 1). Several of these KGRAs, particularly at East Mesa and at the Salton Sea[2] have undergone commercial development during recent decades. East Mesa KGRA is a high temperature (>150 C), liquid-dominated field driven by a major thermal high associated with the southern part of the Imperial Valley (the "Salton Trough", see below). At present, power is generated from a several plants, and further development is planned to exploit the resource.

The work described here was carried out during 1987 to assist in the siting and deviation of geothermal wells for a 40 MW development. The East Mesa geothermal anomaly, which has no surface expression such as geysers or hot springs, occurs in a semi-desert, sparsely vegetated terrain mantled by wind-blown sand and gravel. It has been intensively studied by geophysical methods, and a large body of data, which has been used for detailed modelling of the field, exists in the public domain[3,4,5,6].

The geological structure of the field has received considerable attention in the past, reflecting the potential influence of fracture permeability on reservoir productivity. The remote sensing study described here was conceived as a fresh look, using data acquired by newly available sensors and new processing techniques, for surface data which might contribute to the understanding of geological structure at East Mesa. In attempting this it was considered from the outset that the accurate integration of new surface data with existing sub-surface data would be essential.

REGIONAL SETTING

The Salton Trough is a NW-trending structural depression representing the northern onshore continuation of the Gulf of California, where active sea- floor spreading is occurring in a series of short NE-SW oriented spreading axes offset by dextral transform faults of San Andreas affinity[7,8]. The trough is therefore part of a complex transition zone which links the East Pacific spreading axis to the strike-slip San Andreas Fault (Figure 1). The high heat flow, together with seismic and gravity modelling, suggest that the Salton Trough is underlain by a spreading axis[9], and is therefore a rare example of onland spreading.

The trough is floored and flanked by crystalline rocks of Mesozoic and older age (including PreCambrian), and contains a complex and laterally variable sequence of Tertiary and Quaternary sediments up to 6km thick[10]. Marine, fluviatile and lacustrine facies are all present, reflecting the transient influences of the Gulf of California and the Colorado River. The latter is thought to have been the dominant source of the largely

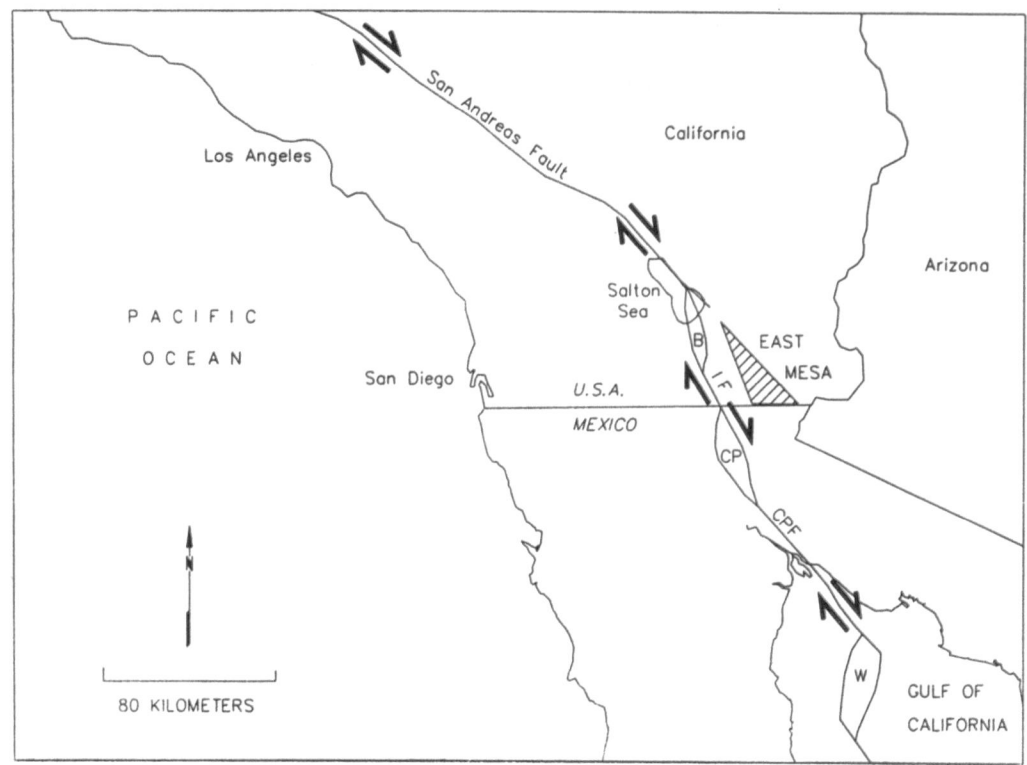

Fig. 1 Location of East Mesa.

B=Brawley Seismic Zone.   IF=Imperial Fault.
CPF=Cerro Prieto Fault.   W=Wagner Basin.
CP=Cerro Prieto.

terrestrial sandstones siltstones and clays which fill the central part of the trough[3,10].

The dominant structural element in and adjacent to the trough is a family of approximately NW-SE dextral strike-slip faults (Figure 2). These include the San Andreas Fault, whose south-easterly continuation is uncertain[10], but which possibly lies some 20km east of the East Mesa KGRA[9]. A similar major strike-slip structure, the San Jacinto Fault, obliquely transects the south-western boundary of the trough and may link south-eastwards with the NNW-SSE striking Imperial Fault[9]. The Imperial Fault, located approximately 10km west of East Mesa KGRA (Figure 2), is thought to be the main tectonic link to the Gulf of California spreading centres. The Imperial Valley is one of the most seismically active parts of the United States, with a catalogue of large earthquakes which, in recent years, have been intensively studied[11]. At least two major events (1940, Ms=7.1, and 1979, Ms=6.9) caused surface rupture on the Imperial Fault. The 1979 event produced a maximum dextral offset of 1.5m on the main fault, plus minor surface deformation on associated splay faults[12].

Seismic and geodetic monitoring[13] continue to reveal active deformation on structures both parallel and transverse to the San Andreas Fault system in the Imperial Valley. Of particular interest to this study is the observation[9,12,14] that transverse faults are an important part of the tectonic framework. An example is provided by the Brawley Seismic Zone, which appears to channel deformation sideways from the north-west end of the Imperial Fault towards the San Andreas Fault (Figure 2), and which contains within it zones of NE-SW striking seismicity. These are interpreted[9] as left-lateral, probably steeply dipping en echelon faults displaying mainly strike- or oblique-slip focal plane solutions, though some have a component of reverse movement. Elsewhere in the Imperial Valley, other field and geophysical observations[9,10,15] have suggested that transverse structures are an integral part of a complex linked fault system, often helping to define areas of extension where the NW-SE dextral faults are offset[16]. These areas of extension are commonly associated with enhanced microseismic activity, enhanced heat flow, volcanism, and geothermal fields[8,15,17].

RESERVOIR GEOLOGY

The East Mesa reservoir has been investigated by geophysical and geochemical methods, and by numerous boreholes drilled in the course of field exploration and development. The available database has been described and interpreted by many authors[3,4,6,18,19] while others have used the data to model reservoir flow characteristics[5,20,21]. This paper provides an outline of the reservoir geology, particularly in regard to structure.

134

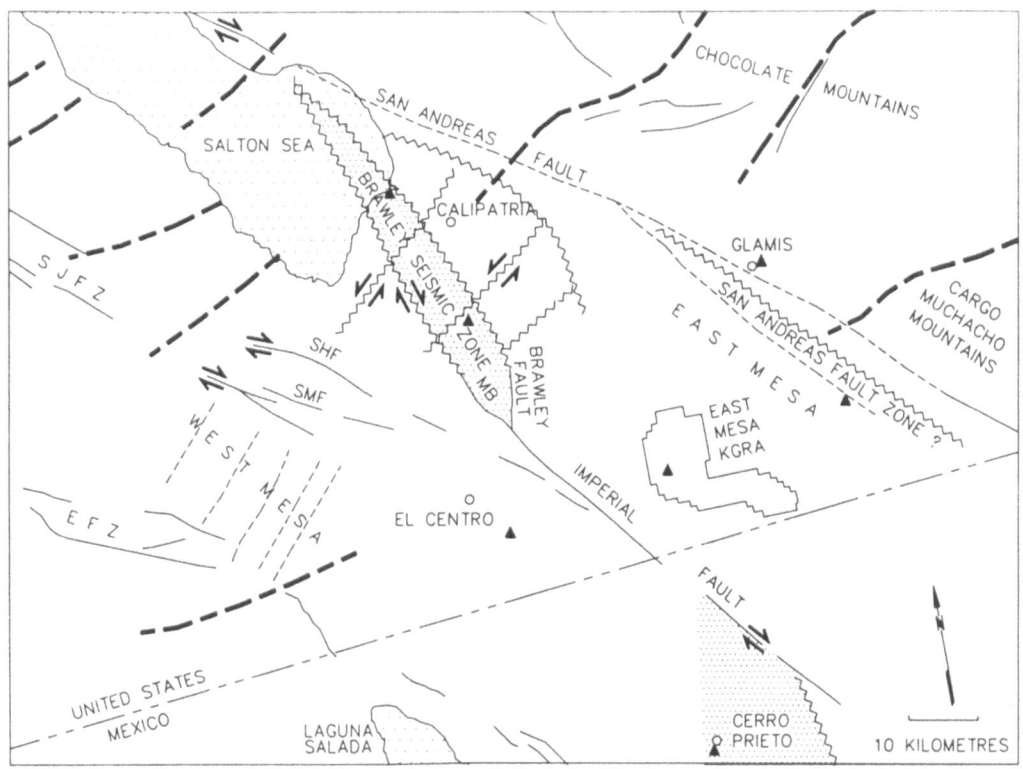

Fig. 2 Tectonic setting of East Mesa KGRA.

EFZ=Elsinore Fault Zone.   SJFZ=San Jacinto
Fault Zone.   SHF=Superstition Hills Fault.
SMF=Superstition Mountain Fault.
Triangles=geothermal fields.   Stippling=areas
of   high   microseismic   activity.   Thin
lines=faults (dashed where uncertain).
Sawtooth lines=seismicity lineaments. Thick
lines=gravity lineaments.

Borehole logging has demonstrated an upper clay-cap from the surface to about 600 to 700m depth, overlying a dominantly arenaceous sequence containing the geothermal reservoir and extending to about 2100m depth. Argillaceous deposits follow and apparently extend to the pre-Tertiary metamorphic and igneous basement at approximately 3000 to 4000m depth[5,19]. Well logs (electric, sonic, density, gamma-ray) show that the reservoir rocks are laterally very variable in thickness and facies, reflecting their deltaic origin. Many of the sandstone bodies have been interpreted as channel fills and bars[19].

Seismic reflection profiles acquired in 1966 and 1977 were used by Howard et al[19] to map reflectors and structure in the KGRA. Reasonably continuous reflectors were identified in the upper part of the sequence, above 1.2 to 1.6 secs TWTT, and tied to wells using time-depth curves computed from sonic logs. Structure contour maps derived from the reflectors defined a NE-SW striking anticline passing through the KGRA[19,22] and were also used to map a system of faults (Figure 3), whose NW-SE strike was presumably estimated from regional considerations. Other geophysical data, in particular microseismic monitoring[18] and self-potential contours[4], have tended to support a generally NW-SE striking pattern of faulting. However, as pointed out by Corwin et al[4], the evidence gleaned from these surveys is equivocal. Although most authors[5,19,21] agree that the reservoir is probably fault charged at depth, and that faults and fractures probably play an important role in enhancing reservoir permeability, a consensus model of the geological structure did not appear to exist.

The seismic profiles also helped to define the form of the geothermal reservoir. Below 1.2 to 1.6 secs TWTT the quality of the seismic record deteriorates, apparently indicating a zone of more dense, fractured sediments that have undergone temperature dependant alteration. Wells that have penetrated this "poorly reflective zone" (PRZ) show that it has lower porosity and permeability than the overlying material, and significantly higher seismic velocities. The PRZ, whose upper surface cuts across reflectors and is therefore not stratigraphically confined, is known to produce high flow rates in geothermal wells, probably due to fracturing. As mapped from the seismic data, the top of the PRZ is aligned approximately NNE-SSW, consistent with the form of the reservoir as defined by temperature contour maps at levels between 600m and 2135m[19]. The reservoir geometry appears therefore to be unrelated to the interpreted NW-SE striking fault system (Figure 3).

Bedrock at East Mesa is largely obscured by wind-blown sands and other superficial deposits associated with the late Quaternary Lake Cahuilla, the predecessor of the Salton Sea[23,24]. It was therefore decided that interactive processing and enhancement of digital Landsat Thematic Mapper (TM) data offered the best hope of displaying linear features related to bedrock faults. Winter imagery, with its low sun-angle at time of acquisition, was chosen to best display subtle topographic effects.

maps. This calibration exercise provided the basis for recognising possible new structural information in the East Mesa region, particularly when integrated with geophysical data sets.

Geological interpretations of the imagery were digitised into a computer mapping system based on a Compaq 386 running Autocad and other routines modified in-house for registering and manipulating 2D and 3D data sets. Lineaments derived from

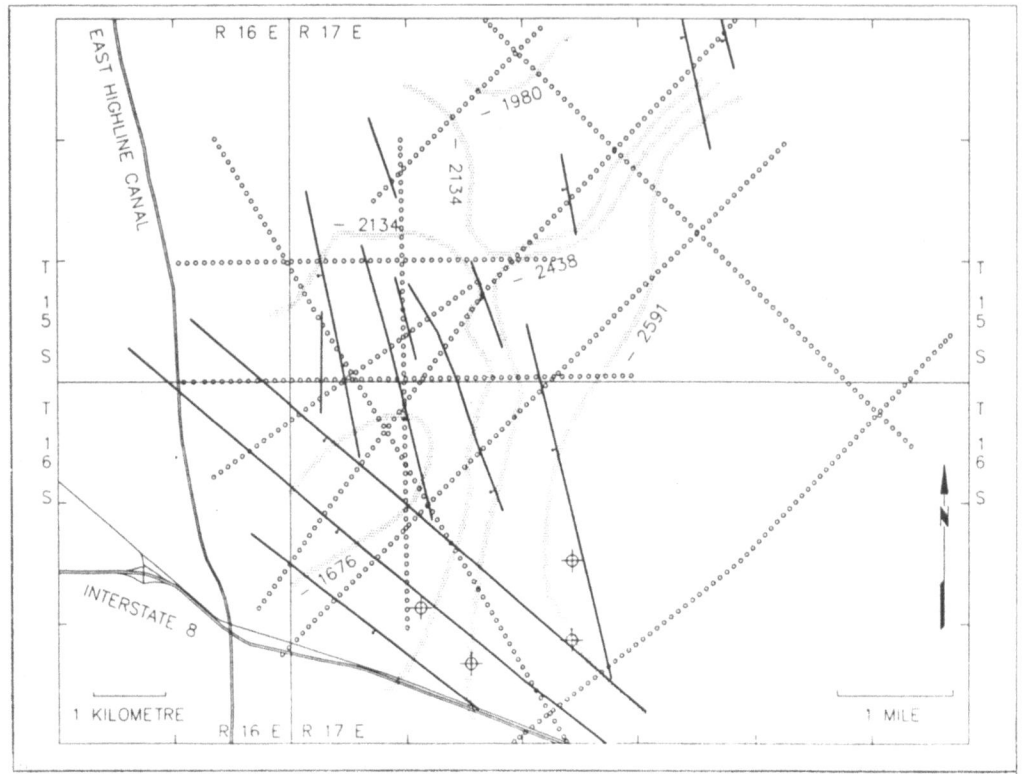

Fig. 3 Previous fault interpretation at
East Mesa KGRA.

Lines with bars=faults, with downthrow side.
Circle lines=seismic profiles. Crossed circles=geothermal wells. Thick lines with figures=contours (m. O.D.) on top PRZ.

Image processing was carried out using a variety of techniques. Of these, principal components analysis (PCA), which compares levels of statistical correlation between wavebands, provided the clearest discrimination between ground cover types, and also displayed linear tonal features (hereafter, "lineaments"). Colour-composite and black and white images were recorded onto 35mm slides for projection and interpretation.

The regional structural context of the KGRA was also studied, using Thematic Mapper Band 5 black and white prints at a scale of 1:250,000, and prints made from Seasat and SIR-A (Shuttle Imaging Radar) radar imagery. Interpretation of these started with the correlation of lineaments on the imagery with known structures shown on published

different sensor sources and enhancements were registered with a base map to build up a best interpretation of the area. In the vicinity of the wells, lineaments were integrated with the extrapolated surface traces of seismically-mapped faults (see below).

During the image interpretation phase of the study a field visit was made to East Mesa to collect relevant ground data, particularly regarding the variations in the gravel/sand ratio, and in vegetation cover and density. Some linear features observed on the imagery were also investigated, such as N-S striking shorelines associated with Lake Cahuilla. This field information contributed to the final interpretation of imagery.

Lineaments interpreted from the Thematic Mapper imagery of East Mesa are shown on Figure 4. Although the NW-SE trend is represented, particularly in the northern part of the KGRA and immediately east of it, relatively short lineaments (1 to 5km) of NE-SW strike dominate the area of interest surrounding the wells. Two members of this set, lineaments A and B (Figure 4), pass close to the wellheads and are clearly seen on the imagery.

coincide with zones of bedrock faulting and also with fracturing in the overlying superficial cover[25]. Surface deformation of this nature may be expected in an area of high seismicity (see above).

Thus, lineaments A and B were assumed to be the surface traces of two faults striking transverse (NE-SW) to those previously mapped from the

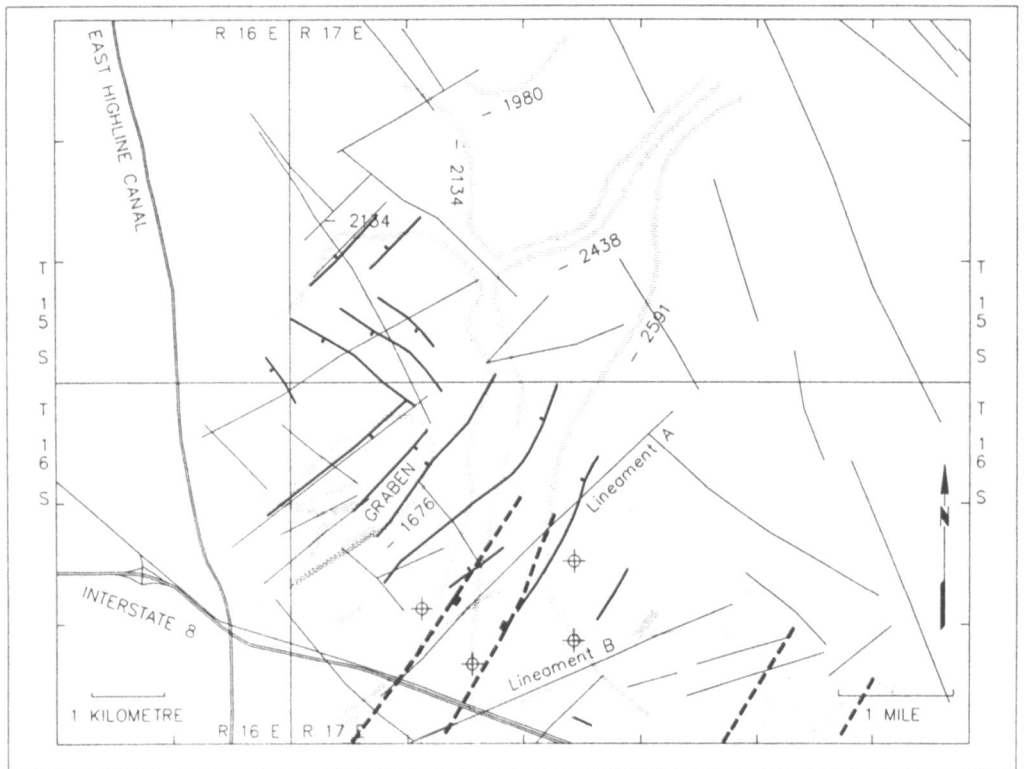

Fig. 4 Revised fault interpretation at
East Mesa KGRA.

Symbols as for Fig. 3, plus:
Thin lines=TM and/or radar lineaments.
Hatched zones=air photograph lineaments.
Dashed lines=faults identified from TDEM data.

During the field visit black and white stereo air photographs (1:12000 scale) of the wellhead area were obtained. Interpretation of these revealed narrow linear zones of slightly denser scrub growth, which were impossible to identify on foot. These features, entered into the computer mapping system, correlated closely with lineaments A and B and with other TM lineaments in the area (Figure 4). In places the TM and airphoto lineaments were found to coincide with breaks of slope, as revealed by correlation with topographic contours entered into the mapping system.

The linear zones of denser scrub were interpreted as the surface representation of faulted bedrock with enhanced groundwater supply. It was not possible to carry out trenching across the lineaments to prove faulted bedrock. However, similar features mapped from air photographs to the north of East Mesa were trenched and shown to

seismic data (Figure 3). On the basis of this finding, it was decided to re-work the original seismic interpretation assuming a NE-SW fault strike rather than a regional NW-SE strike. As a result, lineament A appears to correlate with a normal fault downthrowing NW, whose surface location (Figure 4) is approximate only because of uncertainties in projecting upwards from its highest mapped level on the seismic profiles. Unfortunately, lineament B is difficult to correlate with faults because of the poor density of seismic lines in its vicinity.

A further control on the geometry of the two proposed faults in the sub-surface was provided by dipmeter logs and analysis of cuttings[26] from the geothermal wells. The cuttings yielded evidence of shearing at certain depths, while dipmeter logs, which have been widely used elsewhere in the KGRA to map faults and roll-over structures[22], revealed

137

fault intersections which were used to interpolate fault planes between the surface and the wells. This process of interactive structural modelling was carried out in 3D on the computer mapping system. Fault plane dips of 60 and 80 degrees were tried, as suggested by the seismic profiles, in order to produce a "best-fit" visualisation (Figure 5) of the two faults represented by lineaments A and B. The fault geometries displayed on Figure 5 are consistent with the dipmeter intersections and with the surface fault traces as mapped from TM imagery.

In general, the revised fault pattern for the KGRA (Figure 4) is more closely related to the NE-SW striking anticline[19], and to the form of the geothermal reservoir (see above) than was the old interpretation. In particular, the highest part of the PRZ (marked by the -1676m contour), corresponds to a zone of graben faulting striking NE-SW. These proposed faults may have acted as conduits for rising geothermal fluids, which caused alteration at higher stratigraphic levels. This area of the KGRA also has the highest heat flow[5], and corresponds with a topographic low at the surface.

DISCUSSION

Transverse structural trends are known to occur in the East Mesa region and are generally interpreted as local transfer elements in a linked fault system whose regional elements strike NW-SE. It is suggested that the proposed NE-SW striking faults at East Mesa (Figure 4) represent just such a transfer element, linking regional, NW-SE faults located to west and east. The regional fault to the west would be the Imperial Fault, while that to the east may be a south-easterly extension of the Calipatria Fault (Figure 2)[2], represented by the TM lineaments immediately east of the KGRA (Figure 4). Dextral slip on these boundary faults would cause extension in the intervening area, possibly explaining the presence of the geothermal anomaly at East Mesa. This hypothesis is broadly consistent with the regional tectonic framework and will need further seismic reflection data in the vicinity of the KGRA to test it.

Interpretation of TM imagery at East Mesa KGRA has lead to a new model of the faulting which is more compatible with other data sets and with the

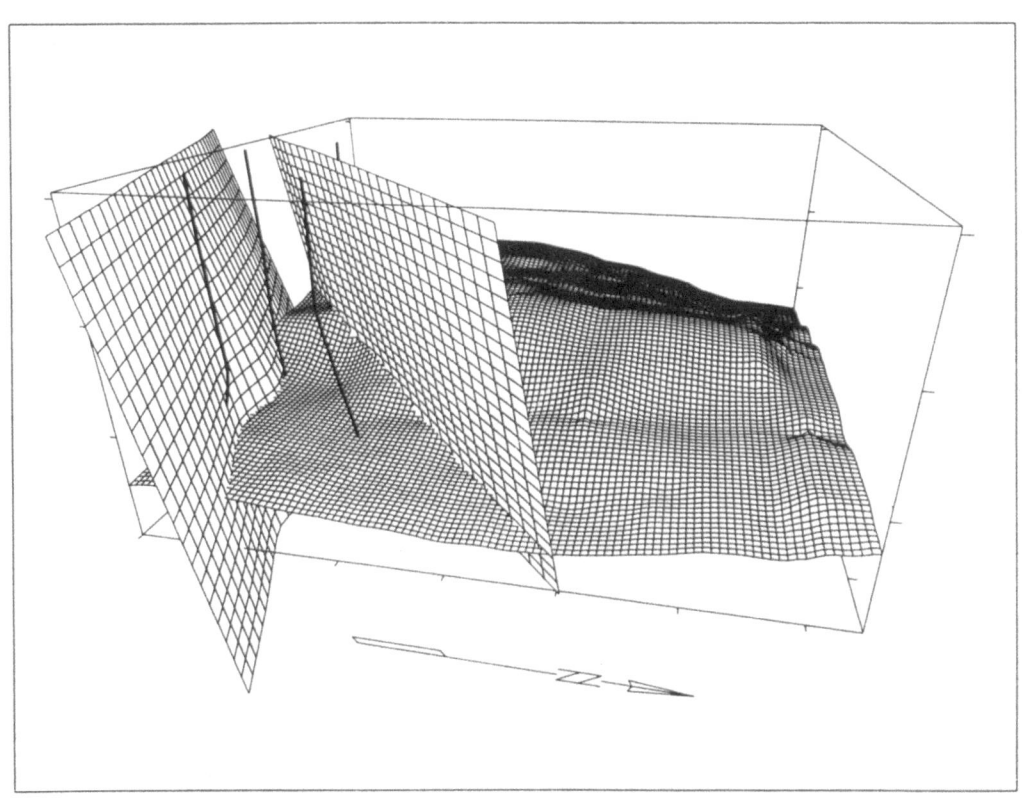

Fig. 5 3D block model of the proposed faults
at lineaments A and B (see Fig. 4).

Flat-lying surface is top PRZ. Thick sub-vertical
lines are wells.

138

geometry of the reservoir. Although it must remain tentative without new seismic reflection data, a system of mainly NE-SW striking faults would be parallel to the main fold structure in the field, and to the geothermal reservoir as defined by altered rocks and temperature contours. The results of the remote sensing study led to a re-appraisal of other geophysical data sets in the KGRA. In particular, re-interpretation of time-domain electromagnetic soundings (TDEM)[27] identified a NE-SW zone of faulting coincident with lineament A (Figure 4). A recent self-potential survey[28] clearly showed the influence of both NW-SE and NE-SW structural trends within the KGRA, and suggested NE-SW faulting in the vicinity of lineament A.

Satellite imagery proved to be a useful source of new information to help with the interpretation of geological structure at East Mesa. Although imagery is often used for regional studies, the results of the East Mesa study indicate the potential value of using high-resolution, interactively enhanced imagery at the local, well-specific scale. As the geothermal plant comes on-line, production data from all wells and from the different producing horizons may indicate a correlation between reservoir flow characteristics and the revised geological structure.

Acknowledgements

Permission to publish from the GEO East Mesa Limited Partnership and the Directors of GeoScience Limited is gratefully acknowledged. The work was reviewed by J.Combs, C.Swanberg and J.Ziagos (of Geothermal Resources International), and R. Whittle and P.Ledingham (of GeoScience). The figures were prepared by H.Lee, M.Dobson and W. Howliston of GeoScience.

References

1. Mohorovich L.M. Geothermal resources of the California desert-a summary review. In: Geology and Mineral Wealth of the Californian Desert. South Coast Geological Society, 1980, p.171-189.

2. Younker L.W., Kasameyer P.W. and Tewhey J.D. Geological, geophysical and thermal characteristics of the Salton Sea geothermal field, California. Journal of Volcanology and Geothermal Research, Vol. 12, 1982, p.221-258.

3. Swanberg C.A. The Mesa geothermal anomaly, Imperial Valley, California: A comparison and evaluation of results obtained from surface geophysics and deep drilling. In: Proceedings, Second United Nations Symposium on the Development and Use of Geothermal Resources, Government Printing Office, Washington D.C., 1975, p.1217-1229.

4. Corwin R.F., DeMoully G.T., Harding R.S. and Morrison H.F. Interpretation of self-potential survey results from East Mesa geothermal field, California. Journal of Geophysical Research, Vol. 86, 1981, p.1841-1848.

5. Goyal K.P. and Kassoy D.R. A plausible two-dimensional model of the East Mesa geothermal field, California. Journal of Geophysical Research, Vol. 86, 1981, p.10719-10733.

6. Goldstein N.E. and Carle S. Faults and gravity anomalies over the East Mesa hydrothermal-geothermal system. Transactions, Geothermal Resources Council, Vol. 10, September 1986, p.223-228.

7. Angelier J., Colletta B., Chorowicz J., Ortlieb L. and Rangin C. Fault tectonics of the Baja California peninsula and the opening of the Sea of Cortez, Mexico. Journal of Structural Geology, Vol. 3, 1981, p.347-357.

8. Silver P.G. and Valette-Silver J.N. A spreading episode at the southern end of the San Andreas fault system. Nature, Vol. 326, 9 April 1987, p.589-593.

9. Fuis G.S., Mooney W.D., Healey J.H., McMechan G.A. and Lutter W.J. Crustal structure of the Imperial Valley region. United States Geological Survey, Professional Paper, 1254, 1982, p.25-49.

10. Sharp R.V. Tectonic setting of the Imperial Valley region. United States Geological Survey, Professional Paper, 1254, 1982, p.5-14.

11. United States Geological Survey. The Imperial Valley, California, Earthquake of October 15, 1979. USGS Professional Paper, 1254, 1982.

12. Reilinger R. and Larsen S. Vertical crustal deformation associated with the 1979 M=6.6 Imperial Valley, California, earthquake: implications for fault behaviour. Journal of Geophysical Research, Vol. 91, 1986, p.14044-14056.

13. Thatcher W. Horizontal crustal deformation from historic geodetic measurements in Southern California. Journal of Geophysical Research, Vol. 84, 1979, p.2351-2370.

14. Nicholson C., Seeber L., Williams P. and Sykes S.R. Seismicity and fault kinematics along the Brawley seismic zone amd adjacent regions. EOS, Vol. 66, No. 46, 1985, p.953.

15. Sibson R.H. Earthquake rupturing as a mineralising agent in hydrothermal systems. Geology, Vol.15, 1987, p.701-704.

16. Sibson R.H. Stopping of earthquake ruptures at dilational fault jogs. Nature, Vol. 316, 18 July 1985, p.248-251.

17. Newmark R.L., Kasameyer P.W. and Younker L.W. Research drilling at the Salton Sea geothermal field, California: the shallow thermal gradient project. EOS, Vol.67, No.39, 1986, p. 698-707.

18. Combs J. and Hadley D. Microearthquake investigation of the Mesa geothermal anomaly, Imperial Valley, California. Geophysics, Vol. 42, No.1, February 1977, p. 17-33.

19. Howard J.H., Apps J.A., Benson S.M., Goldstein N.E., Graf A.N., Haney J.P., Jackson D.D., Juprasert S., Majer E., McEdwards D.G., McEvilly T.V., Narasimhan T.N., Schechter B., Shroeder R.C., Taylor R.W., Van de Kamp P.C. and Wolery T.J. Geothermal resource and reservoir investigations of U.S. Bureau of Reclamation leaseholds at East Mesa, Imperial Valley, California. Lawrence Berkeley Laboratory, Energy and Environment Division, 7094, 1978.

20. Riney T.D., Pritchett J.W. and Rice L.F. Three-dimensional model of the East Mesa hydrothermal system. Transactions, Geothermal Resources Council, Vol. 4, September 1980, p.467-470.

21. Morris C.W. and Campbell D.A. Geothermal reservoir energy recovery-a three dimensional simulation study of the East Mesa field. Journal of Petroleum Geology, Vol. 33, 1981, p.735-742.

22. Davis D.G. and Sanyal S.K. Case history report on East Mesa and Cerro Prieto geothermal fields. Los Alamos Science Laboratory, Vol. 7889, 1979.

23. Norris R.M. and Norris K.S. Algodones Dunes of south-eastern California. Geological Society of America Bulletin, Vol. 72, 1961, p.605-620.

24. Sharp R.P. Intradune flats of the Algodones chain, Imperial Valley, California. Geological Society of America Bulletin, Vol. 90, 1979, p.908-916.

25. Heath E.G. Evidence of faulting along a projection of the San Andreas Fault, south of the Salton Sea. In: Geology and Mineral Wealth of the Californian Desert. South Coast Geological Society, 1980, p.467-474.

26. Waibel A. A review of well site data from Geo East Mesa drill holes 27-8 and 53-17. Confidential report, Columbia Geoscience, 1987.

27. Swanberg C.A. Personal Communication, 1987.

28. Corwin R.F. Self-potential survey East Mesa geothermal field, Imperial County, California. Confidential report, February 1988.

New methods and technologies for forecasting onshore and offshore oil and gas fields

A. L. Stavtsev
O. I. Karasev
'Aerogeologia' Association, Moscow, U.S.S.R.

Remote sensing data are widely used at the present time in the study of oil-gas-bearing basins and hydrocarbon field forecasting.
In addition to satellite image geological and structural interpretation and morphometrical analysis of topographical and bathymetrical maps original methods of oil-gas field forecasting are developed.

1. Remote sensing geochemical study is a principally new approach to earth crust study and mineral deposit forecasting. The method was developed and introduced by the Soviet geologist A.A. Ivanov.

This method is based upon the new scientific idea on the geochemical field in which the geochemical relationships manifest themselves at a distance of tens and hundreds kilometres. Information on the sites of ore, oil or gas accumulations is recorded in geochemical signatures of rocks at great distances from accumulation sites.

Remote sensing geochemical investigation technology included lithochemical sampling, chemical element analysis, special mathematical processing of analysis data, determination of calculated depth of objects forecasted, prognostic geochemical map production.

Lithochemical sampling is conducted every 10-50 m along random traverses (depending on the scale of surveying) or within a grid of individual 10-20 x 10-20 m areas where 20-25 samples are grouped in sets. The sample sets are distributed more or less uniformly over the grid.
The total number of sample sets is 4-5 per 1 sq km of map area independently of its scale (not less, however, than 80-100 sample sets for the whole territory under study).

Samples are collected from loose sediments or hard bedrock. The samples may be collected at equal intervals from borehole core.

The samples are analyzed using any laboratory method, allowing determination of elemental rock composition with accuracy not less than 10^4 - 10^{-5} percent. The number of chemical elements analyzed should not be less than 18-20.

Special mathematical processing of analysis results in the distance from the sample (sample set) collection site to inferred accumulations being determined and their suggested depth.

Anomaly maps are made manually and using computers. The operations consists in construction of circles, radii of which are equal to calculated distances from sampling sites to inferred accumulations. Each circle is a geometric point site favourable for target detection. Local areas of multiple circle intersections are anomalies, corresponding to inferred targets.

The production of a prognostic geochemical map is accomplished. The map shows the anomalies in geological context, possibly corresponding to mineral deposits of inferred mineral type and at the depths indicated.

This method of geochemical investigations has been tested in a number of ore and petroleum-bearing districts of the U.S.S.R. In all cases, the prognostic geochemical maps record in the form of anomalies almost all the known deposits and delineate the promising areas. Preliminary testing has shown positive results for some of them.

This method of mineral deposit prediction is highly effective, both in regard to forecast reliability and in economics. Its effectiveness in coastal shelf studies should be noted specially, since sampling coastal areas allows prognostic geochemical maps to be produced of offshore areas that are situated at distance of some tens and hundreds kilometres from the coast. The forecasting is conducted without ships and any other technical means.

Ivavov's method is currently still in the development stage. There are many problems unsolved both in the assessment of fundamental theory and in the technology.

2. Satellite imagery manipulation
An interesting method of satellite imagery transformation in direct prospecting for petroleum pools is developed by O.I. Karasev. Karasev's method is based on the fact that hdyrocarbon gases migrate from petroleum

accumulations to the land surface through the overlying rocks and affect the vegetation. The results of such exposure are not detectable visually, neither directly in vegetation cover, nor on the satellite imagery. However, combination of the satellite imagery, obtained simultaneously in different bands of the electromagnetic spectra and correlation of those images by a specific manner enable one to distinguish hydrocarbon gases anomalies in vegetative cover. The example of known deposits shows that these anomalies correspond to petroleum accumulations.

Karasev's method is applicable both to anticlinal and non-structural traps (lithological, stratigraphical ones, the traps beneath allochtonous tectonic covers etc.).

3. Satellite imagery in coastal areas

L.T. Shevyrev proposed a method of neotectonic structural studies for deep coastal shelves using satellite images. The method is based on the information extracted from the particularities of suspended matter. These are recorded rather well on images, obtained in blue-green visible bands of the electromagnetic spectrum. Thus, the Shevyrev's method brings into use information which was regarded up to the present times as an obstacle in shelf studies using aerial and satellite survey data.

The method includes photo-tone density measurements on the satellite (or aerial) images and production of a map of equal densities, correlation of this data with information on bottom sediment composition at individual sites, as well as with sea floor topography recorded on bathymetrical charts. Such an integrated analysis allows a map of recent bottom sediments and a map of recent tectonic structure of the sea bed to be produced.

The map of recent tectonic structure of the sea bed shows the large uplifts, troughs, blocks, faults, flexures, as well as local uplifts which may correspond to deep-seated anticlinal traps.

Shevyrev's method has been tested at different depths and favourable results were obtained. During these studies, a 70 percent fit of the results of marine seismic investigations was achieved. The maps produced using Shevyrev's method proved, however, to be more sophisticated than the ones produced using seismic data.

4. Potential petroleum resource from satellite imagery

The method of potential petroleum resource calculation using satellite images is developed by A.I. Antsyforov. The method assesses the degree of known petroleum-bearing structure fit in planar view with the photo-anomalies on satellite images, which corresponds to recent anticlinal uplifts. The correlation of resources of known fields within basins studied (or within analogous fields occuring in similar tectonic conditions) with the areas of photo-anomalies on the satellite images is also

conducted.

Antsyforov's method is based on the observed direct proportional relationship between resources of known fields and areas of photo-anomalies that correspond to appropiate neotectonic and recent structures. Such a relationship was traced during satellite image studies for practically all the main petroleum-bearing basins worldwide, situated in different tectonic conditions -on ancient and young platforms, in marginal troughs and intermontane troughs.

The reliability of potential petroleum resource assessment using satellite images depends largely on geological basin knowledge and known field occurence or absence within them. But even in the most unfavourable conditions such as assessment allows an idea about the inferred hydrocarbon resource range to be established.

5. Structural and geological interpretation of satellite images is based upon the fact that buried fault and fold structures manifest themselves on the earth surface in topography, stream pattern, soil moisture, vegetation and in other landscape components. As a result of this, buried structures are expressed on the satellite images in the form of photo-tone and photo-pattern anomalies. The most prominent structures are those that actively develop at the neotectonic and recent stages of evolution. The studies of vast petroleum basins of different tectonic types have shown that these structures play a most important role in the general distribution of petroleum-bearing structures.

Interpretation is conducted in sequence beginning from small-scale satellite images to large-scale ones. This allows the boundaries of petroleum basins or important parts of them to be determined accurately to delineate potential zones of petroleum migration and accumulation and to detect local anticlinal structures which may serve as the petroleum traps. The large-scale satellite or aerial image analysis results in control and direct guidance to exploration and even exploration operations.

6. Morphometrical analysis consists in the quantitative evaluation of sea bed and adjacent coastal area relief using bathymetrical and topographical maps. Analysis of results obtained and of geological and tectonical maps at hand as well as geophysical data helps develop an idea of the main tectonic features of the sea bed, to locate and to map the most important structural troughs, uplifts, grabens and horst, monoclines, flexures and faults.

7. Integration

The interpretation of shallow shelf satellite images is fulfilled in those cases when the sea bottom is directly seen through the water. Generally, it is possible for the depth down to 10-12m (rarely to 20m) and depends upon the water clarity. The interpretation of sea bed structure is greatly complicated by currents

142

and suspended solids in the water. The reliability of structure delineation increases where satellite images obtained are used in different times.

Of interest is the experience in fault and local uplift delineation in shelf areas based on ice cover pattern. Its comparison with coastal area structures on the satellite images, obtained during multiple coverages, shows that the faults hidden beneath the water are often manifested in ice by linear zones of ice hummocking and clearings. Based on the pattern of small-scale fracturing of ice fields observed satellite images, it is possible to map recent sea bed uplifts. Faults and structures on the satellite images is possibly controlled by the sea bed roughness, even though it is of limited amplitude.

Of interest is also the experience in mapping large fault zones both in continental and offshore areas based on the character of clouds pattern on the satellite images. Not infrequently it is possible to observe the strong subordination of small cloud chains or continuous cloud field boundaries to fault zones in the earth's crust. This may well be explained by the cloud sensitivity to linear electromagnetic field anomalies, generated by some faults.

The integrated analysis of results and geological and geophysical data on hand allows potential zones of petroleum accumulation and potentially petroliferous local structures to be distinguished. Further more this analysis allows local structures to be ranked according to their prospectivity. This differentiation is based upon the analysis of data on the thickness of sedimentary cover, reservoir properties, availability of direct petroleum manifestations and knowledge about the structures in the sedimentary cover that are located by seismic studies of drilling.

Conclusion

Each method of petroleum field forecasting enables potentially petroleum-bearing structures to be predicted more or less reliably. Each of these methods has its own limitations.

The rational method of integration expands greatly the usefulness of remote sensing data and improves the prediction, particularly when the values obtained independently by different methods are similar.

The technology of remote sensing forecasting within the continental petroleum-bearing basins includes the following methods:
-structural and geological interpretation of satellite images;
-transformation of satellite images using Karasev's method;
-integrated analysis of results obtained and of geological and or geophysical data available;
-potential petroleum resource calculation using Antsyforov's method.

The forecast is achieved for any terrain and geological-tectonic conditions, including those in complexes beneath salt as well as beneath tectonic covers in folded mountain margins. Without drilling or geophysical operations, the depth of inferred petroleum accumulation is estimated and the potential petroleum resources are evaluated.

The technology of remote sensing forecast over continental shelves includes the following methods:
-morphometrical analysis;
-shallow shelf satellite image interpretation;
-satellite image processing using Shevyrev's method;
-special geochemical studies following Ivanov's method;
-integrated analysis of results and geological and geophysical data at hand;
-potential petroleum resource calculation using Antsyforov's method.

The forecast is achieved for any offshore area including ice-covered ones. Ships or other expensive technical means are not used. The depths of inferred targets are estimated and potential petroleum resources are evaluated.

The application of the technologies described allows us to plan seismic or drilling profiles and to eliminate large volumes of expensive geophysical and drilling operations. Further more, the duration of regional investigations of potential petroleum-bearing areas is substantially eliminated, particularly in areas situated in inaccessible regions.

143

Groundwater and environmental issues

Location of high-yielding groundwater sites in Zimbabwe by use of remotely sensed data

J. W. Finch M.Sc., Ph.D., M.I.Geol.
Institute of Hydrology, Wallingford, Oxon, England

SYNOPSIS
Landsat TM data covering an area of basement complex "Older Gneiss" in Masvingo province, Zimbabwe, is analyzed for information to help locate higher yielding groundwater sites. Linear features are identified and their proximity to existing wells is investigated to suggest that east-west trending lineations are associated with above average yields. The reflection characteristics of vegetation during the dry season are used to suggest areas where groundwater may be more abundant or nearer the surface.

Water supply is a fundamental requirement for the successful exploitation of mineral reserves. However, it can be difficult to obtain a reliable supply when the reserve is located in arid or semi-arid areas where the geology is dominantly basement complex. In these conditions a sound exploration strategy is required to efficiently assess the water resources of the area. Remote sensing can play a valuable part in this exploration. The work described in this paper covers an investigation to suggest suitable targets for locating wells with above average yields in an area of SE Zimbabwe where economic deposits of gold and nickel occur in the vicinity.

THE STUDY AREA
The area is centred at approximately 31°18'E 20°26'S, to the south-west of Zaka in Masvingo province, Zimbabwe. It consists of the valley of the river Chenyu which flows west-south-west into the river Mtilikwe, Figure 1. Low hills forming the catchment divides occur in the north and south of the area. The altitude ranges from 620m in the west to over 960m at the top of the highest hills.

The area has a semi-arid climate with a mean annual rainfall of around 750mm. Rainfall occurs predominantly in November to March[1] and is associated with the movement of the Inter-Tropical Convergence Zone into the southern hemisphere. The rainfall is very variable and

is heaviest in December or January. It tends to occur as a few days of rain alternating with dry spells of a few days duration. The mean annual temperature is 21°C and, on a monthly basis, there is no excess of precipitation over potential evaporation.

The area is situated on an area of "Older" gneiss which lies between the Limpopo Mobile Belt which was active[2] between 3800 and 2600 Ma and the Zimbabwe batholith, a "Younger" granite emplaced[3] between 2700 and 2600 Ma. The area has been subject to extensive deformation with at least four phases of deformation recognised in the Archaean and early Proterozoic[4]. Weathering has affected the units to depths in excess of 60m.

Ground truth in the area is provided by large scale geological maps and the 1:50000 topographic maps. The British Geological Survey has compiled a well inventory of the area which gives information on the general geological sequence penetrated and the yields and performance of the wells. Aerial photographics at 1:25000 and 1:80000 are available for the area.

IMAGERY AND INITIAL PROCESSING
A Landsat Thematic Mapper scene, imaged on 23rd July 1986, was obtained for the study. The image was geometrically registered to the UTM grid so as to allow direct comparison with map data. Control points, usually river or stream confluences, on the images and the 1:50000 scale topographic maps were identified and used to geometrically correct the remotely sensed data by using

Table I Correlation matrix of TM bands

Band	1	2	3	4	5	7
1	1.000	0.977	0.925	0.948	0.911	0.851
2	0.977	1.000	0.972	0.952	0.954	0.916
3	0 925	0.972	1.000	0.915	0.974	0.962
4	0.948	0.952	0.915	1.000	0.923	0.874
5	0.911	0.954	0.974	0.923	1.000	0.957
7	0.851	0.916	0.962	0.847	0.975	1.000

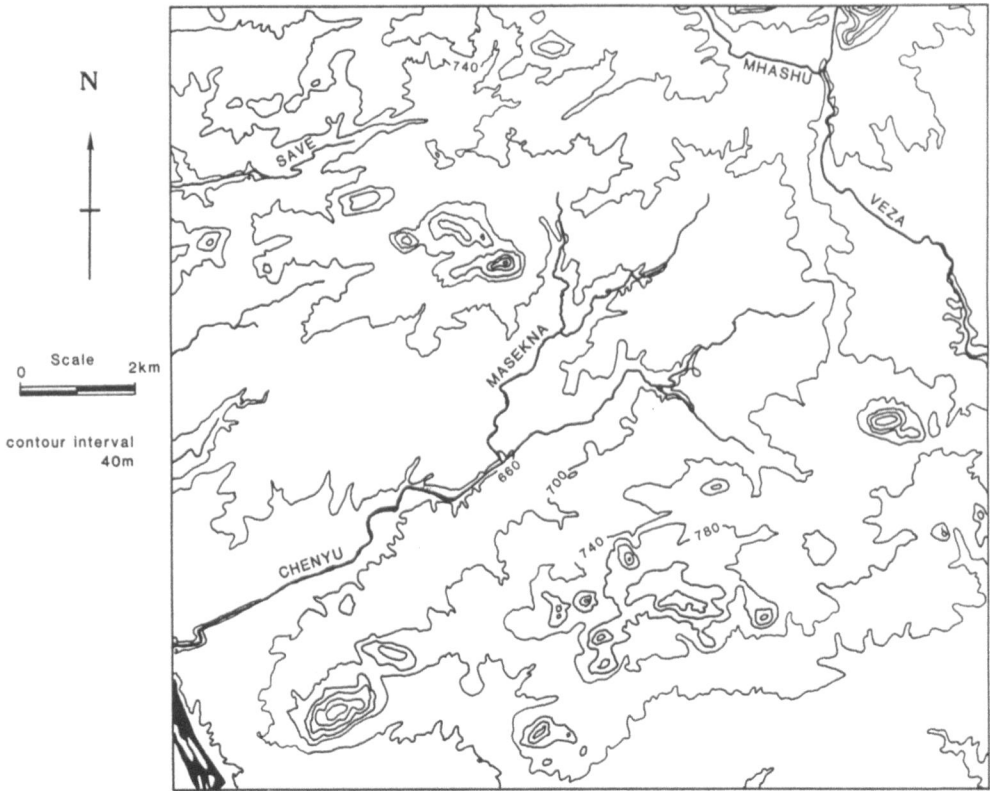

Figure 1 Topography of the study area

nearest neighbour interpolation, resampling to the same pixel size as the original data. The accuracy achieved was to an RMS error for the transformation of 1.6 pixels.

The correlation matrix of the Thematic Mapper (TM) reflected bands, Table I, shows that there is strong correlation between all the bands. One of the intentions of the study was to utilise the occurrence of vegetation to delineate structures with higher than water content and so band 4, near infra-red, was selected as the most important band to be used. It can be seen that band 7 has the least correlation with other bands and in particular with band 4 and so this was also chosen. It is also the band that might allow discrimination between rock and soil types. The visible band with the least correlation to band 4 is 3 and to band 7 is 1. Scattergrams of bands 1 and 2 against bands 4, 5 and 7 showed that they are strongly affected by particles in the atmosphere and so bands 4, 7 and 3 were used to compose a decorrelation stretched false colour composite image. A visual comparison between this and an contrast stretched false colour composite of the same bands suggested that it was easier to recognise features in the decorrelation stretched false colour composite image.

LINEAR FEATURE MAPPING

The image showed clear indications of linear features and, as these may be a source of subsurface water, an analysis of their distribution was carried out. The positions of the linear features were manually interpreted based on several differently processed images. The decorrelation stretched false colour composite of bands 4, 7 and 3, and principle component 1 of all optical bands were found to be the most useful images. The position of the lineations and the specific capacities of wells, are shown in figure 2. These lineations can be recognised on the aerial photographs as soil tones and vegetation alignments. Histograms of the lineation orientations and lengths are shown in figure 3. These are similar to those obtained by other workers[5] using satellite imagery and ground measurements and so are almost certainly fractures. The histogram of orientations shows two broad modes, around 60° and 150°. These correspond to the lines of deformation that could be expected from the four separate systems of semi-ductile and brittle, faulting and shearing from the Late Archaean[6]. There is another, smaller mode around 90°. The histogram of the lineation lengths has a mode around 800m but there are lineations with lengths up to 3200m.

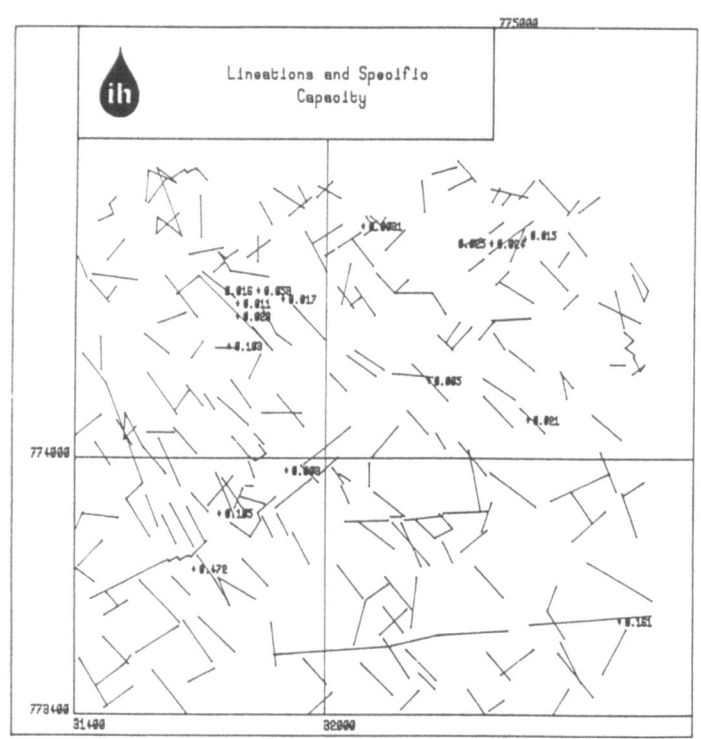

Figure 2 Location of lineations

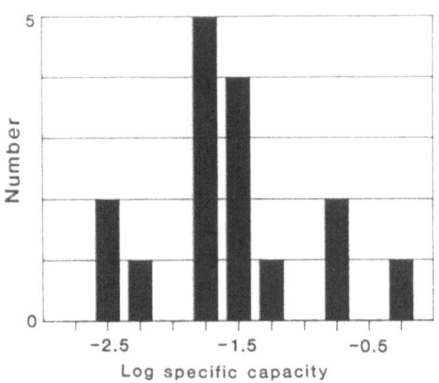

Figure 4 Histogram of well specific capacities

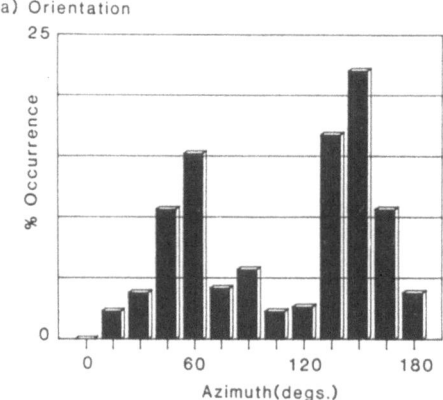

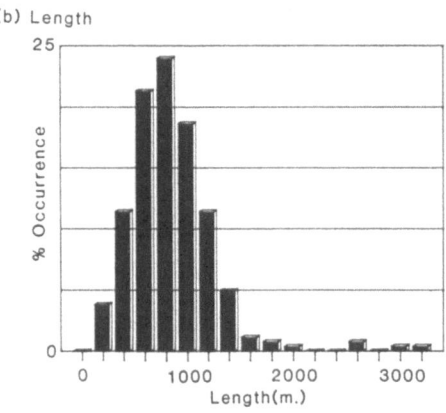

Figure 3 Histograms of lineations

The locations and specific capacities of the wells and the location of lineations were input to a Geographic Information System (GIS) in order to test whether there was any link between high specific capacity and proximity to a lineation. Values for specific capacity are available for 16 wells. The histogram of the specific capacities, Figure 4 shows that 3 wells have high specific capacities, i.e. greater than 0.1 litre/s/metre drawdown, for the area.

There is uncertainty in the exact position of both the wells and lineations so it was decided to assume an error zone of 100m around the locations of both. Thus, wells greater than 200m from a lineation were considered to be definitely not associated with the lineations. Wells within 200m of a lineation might be associated with it. Five of the wells are within 200m of a lineation and 2 of these have a specific capacity greater than 0.1 litre/s/metre drawdown. The lineations associated with these wells are approximately of east-west orientation. These lineations are associated with vegetation which suggests that they are zones of more abundant subsurface water. Although the number of wells is too few to allow a statistical test of significance, the association with vegetation does suggest that east-west lineations might be associated with higher yielding wells.

VEGETATION CHARACTERISTICS

The images also show the distribution of vegetation in the area. In a semi-arid environment this can provide important indications of the distribution of subsurface water. Figure 5 shows some examples of reflection spectra of green vegetation. The strong absorption in the visible portion of the spectra, 400 to 680 nanometres, is due to the presence of chlorophyll. The comparatively high reflection between 680 and 1000 nanometres was recognised[7] as being due to the mesophyll structure of leaves, in particular, to the stomata which are used for transpiration. The ratio between the reflectance in the red portion of the spectrum and that in the near infra-red has been used in the past as a measure of biomass or Leaf Area Index etc. but is now thought[8] to be more a measure of the transpiration capacity of the vegetation. The general increase in absorption at wavelengths greater than 1000 nanometres is ascribed[9] to the presence of water in the leaves. This feature can be used as a measure of the amount of leaf water.

In semi-arid environments these features of the reflection spectra of vegetation can be used to provide information about the amount and distribution of subsurface water that is near enough to the surface to be utilised by vegetation. The effect of rain fed vegetation can be reduced by selecting remotely sensed data that was imaged during the dry season so that all active vegetation is due to the presence of 'permanent' subsurface water. The normalised difference vegetation index[10] calculated

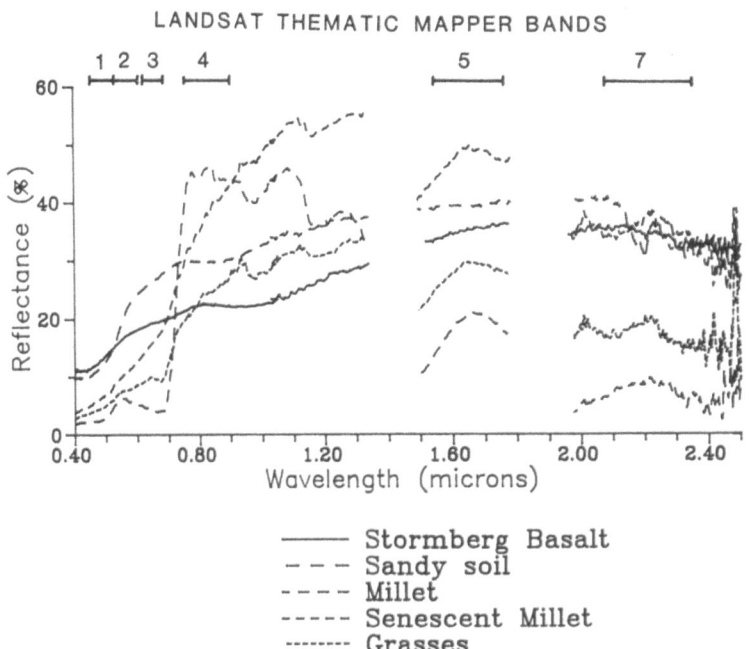

Figure 5 Refelection spectra

150

from bands 3 and 4 of Landsat Thematic Mapper (TM) data, will be a measure of the transpiration of the vegetation. The controlling factor, other than man's activities, on the transpiration is likely to be the amount of water available to the vegetation. Thus, the higher the vegetation index the more subsurface water may be present.

Similarly, the information from the ratio between the radiance recorded by bands 5 or 7 of the Landsat thematic mapper and that of band 4 can be used as a water content index which will give information about the amount of leaf water. However, variations in surface soil water will have similar effects and there are some minerals that have strong absorption bands in this part of their spectrum so care is need in interpreting this index.

The vegetation index image, figure 6, also shows this distribution but picks up areas where vegetation is moderately active, shown by the mid-grey tones. It is these areas where there may be larger supplies of subsurface water. The areas occur on the northern flanks of the hills running WSW across the corner of the area, associated with the rivers. Particularly interesting is the large area in the centre left of the image. This area is almost certainly associated with greater subsurface water as the rivers traverse it and it does not seem to be associated with any other feature. These areas may be a result of human activity, either due to being subject to less grazing pressure, or not exploited by crops.

The water content index, figure 7, shows a similar variation to the vegetation index, although it is less

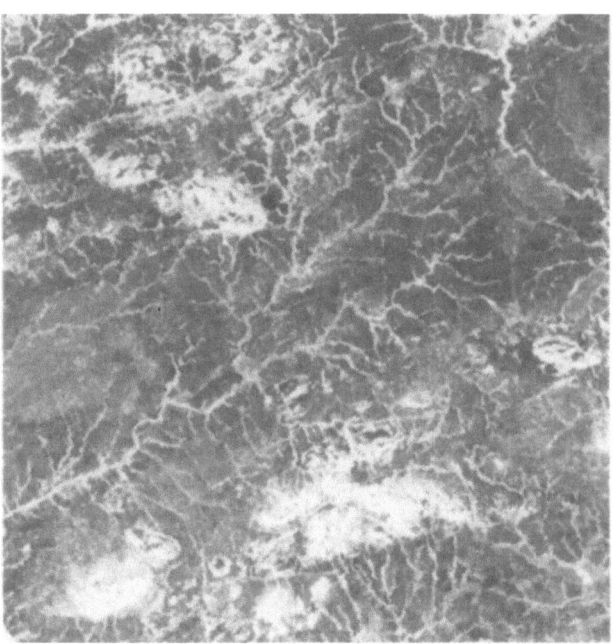

Figure 7 Water content image

obvious. The white areas are due to vegetation associated with the river beds and the hills. However, the same areas that appeared as mid-grey tones on the vegetation index image are also apparent as mid-grey tones in this image. These areas could be due to greater soil moisture in the soil profile close to the surface, but this seems unlikely as the data was acquired during the dry season. It could also be due to changes in the soil or rock types but the coincidence with the features in the thermal band and the vegetation index suggests that an interpretation in terms of vegetation water is correct. Examination of the aerial photographs show that the areas are savannah type vegetation consisting of grasslands with occasional bushes. The outlines of old fields are discernable but there does not seem to be any obvious reason why the fields have been abandoned.

The remotely sensed data shows that there are distinct areas where there is vegetation with a greater amount of leaf water and more transpiration. Some of these areas have little relationship to the river system and their position in relation to the settlement pattern of the area suggest that they are probably not due to human activity. Therefore the most likely explanation is that there is greater availability of subsurface water. This may be due to the water table being nearer to the surface or due to greater storage capacity for water in the soils and rocks of these areas. Unfortunately, none of the wells occur in these areas so there is no direct evidence that these conclusions are correct

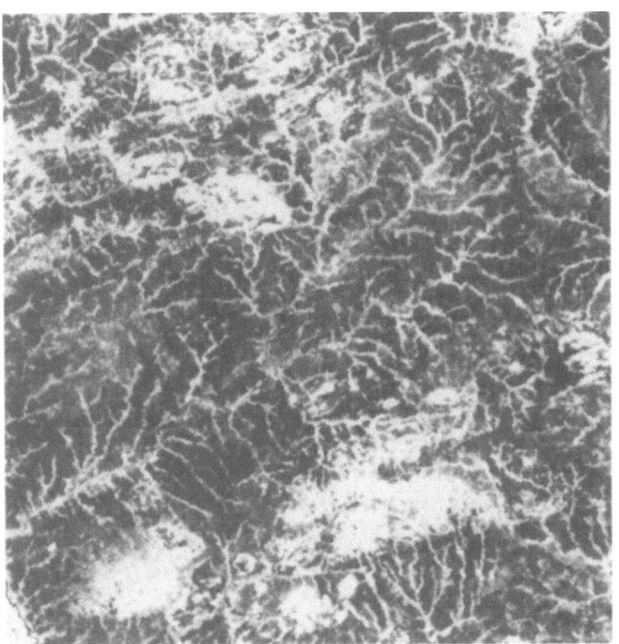

Figure 6 NDVI image

CONCLUSIONS

This study has shown that wells sited on, or close to an east-west fracture are more likely to have a higher yield than other wells. In addition, the distribution of vegetation in the area is likely to give indications of where subsurface water is either more plentiful or nearer the surface. Although these results still have to be confirmed by drilling, it clearly shows that the synoptic overview provided by remotely sensed data can have an important role in assessing the water resources of an area and where more detailed survey work should be concentrated.

REFERENCES

1 World Survey of Climatology Volume 10 -Climates of Africa. J F Griffiths ed., Amsterdam: Elsevier, 1972

2 Barton J.M., 1983, 'Introduction to Limpopo Belt' in Van Biljon W.J. and Legg J.H. (eds.). The Limpopo Belt, Special Publication Geological Society South Africa, 8, 191-203

3 Wilson J.F., 1979, 'A preliminary reappraisal of the Rhodesian Basement Complex', Special Publication Geological Society South Africa, 5, 1-23

4 Coward M.P., 1983, 'Some thoughts on the tectonics of the Limpopo Belt' in Van Biljon W.J. and Legg J.H. (eds.). The Limpopo Belt, Special Publication Geological Society South Africa, 8, 175-180

5 Greenbaum D., 1987, 'Lineation Studies in Masvingo Province, Zimbabwe', British Geological Survey Report WC/87/7

6 Stowe C.W., 1980, 'Wrench Tectonics in the Archaean Rhodesian Craton' Transactions of the Geological Society South Africa, 83, 182-205

7 Gausmann H.W., 1977, 'Reflectance of Leaf Components', Remote Sensing of Environment, 6, 1-9

8 Sellers P.J., 1987, 'Canopy Reflectance, Photosynthesis and Transpiration II The role of Biophysics in the Linearity of their Interdependence', Remote Sensing of Environment, 21, 143-183

9 Gausmann H.W., Escobar D.E., Everitt J.H., Richardson A.J., Rodriguez R.R., 1978, 'Distinguishing Succulent Plants from Crop and Woody Plants', Photogrammetric Engineering and Remote Sensing, 44, 487-491

10 Tucker C.J., 1979, 'Red and Photographic Infrared Linear Combinations for Monitoring Vegetation', Remote Sensing of Environment, 8, 127-150

Structural investigations in advance of underground mining in the Saar Coal Basin, Germany, by use of airborne scanner data

J. Kuhlmann Dr.
K. Rikeit
Interuran/Geoscan, Saarbrücken, Germany

SUMMARY

A multispectral airborne scanner survey using a DAEDALUS AADS 1268 system was performed over 200 sqkm of the Saar Coal Mining District.

This high resolution scanner imagery provides the baseline data for environmental documentation of areas affected by present and future mining operations and for recognition of mining-related surface phenomena with emphasis on prevention of surface damage to buildings and installations in advance of underground mining through identification and mapping of relevant tectonic structures.

This paper presents the structural analysis of scanner imagery by use of a newly developed semi-automatic lineament detector (LINTEC). LINTEC recognizes lineaments of variable length in 5 degree increments of azimuth by analysing in a stepwise operation the spatial orientation of significant radiometric differences between pixels and pixel groups.

The tectonic lineament inventory is correlated to the regional and local structural setting by integration of existing geological data and extrapolation of mine data. Subsequent structural interpretation has also located previously unknown fault zones in future mining areas. Predicted faults were verified in the field by ground geophysics, trenching, and drilling.

This integrated approach is expected to reduce considerably the expenditures necessary for remedial action of surface damage.

INTRODUCTION

Surface damage from underground hard coal mining represent a major environmental problem and an increasing cost factor in densely populated areas. Damage to buildings and surface installations can often be correlated with tectonic faults. The lines of outcrop of these faults usually constitute zones of inhomogeneity where underground mining results in the formation of localized fractures.

The hard coal mining industry is committed to the delineation of these zones in advance of mining to implement appropriate preventive measures to threatened structures.

In the Saar Coal Basin airborne scanner data were employed for the first time to identify relevant tectonic structures at a scale of 1 : 5,000 with subsequent verification in the field.

GEOLOGY AND MINING

The Saar Coal Mining District is located on the northwestern flank of an anticline extending from SW to NE over a distance of 140 km (Fig. 1).

The producing Upper Carboniferous strata dip on this NW anticlinal limb at 40° and flatten to 10° dip towards the northwest. Along the Southern Marginal Overthrust (Südliche Randüberschiebung) the strata are thrusted over the southeastern flank, which is unconformably overlain by a thick sequence of Permian and Triassic sediments.

The northwestern anticlinal limb is subdivided by numerous faults. Some faults trending diagonally to the Southern Marginal Overthrust display vertical throw of up to 600 m. NW-SE trending transverse faults with throws of 200 m subdivide the formation into individual blocks. Regional faults subdivide the mining areas into individual working fields; subsidiary faults usually limit stopes. Therefore faults usually represent stoping boundaries along which surface fractures are formed during the mining phase. Structures located on these predetermined weakness zones are threatened and total loss of buildings is not uncommon.

From todays mining depth of 500 to 1,000 m faults exposed underground cannot be projected to the surface with sufficient precision to be located subsequently by costly trenching. In densely populated areas, permission of property

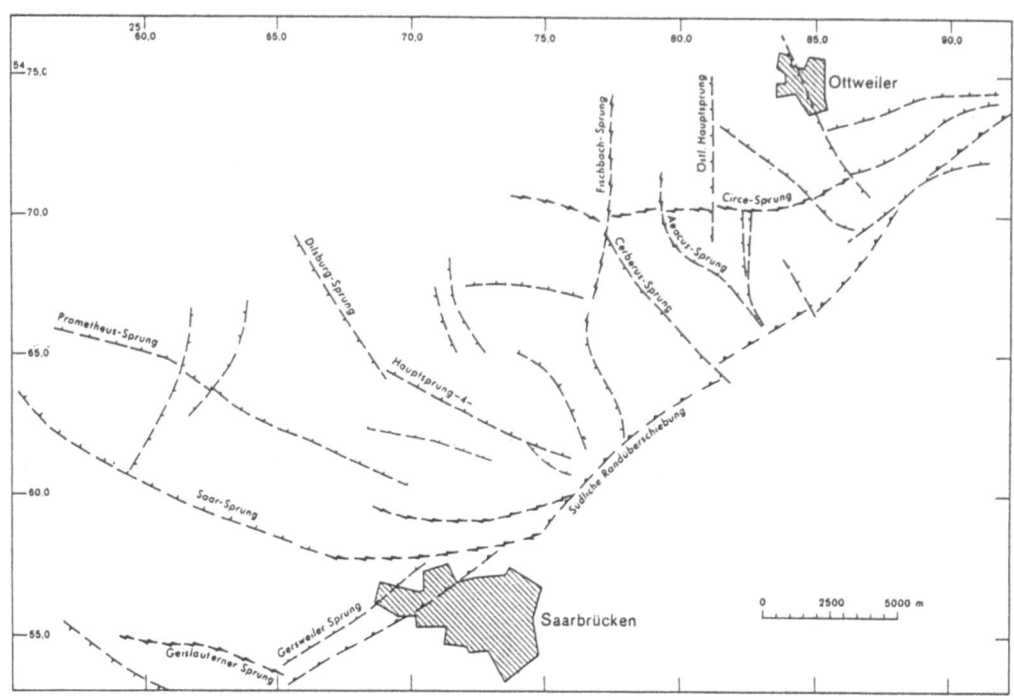

Fig. 1:   Regional Tectonic Framework

Saar Coal Basin

owners for trenching is difficult to obtain. Therefore, technologies for regional fault mapping are necessary with subsequent ground investigations in order to implement stabilization measures to buildings in mining areas and preventive measures in future mining areas.

APPLICATION OF AIRBORNE SCANNER DATA

Objectives

The use of airborne scanner data constitutes an integral part of a long term strategy targeted at improving the environmental acceptance of hard coal mining and cost reduction. The principal objectives can be summarized as:

- Documentation of present surface conditions in mining areas and areas designated for exploitation:
  . data base for long term environmental monitoring
  . recognition of future changes
- Recognition of mining related phenomena at the surface
- Recognition of tectonic faults in advance of underground mining, verification of faults in the field and evaluation of faults with respect to their future damage potential.

Only airborne scanner data to date satisfy the spectral and spatial requirements of this multidisciplinary task, where thematic mapping at a scale of 1 : 5 000 has to be performed in areas with masking soil and/or vegetation cover.

Airborne Scanner Survey Characteristics

The airborne scanner data were acquired June 14, 1988, utilizing the 11 band DAEDALUS AADS 1268 MULTISPECTRAL SCANNER (ATM) owned by Deutsche Forschungsgesellschaft fur Luft- und Raumfahrt e.V. (DLR). A total area of 200 km² was covered with 35 flight lines:

Technical Data

- altitude	1,000 m
- survey orientation	N-S
- line spacing	1,100 m
- scan angle	42.96°
- spatial resolution (of pixel)	2.5 m x 2.5 m
- swath width	1,860 m
- useful swath width	1,280 m
- overlap	> 40 %
- ancillary data acquired	colour infrared photography

DATA PROCESSING

Background

Visual interpretation from digitally enhanced satellite imagery can be considered standard practice today for structural analyses. The discrimination of radiometric differences by the human eye is limited and therefore photogeologic interpretations are subjective and depend largely on the experience of the interpreter. Furthermore, imagery exceeding a scale of 1 : 10,000 is not amenable to visual interpretation techniques.[1] Therefore, image processing for the automatic detection of linear features has been the focus of attention for some time.[2-4]

The airborne scanner data were processed employing a semi-automatic lineament detection technique. This algorithm was developped for Landsat TM data in cooperation with Deutsche Forschungsgesellschaft für Luft- und Raumfahrt e.V. (DLR) for mineral exploration in basement terrain, and for this application has evolved into an operational technique.[5,6]

For the investigations in the Saar Coal Basin, this processing technique was applied, initially without modifications, to airborne scanner data.

Lineament Detector (LINTEC)

This algorithm characterizes a linear feature as a cluster of pixels or pixel groups where individual pixels differ from the neighbouring pixels by a predetermined radiometric difference, and the cluster has significant spectral differences from its surrounding. Radiometrically significant pixels are recorded as a binary product and analysed for spatial distribution and correlation by moving a window over this binary product. The variable size of the window from 3 x 3 pixels to 112 x 3 pixels allows the identification of linears in various lengths in 36 directions in 5° increments of azimuth. A variable threshold within the window analyses the distribution and linear orientation of radiometrically significant pixels. Designated linears are coded according to their significance with the most significant linears being displayed as bright lines. A variable threshold based on geologic criteria eliminates insignificant linears in the final product. With a given set of parameters, window size and threshold definition, identical results are produced. The processing steps are illustrated in Figs. 2 to 5.

Scanner Data Processing

The scanner data were processed with the technique developed for satellite imagery (Landsat TM). The subscene selection utilizes the useful central portion of each swath, resulting from the ground resolution of 2.5 m per pixel and a screen-size of 512 x 512 pixels in an individual processed image size of 1,280 m x 1,280 m.

Based on the work with satellite imagery in basement terrain with comparable vegetation cover, lineament mapping was performed using the data of ATM band 9, a band in the short-wave infrared (1.55 to 1.75 um). The window was set at 16 to 64 pixels with corresponding lineament lengths of 40 to 160 m.

Fig. 2
Airborne Scanner Imagery, Subscene 3
Band 9, Contrast Enhanced

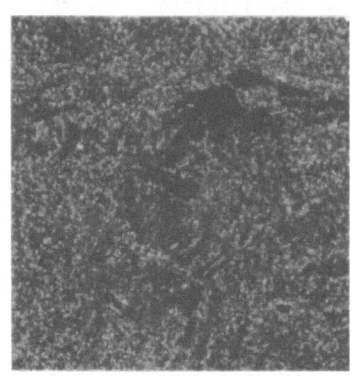

Fig. 3
Intermediate Processing Product
Horizontally and Vertically Processed
Radiometrically Significant Pixels
Threshold Applied

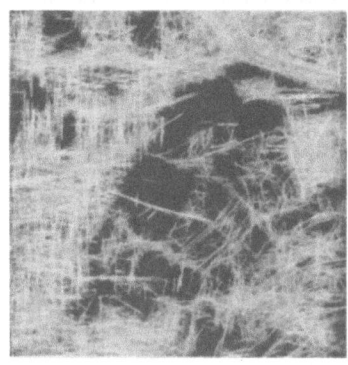

Fig. 4
All Detected Linears
Coded According to Significance

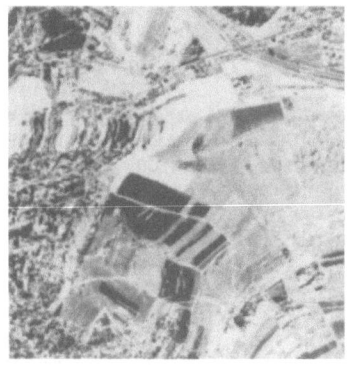

Fig. 5
Significant Linears
Superimposed on ATM Band 9

The lineaments were superimposed on a colour composite of bands 2 (blue), 3 (green) and 4 (red) for topographic reference. Hard copies for interpretation were produced at a scale of 1: 5,000.

EVALUATION AND INTERPRETATION OF LINEAMENT MAPPING

Methodology

The lineament inventory of airborne scanner imagery is composed of geological and non-geological lineaments.

Extraction of geological-structural lineaments is achieved by integration of reference data:

- Topographic maps 1 : 25,000, 1 : 5,000, 1 : 2,000
- Geological maps 1 : 25,000, 1 : 10,000
- Infrared imagery 1 : 6,500
- Mine data (proprietary data at various scales)

Selection criteria for geological-tectonic lineaments relative to location, length, and orientation were based on

- known faults
- stratigraphy and strike directions
- morphology
- fault indicators (springs, wet areas, depressions, etc.)
- anthropogene features (buildings, road systems, land plots, etc.)

Subsequently, correlations between known faults and the lineament inventory were analysed, geologic-tectonic lineaments were defined in length and orientation and transferred to maps at a scale of 1: 5,000. Projected and known faults were summarized in maps at a scale of 1 : 25,000.

Lineament Inventory

Faults, as interpreted from the geological-tectonic lineament inventory, could be traced in places for several kilometers and correlated to known faults from published geological maps (scale 1 : 25,000). Additionally, regional faults were identified which supplement the published data and correspond to the tectonic evolution of the area. To a remarkable extent lineaments were recognized which parallel regional faults at a distance of 10 to 100 m, indicating either wide fault zones or the existence of subsidiary faults.
At a detailed scale of 1 : 5,000 minor displacements between fault traces and mapped lineaments were observed, i.e. the fault trace and the corresponding spectral indication are not coincident, which could be attributed to vegetation and soil condition at the fault trace.

Apart from projected faults, the lineament maps display individual, short lineaments which were identified as geologic/tectonic features and could represent fault segments.

Classification of lineaments as projected faults not only delineates the extension of faults but also attributes a higher priority for subsequent verification in the field.

In areas underlain by sediments of Carboniferous age, faults perpendicular to the strike of the strata can be recognized comfortably, whereas lineaments parallel to the strike of the strata are subtle and reflect lithological features (boundaries of sandstone beds, gravel layers, outcrops of coal seams, etc.). The differentiation of tectonic and lithological features requires in this case the integration of reliable reference data.

To date, the reliability of the lineament data is relative to the land use. In areas used predominantly for agricultural purposes, mapped lineaments and faults correlate on the ground within meters. In forested areas, the high ground resolution results as a function of regional threshold selection in increased lineament density and therefore limits the geologic interpretation to the definition of fault trends within 100 to 200 m range; precise location of faults requires ground investigations.

In populated areas, the lineament inventory represents to a substantial degree anthropogene features (edges of buildings, roads, railways, power and utility lines). Discrimination of faults depends on the distribution and size of open spaces. Also, faults identified in open spaces can be projected into developed areas. The evaluation of the damage potential of faults is, as could be demonstrated in some areas, feasible with a more detailed interpretation at a scale of 1 : 2,000. In this case, verification of the fault mapping in the field is necessary.
It is safe to assume, that a number of local faults are either partially or entirely unrepresented in the lineament inventory. This is principally caused by insufficient radiometric contrast of the fault traces, resulting from:

- Land use (superimposed vegetation lineaments in agricultural areas, masking by anthropogene features such as small-scale land plots with different use of individual plots)
- Thickness of overburden (for example in valley floors)
- Vegetation and soil conditions in the area of the fault trace.

Despite these restriction, experience from the use of this technique in the Saar Coal Basin clearly demonstrates its ability to contribute significantly to the solution of specific problems.

Site Investigations

Site investigtions of faults mapped on scanner

imagery have already demonstrated the potential for time and cost reduction by target oriented deployment of field work:

- In a densely populated area, a ground geophysical survey (resistivity mapping) of an area of 500 x 500 m located a projected fault, trenching to evaluate the damage potential of this fault was avoided. The amount of field work to locate the fault was reduced to 1/5th of the usual time and cost through this progressive approach.

- The licensing procedure for a planned commercial development required identification of faults in view of future mining in this area. Power lines inhibited the use of geoelectrical surveys. Three projected faults as mapped on the scanner imagery were examined in 8 trenches (total length 590 m). Two faults were identified and measures recommended to mitigate potential damage from future mining.

- For an active mining area, subsidence in the range of a few meters was calculated, necessitating the reconstruction of a bridge. At the site, the geological map (1 : 25,000 scale) indicates a regional fault, displacing sediments of the Upper Carboniferous age against Triassic strata. The fault location from the lineament mapping differed from that shown on the geological map. Drilling proved the location of the fault as mapped on the scanner imagery.

CONCLUSIONS

Lineament mapping with the semi-automatic lineament detector represents a geophysical method which characterizes radiometrically anomalous pixels forming part of lineaments. Corroborated by reference data, lineament maps constitute an objective data base for fault identification and mapping.

Compared to conventional visual interpretation of satellite imagery and airphotographs, the semi-automatic lineament mapping features several advantages:

The technique:

- extracts new information not obtainable with other techniques
- results in reproducible products
- is independent of scale and is at present the only remote sensing technique available which can process airborne scanner imagery of a scale of 1 : 5,000. At this scale, only geophysical ground survey could substitute for this technique, for which a time and cost factor (at least 3 to 10) has to be taken into account to achieve comparable results on a regional scale.

This remote sensing technique is designed as a semiautomatic regional lineament mapping tool including subsequent fault interpretation with the objective of restricting the amount of costly field work to relevant targets.

All geophysical techniques are inherently limited. The application of the lineament detector is restricted by geological conditions, i.e. a fault can only be identified within the prerequisite parameters of the technique:

- minimum extension of fault trace along strike
- significant radiometric difference of fault trace

Further development of the technique is directed towards improved discrimination of tectonic structures relevant to the mine operator and integration as an operational technique in the planning and management information system.

References

1. KRONBERG, P. Photogeologie. Eine Einführung in die Grundlagen und Methoden der geologischen Auswertung von Luftbildern. 268 p., Enke Verlag, Stuttgart 1984.

2. VANDERBRUG, C. Line detection in satellite imagery. IEEE Vol. GE-140, 1, p. 37-43, 1976.

3. BURDICK, R.G. and SPEIRER, R.A. Development of a method to detect geologic faults and other linear features from Landsat images. Bureau of Mines, U.S. Dept. of the Interior, RI-8413, 74 p., 1980.

4. CONRADSEN, K., NIELSEN, B.K., PEDERSEN, J.L. Automated analysis of linear features based on satellite imagery compared with geologic setting. Second European Workshop on Remote Sensing in Mineral Exploration, EUR 11317, p. 273-292, 1988.

5. KNÖPFLE, W. Rechnergestützte Detektion linearer Strukturen in digitalen Satellitenbildern. Bildmessung und Luftbildwesen, 56/88 no.1, p. 40-47, Karlsruhe 1988.

6. SCHLICHTER, D., WAEBER, L., KAISER, D., KUHLMANN, L. Untersuchungen an Satellitenbild- und Referenzdaten für explorationsbezogene Aufgabenstellungen. DFVLR-Forschungsbericht 88-46, 43 p., 17 figs., 6 tabs., 1988.

Applications of remote sensing to environmental aspects of surface mining operations in the United Kingdom

Christopher A. Legg

National Remote Sensing Centre, Royal Aerospace Establishment, Farnborough, Hampshire, England

ABSTRACT

Increasing environmental awareness, especially in densely-populated countries such as Britain, imposes new responsibilities on the mining industry in respect of the environmental impact of surface mining operations, and the restoration of land after mining. Satellite remote sensing, with its capability for repetitive digital multispectral imaging of large areas at low cost, can play an important role in environmental impact assessment prior to mining, monitoring the distribution and change in area of different activities during mining, and the evaluation of restoration success and landscape quality after mining.

Examples are presented of the use of remote sensing in combination with map data and digital elevation models to generate perspective views of the results of alternative mine plans in the china clay workings of the west of England. The digital nature of the imagery allows the impact of numerous possible alternative plans to be evaluated rapidly, even on small computers. The use of remote sensing to detect change during mining operations is demonstrated with examples from sand and gravel workings in the south-east of England, and from the china clay districts of Devon and Cornwall. Automated change detection is possible in some cases, but in others a visual interpretation of specially enhanced satellite imagery by people with local knowledge is a more effective technique. Landscape and land quality assessment after mining is demonstrated with examples from open-cast coal mining operations in Dyfed and Northumberland. Remote sensing is shown to be a cost-effective technique for this type of application, as well as permitting observations that are impossible with conventional techniques. The digital nature of the imagery permits quantitative analysis of features such as field size and shape, and direct comparison with pre-mining morphology, while large-area information at infra-red wavelengths permits quantitative comparison of vegetation vigour. The mining industry is urged to make wider use of these relatively simple and cost-effective techniques.

INTRODUCTION

The mineral extractive industries are under increasing pressure to demonstrate their environmental awareness at the planning, production and restoration stages of mining projects. Satellite remote sensing and digital image processing can provide cost-effective assistance to mining organisations, and to the governmental agencies responsible for monitoring their activities. A wide range of different types of satellite data are now available, but for most environmental studies in the mineral industries, the best available spatial and spectral resolution is required in order to discriminate relatively small features on the surface, and to distinguish subtle differences between the surface cover types of interest. In practice, this means using imagery from the Thematic Mapper sensor on the American Landsat satellites, and from the multispectral and panchromatic sensors on the French SPOT satellites[1]. Such imagery is now available globally, and for many areas a time series of imagery covering many years is now available for change-detection studies.

Image processing techniques have become more sophisticated, but at the same time image processing systems have become smaller and cheaper, so that capable systems are now within the reach of most mine and quarry operators, at prices equivalent to the cost of a new pick-up truck. All the operations discussed in this paper could be performed on a state-of-the-art personal computer based image processing system costing less than £20,000.

By their nature, underground mining operations are less amenable to remote sensing studies than surface mines. The mining operations themselves, unless they result in extensive subsidence, do not result in surface changes, and dumps are generally much smaller than for open-casts. This paper will thus concentrate on surface mining operations.

Remote sensing can be useful during the planning stages, either of a new mine or of an extension to an existing mine. It can also be of assistance during mining operations to update thematic maps of the mining district, and to assist in monitoring changes related to mining activities. Restoration after mining normally involves re-vegetation of the mined areas, and the progress and success of such restoration activities can also be assessed with the aid of remote sensing.

MINE PLANNING

A common requirement at the planning stage is the compilation of a detailed database of land-use in the area of the proposed mine. This can be used to assess environmental impact, but also serves as a baseline against which future land restoration after mining can be judged. In a small mining operation, land use can probably be assessed most cost-effectively using low-level air photography in combination with ground surveys. In large mining operations, on the other hand, remote sensing can make an important contribution. Supervised classification of digital imagery based on training areas derived from limited surface surveying can produce a land-use map of sufficient accuracy at a much lower cost than conventional survey techniques, with the additional advantage that quantitative measures of land productivity such as vegetation indices can be compared with similar data obtained after mining and restoration. Image processing techniques can be used to obtain quantitative measures of the proximity of environmentally significant land-cover classes to proposed mine-sites or transport routes. This allows a choice of the best sites for mine dumps or access roads, based on the maximum possible distance between new developments and areas of environmental or ecological importance. The results of a simple proximity analysis study are shown in Figure 1.

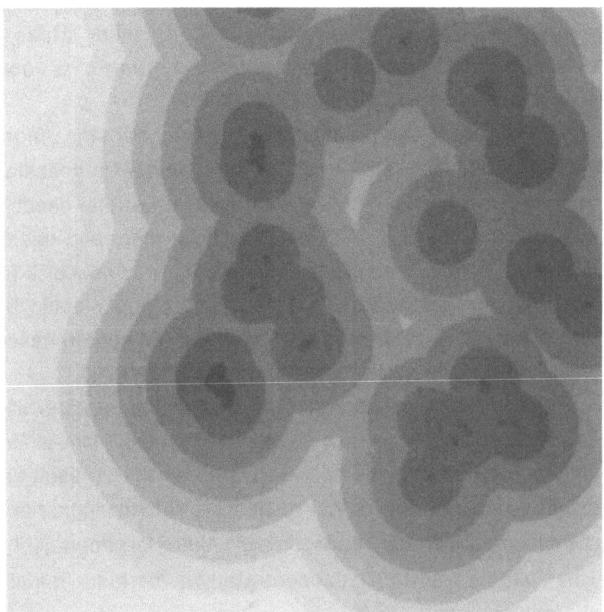

Figure 1. *The result of analysis of proximity to surface water bodies. Water is shown in black on this image and proximity corridors are coded from dark grey (closest) to white (most distant).*

The combination of remotely sensed imagery with digital elevation models on an image processing system presents a very powerful technique for visualising a mine site in three dimensions, and for simulation of the appearance of planned future developments. The digital nature of the imagery and the increasing availability of digital elevation models makes the generation of such views relatively simple on most modern image processing systems. Some computers are becoming fast enough to generate such views in almost real time, opening up the possibility of interactive movement of the view point and angle, and the generation of "fly-by" type moving simulations. The results of all this are very spectacular, but they are also extremely useful in a variety of situations. Mine planners often construct solid three-dimensional models of proposed new mine developments, in order to give a visual impression of the appearance of the the planned developments in a realistic setting. The problem with these models is that they are very expensive to produce, and only illustrate a single possible planning alternative. They also usually only include a relatively small area of the surrounding countryside for reasons of cost, size and scale. A computer model does not have these limitations, except that the "closeness" with which one can approach any feature in the model is limited by the spatial resolution of the data used. Remotely sensed imagery is best suited to examination of the gross appearance of a mine within its setting, from any viewpoint and distance, rather than examination of the details of a pit or dump. Unmodified satellite imagery combined with a digital elevation model of the same date provides a view of the mining operation as it actually appears, but it is also possible to simulate the effects of different planning decisions by modifying both the image and the elevation model. The same combination of digital elevation model and imagery could be used to determine the areas from which this new dump could be visible. A simple iterative process could determine the maximum height of dump permissible within pre-set visibility constraints. This process is illustrated in Figure 2, a satellite image of Lee Moor, showing the areas from which an existing dump is visible.

The generation of three-dimensional views from combinations of satellite imagery with digital elevation models has significant advantages over the more conventional computer-graphics use of "wire models" or "shaded relief models". The actual appearance of the ground surface is more realistically portrayed, and the combination of other image processing operations with model generation allows more flexible interactive analysis of the interrelationships of different landcover types and mining activities. The limitations of remote sensing in this type of application become apparent if one tries to obtain too close or detailed a view of the mining operation, or of any other surface features. Neither the imagery nor the digital elevation models are sufficiently detailed for very close examination, but their strength lies in the capabilities for

Figure 2. *A processed satellite image of the Lee Moor area, showing areas from which a mine dump, situated at the point indicated, would be visible. These areas appear black in the right-hand image.*

portrayal of the mine or other feature of interest in the context of its surroundings. Views of this type can now be produced fairly rapidly even on relatively small PC-based image processing systems, and no longer require the large and expensive image processors that were once considered necessary. The use of perspective views in the mineral industry is not restricted to the examination of mining sites. The choice between alternative transport routes, for example, may often be assisted by the chance to view these alternatives in three dimensions in a natural-appearing setting.

MONITORING AND MAP UPDATING

While surface mining operations are in progress there is often a need for both mine operators and relevant planning authorities to monitor activities within sites. A major task of local government mineral officers is to regulate mining operations within agreed planning permissions and to measure and calculate over their districts the overall areas affected by mining activity. General surveys of sites are often undertaken using aerial photography, which may be out of date, and by reconnaissance field visits. For relatively large sites and for studies involving numerous sites over large administrative districts, satellite remote sensing data can provide a cost effective tool for such survey work. In studies undertaken in the UK it has been found that the use of appropriate combinations of high resolution satellite imagery, in particular SPOT PAN with Landsat TM, allows operators and planning officers to measure and monitor activities within surface workings. Besides its use for general surveying of sites this type of imagery can also provide an important overview of surface workings and

alert operators and planning officers to areas which may require more detailed field investigation. For a county-sized area, composite imagery of this type can be generated for less than one-quarter the cost of aerial photography[2].

In some mining districts, fragmentation of ownership between numerous separate operating companies makes preparation of any district-wide maps, and assessment of changes in land-cover, a difficult process. Each company may map its areas of operation at a different scale, or use a different classification system for land-cover types. There is often a requirement, either from government agencies or for public relations purposes, to produce maps showing current land-cover as well as changes in land-cover, every two or five years. If the final product is to be at scales of between 1:25,000 and 1:100,000, this task can readily be undertaken using remote sensing. In studies of the china-clay mining districts of the south-west of England a methodology has been developed which could be applied in major mining districts elsewhere in the world to produce updated land-cover maps, and derivative change maps, at a relatively low cost using satellite remote sensing.

It is often forgotten that china clay is Britain's most valuable mineral export after oil[3], the most extensive workings being in the St. Austell district of Cornwall and in the Lee Moor area of Devon. Hydrothermally altered granite is excavated by mainly hydraulic means, using powerful water jets or monitors, in large open-pits. Kaolin is extracted in a series of settling tanks. The residual slurry of fine-grained muscovite with minor kaolinite is disposed of in "mica dams". The large waste tips so characteristic of the china clay districts are heaps of sand-sized waste quartz with minor feldspar, together with coarse rubble of partially decomposed granite. Old spoil heaps have become naturally vegetated with heather, grass and rhodedendron thickets, while the sides of new flat-topped heaps are commonly grassed to prevent erosion and lessen their environmental impact.

The main objective of the NRSC study[4] was to develop a technique for rapid low-cost production of thematic maps of the mineral district, on a regular basis, using available imagery irrespective of season. Detailed studies of the Lee Moor workings in Devon indicated that conventional classification techniques did not produce reliable results, particularly at low solar elevations, with a large percentage of unclassified surfaces. Within the china clay workings, four types of surface, spectrally distinct in Landsat TM imagery, have been identified. Water in active pits is often extremely turbid, with high reflectance in the three visible TM bands but low reflectance in the infrared. "Mica" (mainly muscovite, with subordinate lepidolite and fine-grained kaolin), has a relatively high reflectance in the NIR and low reflectances in the MIR. "Clay" (mainly kaolinite, with subordinate feldspar and quartz) has roughly equal reflectance in bands 4 and 5, with a marked decrease in band 7. "Sand" (mainly quartz with subordinate feldspar) has roughly equal reflectance in bands 4 and 7, and markedly higher values in band 5. Studies of the flat and mineralogically simple surfaces of mica dams indicated that the use of band ratios effectively eliminates the effects of solar elevation, remaining differences between different image dates due to atmospheric effects being minor in relation to the variation in response between different types of surface.

Experimentation with a variety of ratios resulted in the development of a sequential ratio thresholding scheme to segment the image into six broad surface cover types. Mining areas without significant vegetation cover are distinguished from surrounding land using a ratio of Band 4 / Band 2. Very low values of this ratio are thresholded to separate water from all other cover types. Thresholding intermediate values of this ratio was found to be a reliable technique for discriminating mineral workings, irrespective of solar elevation, although some bare fields and built-up areas are confused with mineral workings and

require manual editing by reference to suitably contrast-stretched imagery. A ratio of band 4 / band 7 is then used to separate sand from other cover types within the mineral workings, since this ratio is lower for sand than for mica or clay. The next stage is the separation of mica from clay using a ratio of band 5 / band 4, in which mica values are relatively low. The final stage is the subdivision of "clay" areas using a ratio of Band 5 and band 7. High values of this ratio map mainly mixed dumps composed of "stent" (partially kaolinised granite), as well as rock outcrops in the pits, while low values map active clay pits as well as the clay-rich outer areas of mica dams. The series of five masks generated by the segmentation process are then added together to produce a digital thematic map, which can then be coloured by density slicing and added to a colour composite image of the mining district. Maps of this type prepared for different years can be compared digitally to highlight areas of change. A monochrome example of such a change image is shown in Figure 3.

RESTORATION QUALITY ASSESSMENT

Once mining is complete, there is commonly a requirement to restore the land surface to its pre-mining use, which in the United Kingdom usually means restoration to agricultural land. It is desirable to achieve similar agricultural productivity on restored land as on land never mined, and there is an increasing requirement for the scenic quality of the land to be retained. There may even be opportunities for scenic improvement during restoration, with the addition, for example, of small hills in a previously featureless landscape, or the provision of ponds and small lakes. Restoration planning may involve conflict between farmers, who wish to maximise productivity by replanning the field layout, replacing original small irregular fields by large rectangular fields, and conservationists, who would like to retain the network of hedgerows in the old field pattern.

Remote sensing can play an important and cost-effective role in assessing the quality of restored land, as well as in comparing the size and shapes of fields in restored and

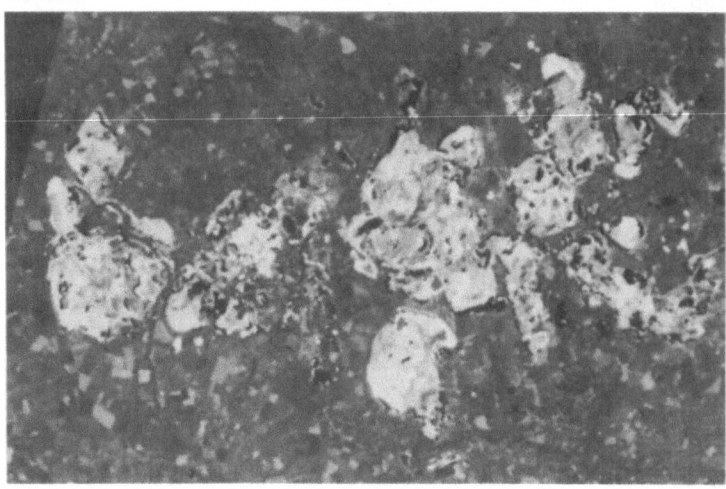

Figure 3. *Change map (Sept.'85-May'88) of china clay workings in the St. Austell area of Cornwall, prepared from Landsat Thematic Mapper imagery. White - increase in area of workings and dumps. Black - restoration and re-vegetation.*

unmined areas[5] A remote sensing study of an area of open-cast coal mining in Northumberland in the north-east of England, has demonstrated the use of vegetation indices to compare productivity of restored and unmined land[6]

The coastal area of southern Northumberland has a long history of coal mining. Most production in the past has been from underground or "deep" mines, but only a few of these still operate. Open-cast production started in the early 1940's, but it was not until 1956 that large-scale open cast mining commenced. A total of thirteen open-casts have operated in the area since then, five being in production in the late 1980's. All mined areas are restored to agricultural or recreational use after mining. The main agricultural activity in the area is sheep grazing although minor acreages of grain crops, potatoes and oil-seed rape are grown. Most of the restored land is under pasture, with strips of woodland to act as windbreaks and to improve the scenic quality. Restoration is usually a continuous process, concurrent with mining. Topsoil, subsoil and waste rock from the initial cut of an open pit are stored next to the site. As mining proceeds laterally from this initial cut, waste rock is moved, usually by dragline, from above the coal seams and dumped directly in the worked-out part of the pit. Topsoil and subsoil are removed from ahead of the advancing face of the pit, and placed directly on rock-filled mined-out areas. Once mining is complete, the final pit is backfilled with stored material from the initial cut. The process of restoration results in an homogenisation of soils and, where the ratio of coal to waste is small, may leave the restored land at a slightly higher mean final elevation than the original surface due to imperfect compaction of waste rock. In cases where the ratio of coal to waste is high, the final cut has been inadequately filled with material at the end of mining resulting in a lower than average land surface and even some open water-filled voids. Success of restoration is reported to have increased as the mine operators and British Coal gained experience. More

consideration has been given to aesthetics in recent restoration, with trees being planted in less geometric patterns, and excess backfill material being used to produce undulating surfaces[7] Drainage can be a problem, particularly in low-lying coastal sites. Excessive compaction of the sub-soil may occur due to heavy machinery used for transporting and dumping the soil, reducing soil porosity and sometimes causing waterlogging of topsoil. Most restored areas are planted with grass, and the farms are ultimately either handed back to their original owners, or if purchased prior to mining by British Coal, sold to new farmers[8]

Landsat MSS imagery from 1973, 1976, 1977, 1979 and 1981 was co-registered with Thematic Mapper imagery for 1984, 1985 and 1986, and images of the near infrared to red ratio (MSS bands 7/5, TM bands 4/3) were prepared for each date. Values of this ratio were extracted for three coal mine sites, restored to agriculture in the mid 1970's, for all dates of imagery, and were compared with ratios from three control areas of similar agriculture on unmined land. Mined and unmined sites were all in excess of one thousand pixels in area, and a crude atmospheric correction was applied to the NIR/Red ratios based on the assumption that, for each date, the value of this ratio over open sea should be zero. No corrections were applied for solar elevation since the use of ratios is assumed to largely eliminate illumination differences. The mean ratios for mined and unmined land were then plotted on a graph of time against the NIR/Red ratio (figure 4), where time is the Julian day of image acquisition, irrespective of year.

With the exception of the July (1979) scene, the ratios lie very close to a generalised curve of grassland vegetation index (after Curran[9]). During spring, summer and autumn the ratios for restored land are slightly lower than for unmined land, the difference being greatest in the spring. In winter, on the other hand, restored land ratios are higher than those for unmined land. This study showed that, in

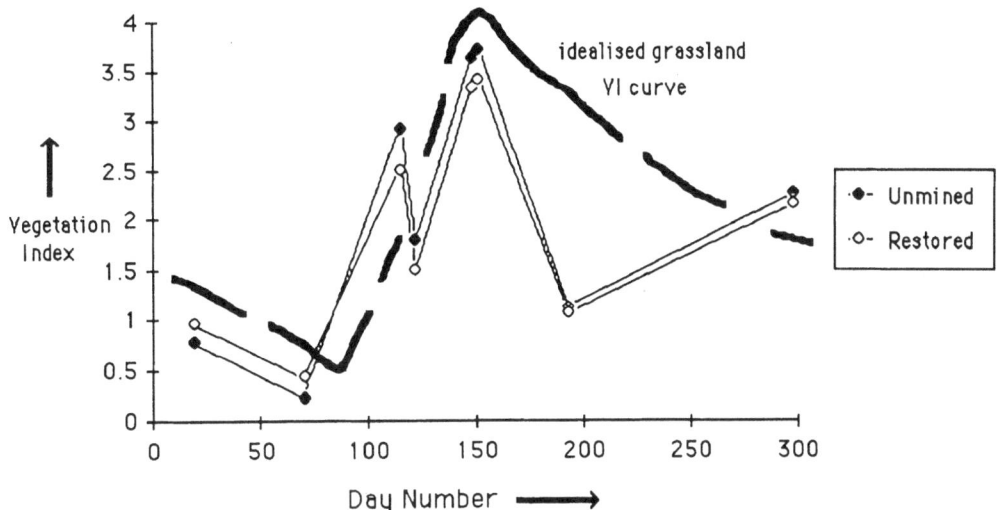

Figure 4. *Vegetation indices for restored and unmined land, Druridge Bay area, Northumberland.*

general, pasture on restored land has almost the same vigour as that on unmined land. Larger differences in spring suggest that grass on restored land starts its most active growth phase after grass on unmined land. This could be due to lower porosities of soil and subsoil on restored land, resulting from compaction by heavy machinery during restoration. Quantitative information of this type is not obtainable by air photography, and would require expensive field work at many different dates to collect on the ground.

Comparisons between field patterns in restored and unmined land can be made visually on the ground or by using aerial photography. These comparisons are usually qualitative, since direct measurement of boundary lengths and angles would be very time-consuming. The combination of land-cover information from digital imagery with digital maps of field boundaries, derived either from recent topographic maps or by the use of automated edge-detectors on satellite imagery, allows direct measurement of quantitative aspects of field size and shape, and thus permits quantitative comparison between different land-use patterns. Many image processing systems allow size and shape analysis of pre-determined features or groups of features, allowing relatively rapid quantitative comparison of field patterns.

CONCLUSIONS

Satellite remote sensing has a cost-effective role to play in studies of the environmental aspects of surface mining operations in the United Kingdom and similar densely-populated countries with temperate climates. While remote sensing cannot replace conventional ground survey and airphoto techniques for detailed studies, it can provide a unique overview of entire mining districts in a digital form and in important parts of the spectrum which are not accessible with photography or human observation. Modern image processing techniques can derive quantitative information on spatial and temporal relationships of land-cover and its changes, and can be used to integrate satellite imagery with other geocoded data sets such as digital elevation models. An increasing use of remote sensing within the minerals industry can be anticipated as a wider range of satellite imagery at appropriate resolutions becomes available, repeated during useful seasons and in successive years, and as the cost of image processing systems decreases dramatically.

ACKNOWLEDGEMENTS

The assistance of British Coal Opencast Division, Taylor Woodrow, English China Clays International, Watts, Blake and Byrne, Tarmac, ARC and other operators of mineral workings visited during the course of this study is gratefully acknowledged. Dr. Norman Jackson and his students at Kingston Polytechnic assisted with the Lee Moor study. Work on sand and gravel workings reported in this paper was carried out by Anthony Harding at the National Remote Sensing Centre, and the author is indebted to all members of the Applications Support Team at the NRSC for their assistance and encouragement.

References

1. Irons, J. R. and Kennard, R. L. The utility of thematic mapper sensor characteristics for surface mine monitoring. Photogrammetric Engineering and Remote Sensing, vol. 52, No. 3, pp. 389-396.1986.

2. Harding, A. E. Monitoring surface mineral workings using TM and SPOT. Proceedings IGARSS '88 Symposium, Edinburgh, pp. 1671-1673. 1988.

3. Bristow, C. M. and Exley, C. S. Kaolin deposits of the United Kingdom. In Genesis of World Kaolin Deposits. Murray, H. and Storr, M. (editors) Springer Verlag, Berlin.1989.

4. Legg, C. A. Updating thematic maps of mining districts. An operational demonstration of remote sensing in the south-west of England. Proceedings RSS conference "Remote Sensing for Operational Applications", Bristol, pp. 243-248, 1989.

5. Anderson, A. T. Evaluating the environmental effects of past and present surface mining; a remote sensing applied research review. Proceedings. 14th International Syposium Remote Sensing of Environment, San Jose, pp. 275-278.1980.

6. Legg, C. A. Monitoring of open-cast coal mining and reclamation works in the United Kingdom using MSS and TM imagery. Proceedings 21st ERIM Syposium on Remote Sensing of Environment, Nairobi, pp. 931-941.1986.

7. Grimshaw, P. N., 1986. Environmental benefits from surface mining for coal in Great Britain. Mining Magazine, December 1986, pp. 581-585

8. Paice, C. How British Coal cares for farmland. Farmers Weekly, Nov. 21, 1986, pp. 50-51.1986.

9. Curran, P. Multispectral photographic remote sensing of vegetation amount and productivity. Proceedings. 14th International Syposium Remote Sensing of Environment, San Jose, pp. 623-637.1980.

Use of SIR-A interpretation for underground water prospecting in southern Iraq

A. A. Omar B.Sc., M.Sc.
Department of Geology, College of Science, Salahaddin University, Arbil, Iraq

SYNOPSIS

SIR-A imagery was used to study the hydrogeology of an area in southern Iraq, covering 15000 Km^2. Geological, structural lineament and drainage density maps were constructed and the significant hydrogeological features extracted from them in order to locate areas of high underground water content.

INTRODUCTION

Among the recent techniques contributing to grate understanding of the environment surrounding us, is the L-band radar known as Shuttle Imaging Radar (SIR-A) carried aboard the space shuttle Columbia in November 1981. The objective of this study was to use SIR-A images as a tool for locating areas of underground water potential in southern Iraq. Sabin[1], King[2], and Harrison[3] have conducted geological interpretation of SIR-A images of rock types and structures. Abdullah and Zwain[4] carried out an assent of the capability of SIR-A imagery in revealing facts about ground cover categories and some sub-surface phenomenon and compared SIR-A to the well known products of the Landsat MSS for the area in southern Iraq

The study area is located in southern Iraq and bounded by latitude 32° 00´ - 30° 00´ and longitude 43° 00´ - 45° 30´ , covering 1500Km^2 (Fig.1). This is an example of a case study of a semi-arid region. The natural vegitation is represented by scatterd bushes in some wadies. Cultivated land is concentrated in the Euphrates valley zone where the fertile land is irrigated by water from the Euphrates river. Underground water is very important because of the role it can play in the development of horizontal expansion projects area from the river for cultivated lands in Iraq.

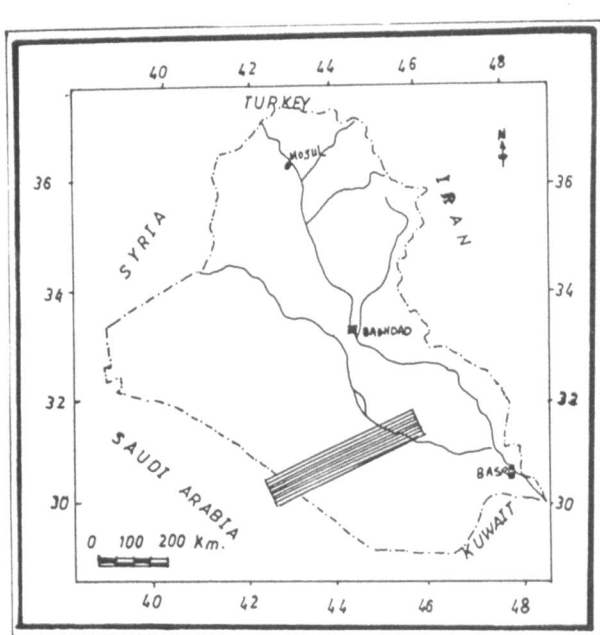

Fig.1: Location map of the study area.

165

HYDROGEOLOGICAL OUTLINES

The hydrogeological conditions in the study area are not well known.However, two main aquifers can be identified. The first and deeper one is associated with the Lower Damam Formation which represent the oldest unit and is about 400 meter thick.It is a confind to semi-confined aquifer,the thick massive and shelly limestone of Middle Damam Formation overlying it,representing the confining horizon[5].

The second aquifer, a shallow one,consist of thick sand,gravels and rock fragments present in the main wadis. This aquifer is unconfined and the water level ranges from 1-10 meter below ground level; the amount of water in this aquifer is often small.

GEOLOGICAL SETTING

The main geological units in the area under consideration were identified on the basis of variations in surface roughness,tone,texture[1],and Complex Dielectric Constant (CDC)[6]. These units were compared (Table I) with the previous geological study and a geological map has been constructed (Fig.2) .

The nature and mode of occurrence of the sedimentary units are of great importance in underground water studies.The oldest sedimentary formation is the Ummer Radhuma Formation,which is of Upper Paleocene age. It is composed of thickly bedded to massive recrystallized limestone. This formation is overlaid by the Damam Formation of Eocene age,which consists of several rock units[7],and forms a number elevated structural plateaux characterized by steep irregular scarps generally facing northeast. These scarps produced bright signatures on the SIR-A images. The clastic formations of Euphrates and Zahra (Miocene) overlay the Damam Formation. Their rocks display intermediate

tones in SIR-A images. Younger sediments are represented by wadi deposits, gravel terraces and rock fragments and aeolian deposits,and display a uniform dark tone in SIR-A images because of the absence of strong reflections from sensor facing slopes. These sediments range in age from Pliocene to Holocene, and their thickness varies from a few centimeters to about 50 meters.

STRUCTURAL LINEAMENTS

Structural lineaments,especially major fractures and faults, are of grat importance in prospecting for underground water[8].They play an important role in controlling the solution openings. Morecver, large faults can act as traps for underground water. In the arid parts of the Colorado Plateau where water resources are scarce, mapping of fractures observed in ERTS imagery guided the selection of drilling sites for water production[9].

The structural lineaments (Fig.3) in the studied area are delineated and the significant lineaments extended to the centres of the valleys[10]. The recorded faults represent two main faulting sets:NNE-SSW andNE-SW. These faults are mainly of gravity types and are responsible for the development of the horst and graben structures in the studied area[11].

DRAINAGE LINES

Drainage lines are one of the important surface indicators of hydrogeological features[12]. It is known that, as terrain transmissibility decreases,a progressive increase in surface flow will occur. Moreover it is found that terrian transmissibility is inversely proportional to the square of drainage density. The drainage lines and the master streams are obtained from SIR-A images and a density

166

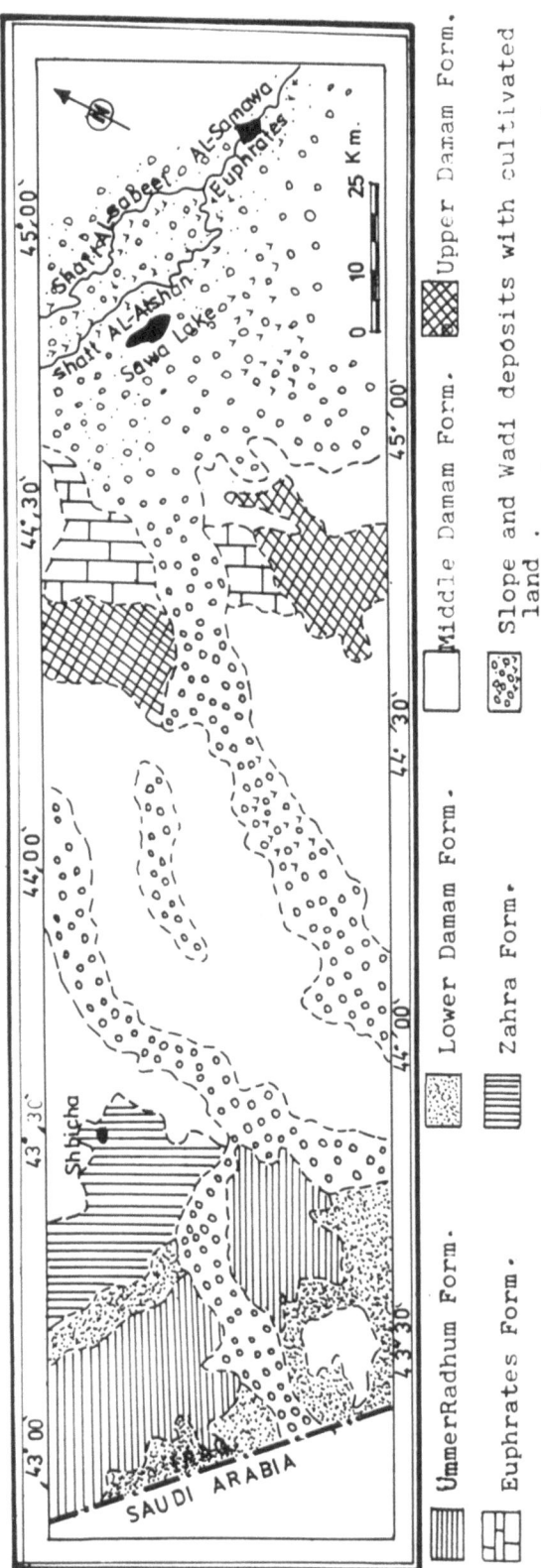

Fig.2 Geological map showing the main lithological units in study area(based on SIR-A image).

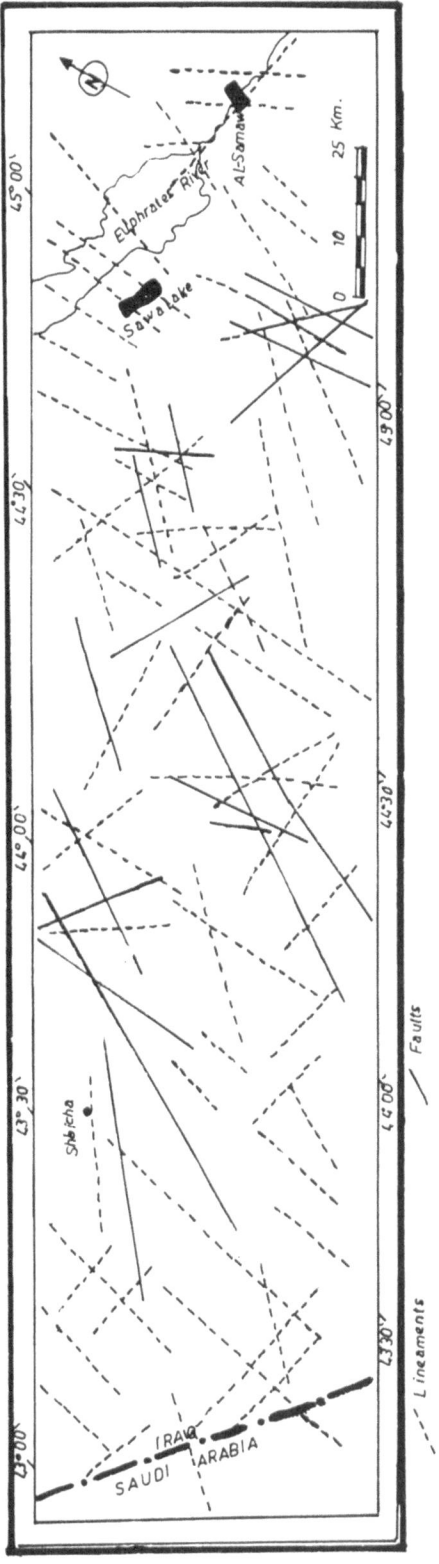

Fig.3 Structural lineament map of the study area (based on SIR-A image interpretation)..

167

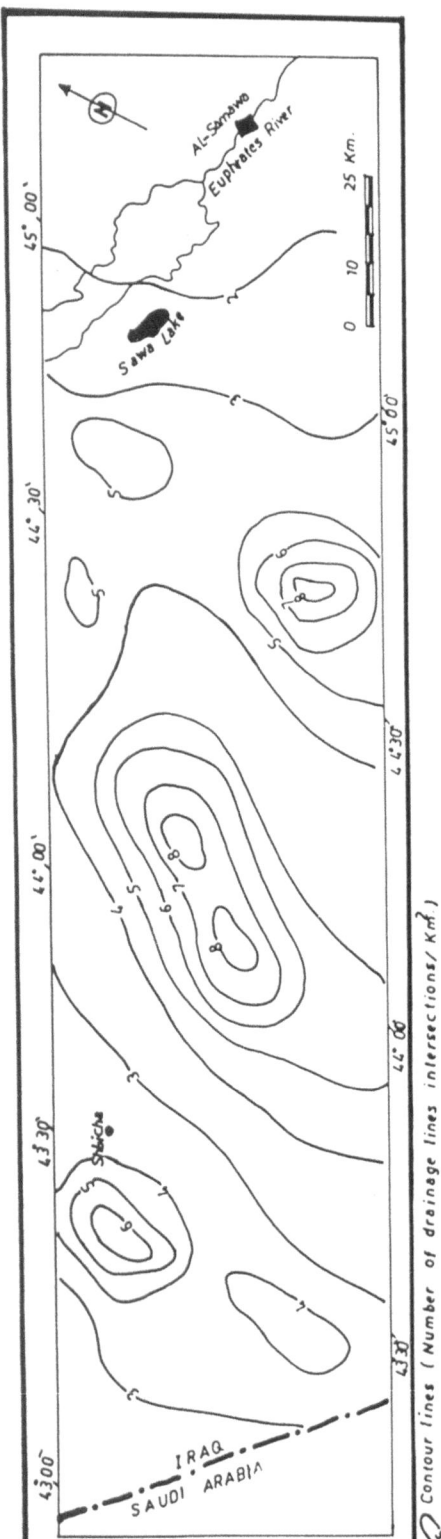

Contour lines (Number of drainage lines intersections/ Km²)

Fig.4 Drainage density map of the study area(based on SIR-A image interpretation).

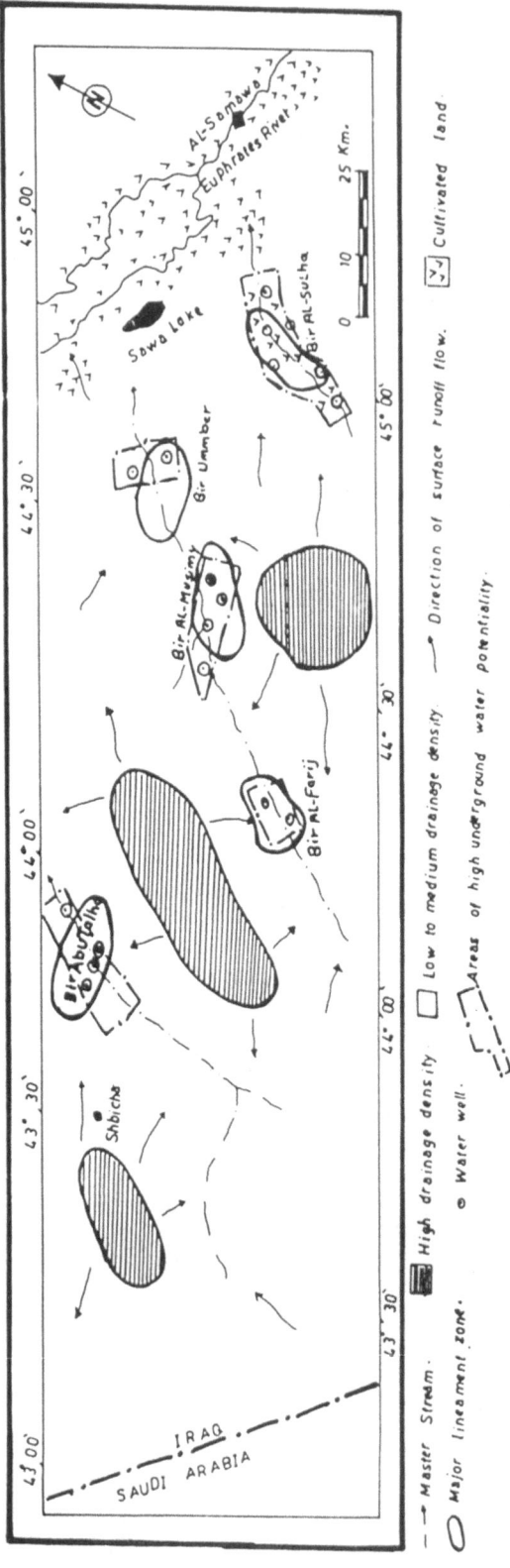

→ Master Stream. ▦ High drainage density □ Low to medium drainage density → Direction of surface runoff flow. ⊻ Cultivated land.

⊙ Water well. ⬭ Areas of high underground water potentiality.

— · — Major lineament zone.

Fig.5 Map of ground water exploration areas,(based on SIR-A images).

168

Table I : Comparison between the main Lithostratigraphic units in the
present work and the previous geological map.

← Previous geological map[7] (Main lithostratigraphic unit.) →				← Corresponding image features →
Quternary	Holocene		Slope sediment	wadi deposits , gravel terraces and slope sediments.	Dark tone because of absence of strong reflections from sensor facing slopes. Terraces are slightly drained.Coarse texture, form gravel wadi fill.
Neogene	Pliocene		Zahra	Red claystone, green marl, Sandstone, Limestone.	Grey tone, characterized by undeveloped drainage system and dry lakes.
	Miocene	Lower Upper	Euphrate	Fossiliferous limestone	Light greyish, dendritic drainage pattern, midium texture.
Palaeogene	Eocene	Upper	Upper Damam	Basal breccia and recrystalized limestone.	Grey to light gray, dendritic to parallel drainage pattern.
		Middle	Middle Damam	Massive limestone Massive recrystalized limestone with thin beds of clay.	Light grey,moderately drained with deep vally,medium to coarse texture, form dissected ridges and hills.
		Lower	Lower Damam	Marl,sandstone marly limestone	Light greyish, weakly drained with wide shallow drainage lines,coarse granular texture,form ridges and low relief hills.
	Paleocene	Upper	Ummer Radhuma	Massive recrystalized fossiliferous limestone.	Blackish to dark grey tone,moderately drained,with dendritic to trellis pattern,fine to medium texture.

map is constructed (Fig.4). It may be
seen from this map that some of the lithological units are of low drainage
density.

DISCUSSION

It is evident that one of the main characteristics of remote sensing images is
a broad regional view of the landscape.
The main characteristic features of the
major lithostratigraphic units were extracted from the regional geological map
(Fig.2). The type of these sedimentary
units,their distribution and the litho-
stratigraphic relations are of grat
importance in prospecting for underground water.
The lower part of Damam units exposed in
the studied area,overlays thickly bedded
massive limestone of the Ummer Radhuma
Formation. The beds generally strike to
the north-west and show gentle inclination to the north-east. This unit represents a remarkable underground water
aquifer which is relatively deep and of
a confined to semiconfined nature. The
Lower Damam unit is exposed in the form
of low to moderate relief,dissected hills
and plateaux with coarse drainage texture
and low drainage density (Fig.2,4).

The second important unit,from a hydrogeological point of view,is composed of
the wadi deposits and gravels. This unit
is significant,because it represents a
shallow underground water aquifer. Most
of the water wells (Fig.5 and Table II)
are present in this unit.
The wadi deposits,gravels and master
stream, in the study area,have been
traced very readily from SIR-A images on
the basis of shape or from pattern,tone
and texture.[13]
It is reported by many authors (for
example, Moore[13], Gelnett and Gardner[14])
that structural lineaments (fractures
including faults and joints) play an
important role in controlling the loca-

lization of underground water.Moreover
many of these structural lineaments were
identified in the field[15],and compared
with the previously available tectonic
map[16].

Table II: List of permanent water supplies around Samawa[17].

Name	Quality	Remark
Bir Abu-Talha	Slightly brackish	abundant,3meter deep.
Bir AL-Farij	Salt	P o o r
Bir AL-Musimy	Brackish	Large supply, 20 meter deep.
Bir Ummber	Fresh	Small supply, unreliable water after rain.
Bir AL-Sulha	Slightly brackish	Unlimited supply, many wells,some Cultivations.

The interpreted major fractures in the
area play an important role in the loca-
lization of underground water.It is
noticed that most of these structural
lineaments are major gravity faults[11].
These faults control the course of many
master stream and are characterized by
relatively wide fracture zones which are
often filled with thick sands and gravel.[15]
This association is very promising since
a remarkable amount of groundwater is
found where alluvial deposits are relatively thick and coarse-grained sediment
occur at depth[7].These fracture zones can
act as suitable channels for migration
of underground water from the deep
aquifer (Lower Damam) to the shallow
aqifer(Wadi deposits and gravels).

170

This assumption is supported at the AL-Sulha locality,where many wells with unlimited water supply are present (Table II).In this example several major faults were interpreted from SIR-A images (Fig.3)and were found to play an important role in controlling underground water in the locality.Moreover, in the same locality a recently drilled deep water well gave an artesian water flow from the Lower Damam unit aquifer[17].

A groundwater exploration map (Fig.5) for the area under consideration has been constructed according to the above SIR-A image interpretation and discussion. On this map master streams, major fracture zones,directions of surface run-off and zones of low to medium drainage density are represented,moreover, localities of high underground water potential are predicted. These localities can be considered as very suitable for more detailed investigations and test wells for underground water.

CONCLUSION

SIR-A imagerinterpretation can be applied to recognize features that are favourable for underground water occurrence. Each image provides a useful tool for the geologist or hydrogeologist, which permits him to bring the field into the office and hence to become rapidly acquainted with a very broad area. This synoptic view greatly increases his interpretative potential.

The most effective features which can be interpreted from SIR-A images covering Southern Iraq are,land form patterns, drainage characteristics,the main litho-stratigraphic units and major structural lineaments.Moreover,direction of surface run-off,soil moisture,cultivation and natural vegetation anomalies can also be recognized. The correlation of these features,in addition to the available

underground water information,can lead to predic ion of locations of high underground water potential. In these predicted locations detailed exploration methods such as interpreting laerial photographs,hydrological mapping,geological mapping, electrical,seismic prospecting, and shallow and deep test drilling can be applied.

REFERENCES

1. Sabin,F.E., Geologic interpretation of space Shuttle Radar images of Indonesia,A.A.P.G.Bull,Vol.67,no.11,1983, P.2076-2099.
2. King,P.B., Comparison of SLAR,SIR and LANDSAT imagery for mapping land system in Kalimantan,Indonesia.Proc.Int. Conf. on advanced Technology of Monitoring and Processing Global Environment Data, London,10-12 Sept., 1985,P .319-329.
3. Harrison,P.G., Development with Multispectral Thermal-IR and Active Microwave System,Frontier for geological remote sensing space.Fouth Gosat Work shop,Flags taff,Arizona,P.49-64.
4. Abdullah,A.Q.,and Zwain,J.A.,Majeed, T.A., Comparison of shuttle imaging radar and Landsat imagery in mapping natural features in southern Iraq,Proc. of the ASCM-ASPRS,Vol.4,Santlous,USA, 1988.
5. Parsons,R.M., Ground water Resources of Iraq,Vol.7,South western desert, 1957,P.157.
6. Lillesand,T.M. and Kiefer,R.W.,Remote Sensing and Image Interpretation,2ed, John Wily and Sons,NewYork.,1987,P.245
7.Mehadi,H.A., Report on the geological mapping of the mesopotamian plain (General Iraq Sheet) using Remote sensing Technique.,SOM Library,Baghdad, 1978.
8. EL-Shazly,E.M.,Abdelhady,M.A., and EL-Shazly,M.M., Groundwater studies in arid areas in Egypt using Landsat

Satellite image, Proc.of the Eleventh
International Syposium on Remote Sensing
of the Environment, Michigan,USA, 1977,
P.1-4.

9. Abdel-Gawad,M., Use of satellite
photography for mineral exploration and
ground water studies, Proc. of the UN/FA
Regional Seminar on Remote Sensing of
Earth Resources and Environment,1974,
P.431-438.

10. Taranik,J.V.,Moore,G.K., Targeting
ground water exploration in south central Arizona using Landsat imagery,
EROS 6th Workshop on Remote sensing,
Sioux Falls,South Dakota,USA,1976,P.1-10

11. Salomy,G.T.,and AL-Khatib,H.H.,
Basment tectonics in AL-Salman area
Southwestern desert,Iraq,IJour.of the
Geol.Soc.Iraq,Vol.15,no.1,1982,P.3-8.

12. Charon,J.B., Hydrogeological applications of ERTS satellite imagery.Procc.
of the UN/FAO Regional Seminar on Remote
Sensing of Earth Resource and Environment, 1974,P439-456.

13. Moore,G.K., Prospecting for ground
water with Landsat images. Procc.of the
word water Conf., Argentina, 1977,P.1-16

14. Gelnett,R.H., and Gardner,J.V, Use
of radar for ground water exploration
in Nigeria,West Africa,Motorola Aerial
Remote Sensing,Inc.,4350 East Camelback,
Phoenix,Arizoa,USA, 1979,P.1-10.

15. AL-Azawi,T.M., The tectonic of the
Land western Euphrates River using
Satellite imageries and geological info-
rmations,Unpub. M.Sc. thesis,Univ. of
Baghdad, 1988, P.108,(in Arabic).

16. Buday,T., and Jassim,S.Z., Tectonic
map of Iraq, 1984.

17. Said,T.M.,and Kadhum,K., Stage report on the geology and the hydrgeological
exploration in Ash-Shabaka,AL-Salman
AL-Thakhadid area,Unpub.Report, SOM
Library,Baghdad, 1980,P.145.

Gold exploration

Turbidite-hosted Proterozoic gold exploration in the Kibaran Belt, Burundi, Central Africa

R. L. Bedell B.A., M.Sc., F.G.S., M.I.M.M.
Burundi Mining Company, Bujumbura, Burundi
M. Odinga B.Sc., M.I.M.M., C.Eng.
London, England
S. Niyondezo B.Sc.
L. Nkurikiye B.Sc.
Burundi Mining Company, Bujumbura, Burundi
M. Fernandez-Alonso B.Sc., Ph.D.
Ph. Trefois B.Sc., C.Eng.
Royal Museum for Central Africa, Tervuren, Belgium

SYNOPSIS

Gold-bearing quartz veins hosted by quartzites in NE Burundi belong to a thrusted turbidite sequence in the external zone of the Kibaran Belt. Remote sensing data combined with computerized field investigations are the basis of exploration methodology. Analogue Thematic Mapper band 4 imagery were used in the initial stages of the project to test the relative capability of the data in this heavily vegetated terrain. Subsequent processing of digital TM data resulted in one main colour product written to film (TM bands 3,4,5 B,G,R with decorrelation stretch and 31x31 high pass filter applied to all bands). Regional interpretation at approximately 1:150,000 scale provides quality data on the distribution of major thrust faults and more recent faulting associated with splays of the East African Rift System. Local interpretation on TM imagery enlarged to approximately 1:25,000 scale provide additional and complimentary information as well as a mosaic base to stereo monochrome airphotos at 1:50,000 scale. Airphotos are enlarged up to approximately 1:12,000 scale for local mapping. Regional and local remote sensing products are integrated with airborne geophysical surveys and stream sediment geochemical surveys provided by aid projects in the past. All data are correlated with local prospect development to provide a basis for further detailed exploration in a given area. All data are vector digitized and converted into standard CAD format (DXF) for use on PC's in Burundi. Data at different scales and projections are warped to TM imagery and airphotos for inspection as rectified overlays using comprehensive "rubber sheeting" techniques on the PC. Ultimately all data are rectified to UTM projections and tied in with ground surveys. Field data are typed into inexpensive laptops in the field and then read into a comprehensive database integrated with the digital maps on a PC.

INTRODUCTION

This study was instigated by a comprehensive gold exploration program undertaken by the Burundi Mining Company (BUMINCO). Collaboration with the Royal Museum for Central Africa in Belgium has involved the joint development of tectonic and integrated digital mapping and database studies. Until recently, most detailed survey work has not concentrated on NE Burundi and therefore some important contributions to the regional geology have been made during gold exploration (Bedell et al. in press). This contribution will focus on the details of the remote sensing and digital mapping techniques employed. We hope to explain the use of certain data during the exploration program and the development of tools to examine the data driven by the ever-changing daily demands of an on going exploration program. Most of the work is PC driven and done in Burundi. It is hoped that the techniques described herein will help to speed the inevitable trend away from centralized data processing and into the hands of the geologists in the field where timely and more cost effective demands are met during exploration.

A preliminary visit to Burundi was made in 1987 to examine the terrain, attempt to establish the nature of the primary mineralization and evaluate the overall gold potential. In April of 1989 the Burundi Mining Company was formed. This contribution was written near the first anniversary of the company, and during the first year an exploration team was established and a vast amount of data have been synthesized. Most field based efforts have been involved in bulk sampling-pilot mining around historic eluvial gold diggings in order to obtain some cash flow as quickly as possible. The expense of sending samples for geochemical analysis to overseas laboratories dictates that processing must be largely done locally. Detailed excavations at a number of sites have therefore provided methods for the rapid processing and evaluation of samples during regional exploration.

Local processing to date has included sluicing and then running this material through a small trommel, or running material through a large trommel and Knelson concentrator. Oversize and undersize are checked by assay. Presently plans are being made to establish a more appropriate 120 ton per day pilot mining plant.

In summary, exploration demands have required detailed work at a number of sites with regional exploration a lower priority during the first year of operation. This method may be somewhat contradictory to ideal exploration philosophy, but helps to boost investor and political confidence. In the second financial year BUMINCO now has a well established infrastructure and is able to investigate other exploration targets in the region.

REGIONAL GEOLOGY

The Kibaran Belt is an intracratonic mobile belt with an extensive history of orogenesis from 1350 - 1100 m.y. in Burundi (Figure 1a). Burundi is located at the northeastern tip of Lake Tanganyika and can roughly be divided into the western internal zone, and the eastern external zone (Figure 1b). The Kibaran in Burundi has been described as a thrust - fold belt (Theunissen, 1988), and has a protracted history of granitoid emplacement (Fernandez-Alonso, 1986 and Klerkx et al., 1987). Early orogenesis resulted in granitoid emplacement (variable in composition and deformation) and well developed S-C fabrics in the internal zone (Theunissen, 1988). Later deformation occurred at around 1100 m.y. along the Cene shear zone, a left lateral shear zone roughly dividing the country along a north south trending line. This late shear event hosts a distinct magmatism of alkalic composition (Tack et al. in press). Ultramafic and mafic bodies found generally east of the CENE also are interpreted as being derived from a deep seated source emplaced along a major intracratonic fracture system (Lavreau, et al., 1989). Thrusting in eastern Burundi has been interpreted as a late Kibaran event and a kinematic model to explain thrusting within a strike-slip regime has been proposed (Theunissen, 1989).

The sediments of NE Burundi consist of a monotonous sequence of Mid Proterozoic quartzites and argillites generally trending NE-NNE and dipping NW. Some tuffs have been found in the argillites although weathering obscures identification. Detailed stratigraphic work along strike in Rwanda suggests that the sediments are relatively shallow water turbidites with an eastern source (Baudet, et al., 1988).

GEOGRAPHY

Burundi is a mountainous country with a dense vegetation cover. The original vegetation was largely tropical rain forest to densely wooded savanah. Burundi is also one of the most densley populated parts of Africa with people divided into small communes. Small agricultural plots dominate the present landscape containing a variety of crops. The main dry season lasts from June through August.

Figure 1a: Kibaran and Panafrican mobile belts. (After Pohl, 1987).

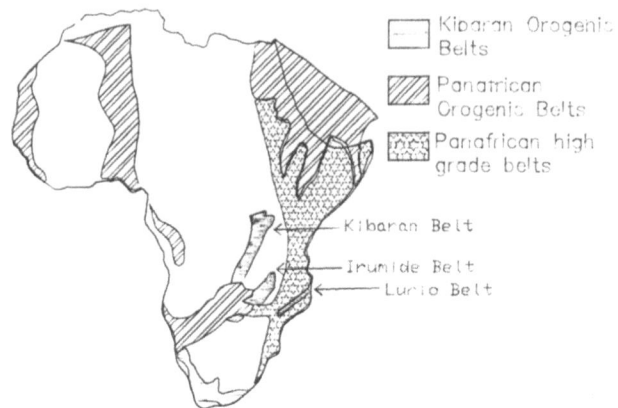

Figure 1b: Generalized map of Burundi (after Klerkx et al. 1980).

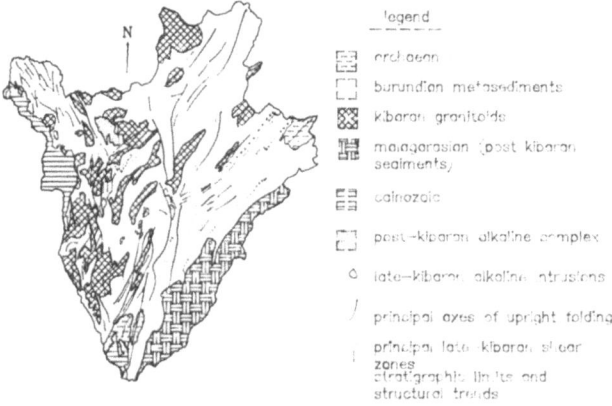

DATA SOURCES

A variety of data are available useful for both small scale and large scale mapping and exploration.

AERIAL PHOTOS: monochrome at 1:50,000. Some of these have been enlarged up to about 1:12,000 scale for detailed mapping.

GEOPHYSICS: are available as magnetics and radiometrics (Bundesanstalt fur geowissen- schaften und Rohstoffe (1981) and Huntings (1974)). These data have been converted to PC format and are processed in Burundi. This allows detailed investigations over specific areas as the need arises.

GEOCHEMISTRY: a full stream sediment survey was made by BRGM in 1981.
TOPOGRAPHIC MAPS: all the areas are available at 1:50,000 scale. Some of the areas are available at 1:25,000 scale.
GEOLOGICAL MAPS: are available at 1:100,000 scale produced in 1986.

In addition, to the above data remote sensing imagery were obtained.

REMOTE SENSING

After some consideration LANDSAT THEMATIC MAPPER satellite data, scene 172-62 acquired 12/6/86 was purchased. This data was chosen over SPOT for a number of reasons. Firstly, relatively cloud free imagery were available over the whole concession area on one TM scene in the dry season. The SPOT archive had only one image over a small part of the concession area. Secondly, SPOT does not have the longer wavelength bands which penetrate the ground haze. Digital processing of the TM demonstrated that a standard false color composite (FCC) of SPOT would have contained significant haze. Thirdly, 1:50,000 scale airphotos were available and the data overlap of higher resolution SPOT would be an unnecessary expense. Digital processing and subsequent photographic enlargement of TM imagery in this study have provided 1:25,000 scale imagery over areas of high interest which is twice the resolution of the unenlarged airphotos.

In summary, TM data were chosen due to data availability, preferred spectral advantages, and sufficient spatial resolution.

The TM data were tested by purchasing a monochrome TM band 4 negative and making approximately 1:200,000 scale enlargements for the initial logisitics visit. Much of the geology is outlined by quartzite ridges which are texturally smooth and have low reflectance in band 4 (due to dry vegetation) during the dry season relative to the valley floors (Figure 2). The simple single band monochrome image was considered to be succesful, but further improvements were thought to be gained by purchasing the digital data. Firstly, by custom processing, a higher resolution image could be obtained. Secondly, increased spectral discrimination could be obtained using the additional spectral bands and customized contrast stretching. Thirdly, a non-spectral product might be advantageous in lineament mapping. The digital data were therefore purchased.

TM PROCESSING

A variety of images were produced during image processing and archived on 35mm film directly from the VDU. Ultimately, two images have been processed and digitally written to large film negatives. The first film product is a false colour composite of TM bands 3,4,5 to blue, green, and red (Figure 2) with a decorrelation stretch and 31x31 high pass filter (Figure 3). The decorrelation stretch is now a standard processing technique which uses principal components analysis to find orthogonal axes of maximum variance among the spectral bands. Linear stretching along the principal component axes (which are combinations of three spectral bands) decorrelates the data for greater colour variation. The technique differs from standard principal component analysis by rotating the data back so that the colours represent specifc spectral bands and not unknown combinations. This method gave a qualitatively superior product to a variety of linear and non-linear stretches as well as Hue-Saturation-Intensity (HSI) transforms. The 31x31 high pass filter is visually a superior edge enhancement, particularly when looking in detail. Theoretically this is also near the limit of the frequency detail obtainable by the Human Visual System (e.g. Drury, 1986). Be warned however that such large filters are very CPU intensive.

The second film product is a monochrome 3x3 edge enhancement of band 5. This method smooths the image and then subtracts the smoothed image from the original, placing most of the image in mid gray tones while the edges (e.g. faults and fractures) are black and white. Photographic enlargements have been produced at about 1:150,000 scale.

Figure 2: Spectral bands of TM and SPOT displayed with approximate reflectance curves for vegetation and dry soil. The false colour composite used in this study used TM band 3 as blue (visible red), TM band 4 as green (very sensitive to vegetation reflectance in the infra red), and TM band 5 as red (sensitive to vegetation and soil moisture).

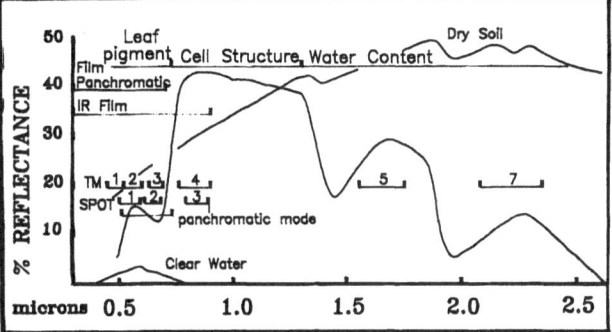

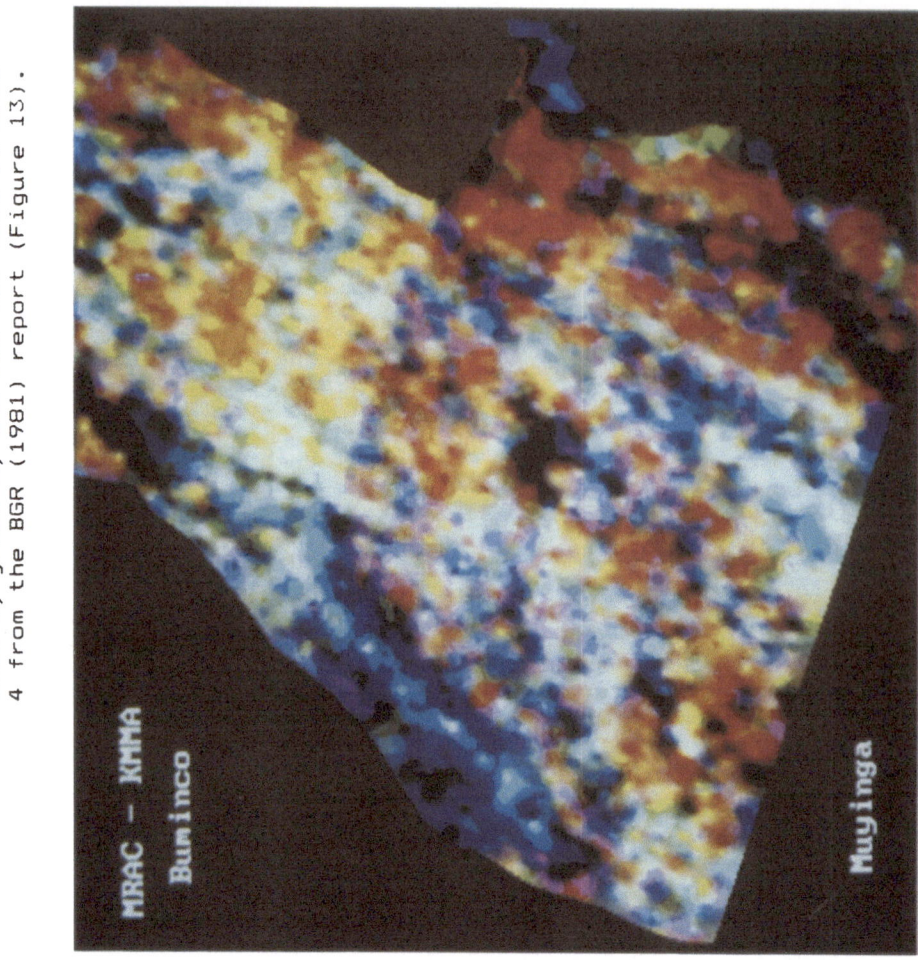

Figure 12: False colour composite of airborne radiometrics. The colours are
blue=K, green=U, and red=Th. This is area
4 from the BGR (1981) report (Figure 13).

FIGURE 3: Landsat Thematic Mapper false
colour composite processed with a
decorrelation stretch of TM bands 3
(blue), 4 (green) and 5 (red). All bands
have a 31x31 high pass filter applied.
Compare electromagnetic spectrum and reflectance curves in Figure 2. Image is
approximately 115 Km across.

178

Figure 5: Generalized Thematic Mapper interpretation of NE Burundi. Similar to
Figure 4, but here the lineaments are
added. Most lineaments have been reactivated during East African Rifting. The
dominant sets of lineaments are E-W and
NW-SE. Also notice the shorter lineaments
near the N-S trend. Also notice the ductile fabrics, suggesting left lateral
shear, associated with the area of the
Cene shear zone in the lower left of the
figure.

Figure 4: Generalized Thematic Mapper interpretation of NE Burundi. Lithology and
fabric is dominated by quartzite distribution. A number of thrusts are outlined
by thick lines where they can be interpreted from the satellite image.

Figure 6: Rosenet of
lineaments in Figure 5.
The top Rosenet is by
number and the bottom
Rosenet is by length.
Note the dominant E-W and
NW-SE lineaments reacti-
vated by recent rift tec-
tonics. Notice the near N
population which have
little length (as they
were not reactivated by
the rift), but are numer-
ous.

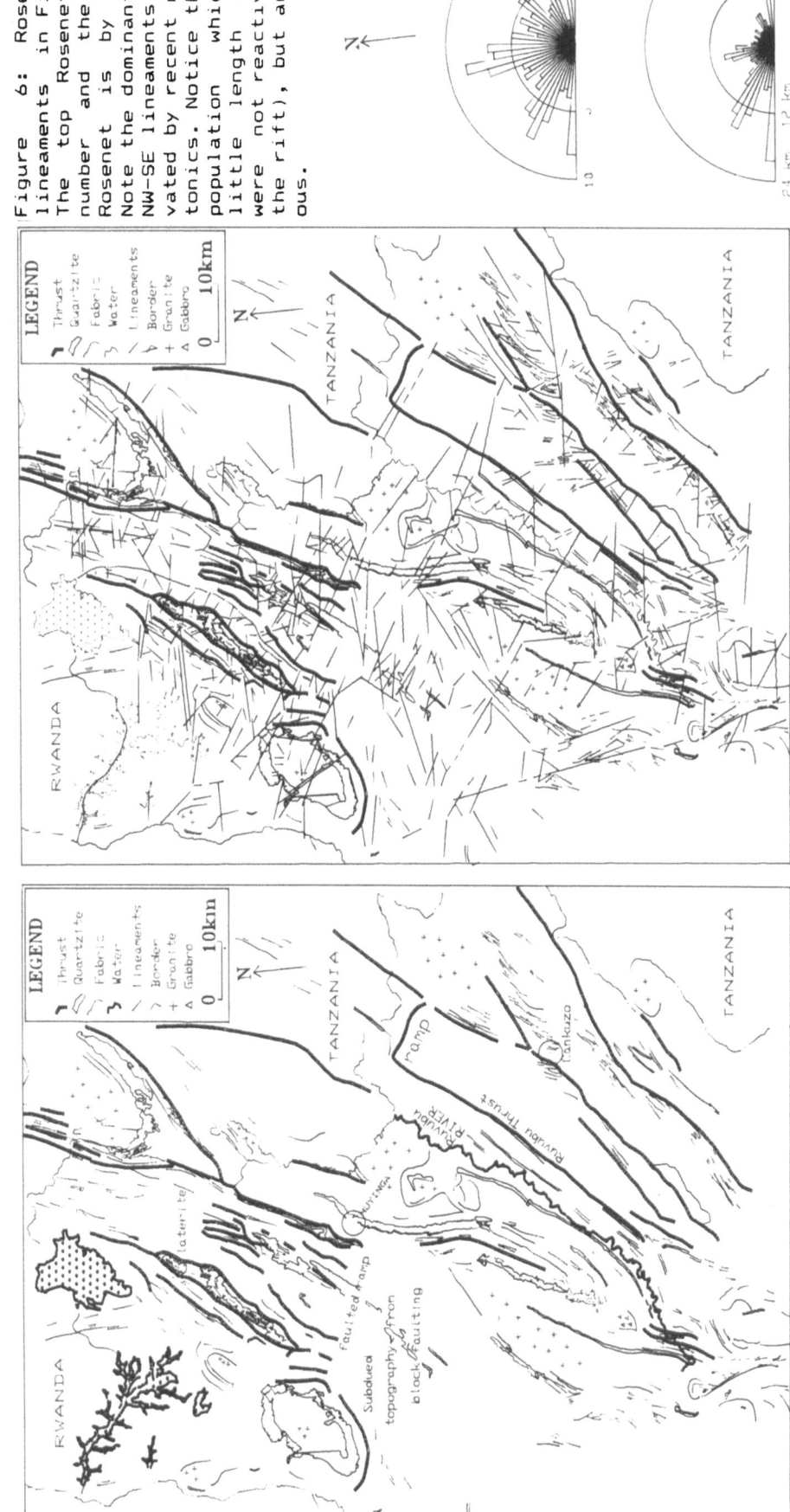

179

REGIONAL TM INTERPRETATION

Most of the comprehensive interpretation has focused on the FCC generated with the decorrelation stretch and using the monochrome edge enhancement as a backup for resolving specific lineaments. A regional geological interpretation has been produced of NE Burundi using the Thematic Mapper Imagery described above. It is important to note that this work has led to a fundamental re- interpretation of the geology. Regional mapping in the past has outlined a number of large fold structures dominated by upright folding. This work produces a different interpretation based on SE directed overthrusting (Bedell et al. in press).

Figure 4 is a reduced overlay of this work showing the major features of NE Burundi (compare colour image in Figure 3). The dominant trend of the belt is NE trending in the south east of the image and more NNE trending in the north east of the image. Major and minor thrusts are delineated on the TM interpretation with similar NE to NNE trends to the regional lithology. A major thrust is viewed just SE of the Ruvubu River (Southern most source of the Nile). Near the Tanzania border the thrust displays a right lateral ramp structure. Throughout the rest of the image a variety of smaller thrusts are demarcated. Toes of the thrusts are extending largely to the SE-ESE suggesting direction of transport. The transport direction is verified in the field by numerous drag folds and slickensides.

Figure 5 describes the lineaments interpreted from the TM image; note that the north arrow is not vertical. Dominant lineament trends (lengthy, with apparent spectral and topographic expression) are NW-SE and E-W (Figure 6). Both of these trends are reactivated during East African Rifting. Both trends show significant topographic breaks and drainage realignment. The NW-SE trend cross cuts the belt at a high angle and is considered to be a combination of reactivated lateral ramps and simple cross fractures. The E-W trend has been considered as being reactivated Malagarasian or Pan African (approximately. 700-800m.y.) (Theunissen, personal communication 1990).

Less dominant lineaments cut the belt at a lower angle and are found trending NNW to N (Figure 5 and 6). These lineaments are relatively short and are terminated by the more recently activated East African lineaments. In the NW part of the image sharp changes in metamorphic grade may be found by crossing these NNW to N trending lineaments. Argillitic rocks are found on the eastern side while garnet bearing schists can be found on the western side. This tectonic and metamorphic zone has been interpreted as a back thrust along strike in Rwanda (Theunissen, personal communication

1990). SE dipping beds can be found locally along this structure and therefore do not contradict the possibility of this backthrust continuing through into Burundi.

A major group of subparallel linear features is the Cene shear zone running NNW through the left hand side of Figures 3-5. This is a major left lateral shear zone, for which the sense of movement can be interpreted from the TM imagery by second order ductile structures (Figure 5, lower left). This is a major shear zone with mylonites and associated alkalic magmatism dated as Late Kibaran (1137m.y. Tack et al. in press). The CENE has been interpreted as a bounding strike-slip fault to the assymmetrical package of Burundian sediments (Theunissen, 1989).

Spectrally the image gives superior discrimination between the bright red (dry vegetation) quartzite ridges and the bright green (healthy vegetation) argillite valleys (compare Figures 2 and 3). Blue is highlighting water and exposed soil and rock. The lateritic soil is visible red but note that rock in general will give greater reflectance over vegetation in band 3 (Figure 2). This is particulalrly useful as it can lead one to exposures throughout the belt and can even pickout white quartzite boulders in a field of metre high grass in the images viewed at 1:150,000 scale.

The spectral signature of the quartzite ridges is so vivid it could probably be mapped with reasonable accuracy using supervised classification techniques. Acceptions to a complete classification will be small eucalyptus forests on the quartzite ridges. Laterite caps also occur over quartzite and will give erroneous classification results if one is attempting to map bedrock. Laterite caps are texturally rougher than the smooth quartzite ridges and have spotty spectral vegetation anomalies. This is probably due to trapped ground water relative to the well fractured quartzites. In any event the vivid quartzite ridge distribution could produce a reasonable map that in a number of areas would be more reliable than the 1:100,000 scale geological maps produced in 1986 using field traverses air photography and monochrome TM imagery.

The other major lithology in the belt besides quartzite, argillite, and laterite is granite. Granites are demarcated by the distribution of bounding quartzite ridges, their internal drainage systems, and subdued topography.

AIR PHOTO INTERPRETATION

Monochrome airphotos can give more detailed data on the quartzites. Although the quartzites are less readily apparent without the strong spectral signatures, the smooth texture and particularly the topographic expression of the quartzites is apparent with stereo and the higher resolution which can be provided (enlargements upt 1:12,500 scale have been produced). On some ridges bedding strike and dip can be mapped by the blocky outcrops and linear grass patterns. Dip can also be mapped by the smooth slopes. Drainage incision is most apparent in the quartzite footwall contact with the underlying argillites. This type of geomorphology is known as a strike ridge and cuesta pattern. The indented pattern (cuestas), created by drainage incision into the soft argillites below the resistant quartzite, allow direct interpretation of the direction of dip. This has been a particularly useful tool for discriminating against some of the detailed fold structures that were previously mapped. Lithological contacts for areas of detailed mapping are constructed using the combined TM and air photos.

DATABASE

In addition to the spatial data, tabular datasets are used to record field observations using inexpensive laptop computers. The database described here is used by BUMINCO. The Royal Museum for Central Africa in Brussels uses the UNESCO sponsored CDS-ISIS system.

The data are originally input using a BASIC program which is small enough to use even on hand held computers. Programs with built in error checking include input for hardrock, eluvials, alluvials, and samples. For profiles, thicknesses are calculated. These data are then read into dBASE IV using standard Sequential Data Files (SDF) for manipulation. dBASE IV requires a computer with a hard disk.

A drawback of dBASE IV is that most fields are fixed, however memo fields are of variable length. The two types of fields appear to work fairly efficiently because necessary data such as dip and dip direction do not take much storage space and are of a predictable length. The memo fields however offer the geologist up to 64 kb of waffle to describe the feature in any amount of detail.

A major advantage of the dBASE IV is the control center programming and built in compiler. This allows you to make database applications ad infinitum and compiled so that field geologists can be presented with a custom system for their specific task. At the time of this writing dBASE IV is running under a pseudo compiler and still requires a hard disk system to run the 1 mb of runtime files. The full compiler promises to run small applications on a computer without hard disk and should eliminate the need for

using an intial BASIC program for data input on most small computers.

Presently, work is pursuing a link between the digital mapping based on AutoCAD and the dBASE. Commercial links are available providing a GIS environment between dBASE and AutoCAD, but there are a number of books available which can show the dedicated user how to accomplish this without additional software.

DATA RECTIFICATION

All field based geological observations are being input into the database using the recent 1:50,000 scale map sheets in UTM coordinates. The UTM coordinate system is the most preferable projection for many reasons (Merrill, 1986), but most of all because it is metric and the most ammenable to direct computations for distance, area, and volume calculations in metres.

Comprehensive rectification are required for all other datasets. This is accomplished using another off-the-shelf software package called WARP (c) 1988 Terra Investigations & Imaging Ltd. which uses a variety of polynomial transforms to provide non-linear distortions or rectification to a computer drawing. For comprehensive rectification there are 3 problems. Firstly, the original survey may have non-linear distortions due to a lack of control. Secondly, if you are digitizing analogue data, these data are invariably distorted in a non-linear way due to paper shrinkage. Thirdly, the actual change in projection, which in the case of air photos may have to remove radial lens distortions and perspective. A discussion of problems faced by working geologists with digital maps is found in Bedell (1989).

An example of detailed rectification over an area with a high gold occurrence density will give you some idea of the accuracy which can be achieved.

Butihinda Block
The Butihinda Block is an uplifted, structurally anomalous area containing a high density of gold occurrences (Figure 7). Most of the area consists of NW dipping quartzite units with evidence of thrusting throughout. The northern part of the block with anomalous tectonic thickening is interpreted as a complex antiformal stack with a NE plunge (Figure 8). The geology (quartzite and laterite distribution) have been mapped with a combination of TM and air photos. Field checking has verified some quartzite outcrop, but the contacts are best described on the remote sensing imagery.

Figure 9 shows the combination of lineaments rectified to the UTM projec-

Figure 7: Map of the Butihinda Block digitized from two 1:50,000 scale topographic maps. The broad geology was rectified using the computer from Thematic Mapper imagery and stereo airphotos with errors of less than 30 metres. Gold occurrences are marked. Note the complex distribution of the quartzites. It is probably no mistake that the structurally complex Butihinda Block hosts a large number of gold occurrences.

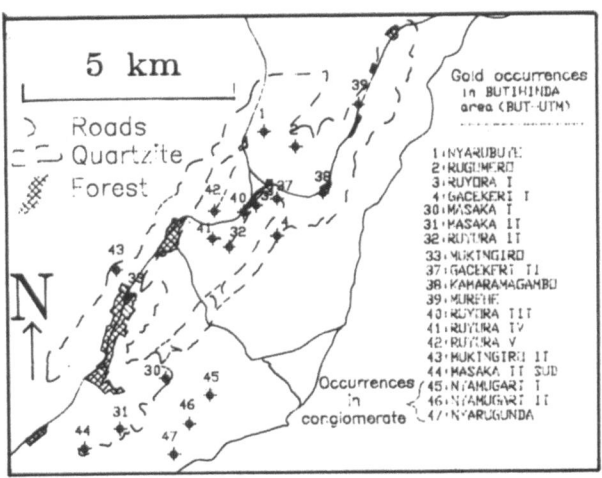

Figure 8: The Butihinda Block geology is dominated by a complex pattern of quartzites (dotted lines). Duricrust caps are found along the NW limbs of the quartzite ridges. The beds are dominantly dipping to the NW and a number of reverse fault structures have been found in the bedding plane. The north tip of the Butihinda Block shows an anomalous thickness of quartzites. Topographically this area is also the highest point in NE Burundi. This region is bordered to the south by a major cross fault along which a marked topographic break occurs. This zone may have been some kind of ramp structure and been modified by more recent block faulting. The Northern end of the Butihinda Block is undoubtedly created by some tectonic thickening and is best described as an antiformal stack.

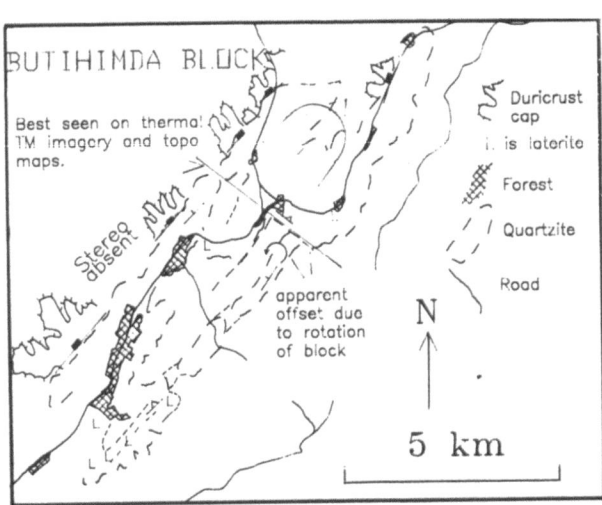

Figure 9a: Lineaments over the Butihinda Block. Lineaments interpreted from the TM image are thick solid lines and lineaments interpreted from the airphotos are in thick dashed lines. The quartzite distribution is in dots and drainages are in thin lines. Forests are hatched. North is vertical. Note the relative N-S bias of the airphoto interpretation and the relative NW-SE bias of the TM imagery.

Also note the drainage density relative to the quartzites. One could almost map the quartzites by contouring areas of low drainage density.

Figure 9b: Rosenet of the lineaments in Figure 9a. Notice the difference in the N-S bias of lineaments in the airphotos relative to the TM.

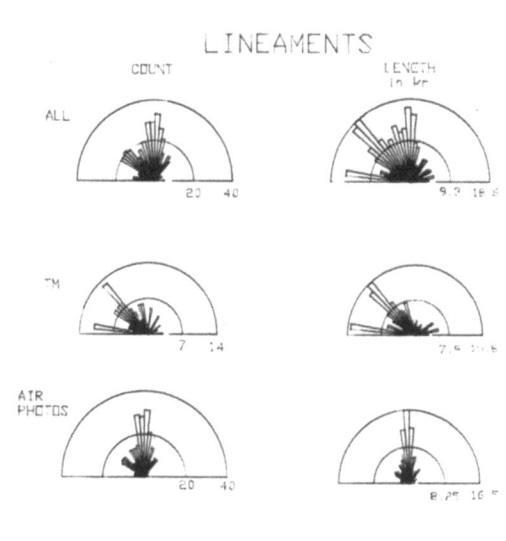

182

tion. The most complex rectification is with air photos which have radial as well as perspective distortions. Errors were minimized using a fourth order polynomial transformation with 16 control points (Table I). The highest error encountered in one of the airphotos was 37 metres although most errors are significantly less. Remember 37 metres on a 1:50,000 scale map base is only 0.74mm which could be digitizing errors.

Table I: Errors in warping airphotos over the Butihinda Block

```
-------------------------------
AIR PHOTO 4-12
WARP by 4th order polynomial
-------------------------------
control ERROR in metres
point   X        Y
-------------------------------
1     -3.79    28.03
2     -1.10     5.45
3      0.58    -7.44
4     -1.15     2.86
5     -1.79     7.41
6     16.80   -35.00
7      5.91   -25.48
8    -10.27     5.18
9     -0.76     3.68
10     2.75    -8.76
11    -5.69    -6.06
12    -8.04    37.44
13     3.35   -14.43
14     3.92   -19.15
15     2.23     4.62
16    -2.95     9.53
-------------------------------
AIR PHOTO 3-74
WARP by 4th order polynomial
-------------------------------
control ERROR in metres
point   X        Y
-------------------------------
1      0.37    -0.47
2     -0.51    -4.01
3     -9.32    12.21
4      0.53    -0.49
5     -0.19    -0.13
6     -0.17     0.17
7      0.42    -0.26
8     -0.73     0.62
9      9.17   -10.37
10     5.72    -2.03
11    -1.44     1.56
12    -2.82     2.80
13    -4.43     6.92
14    -0.14     0.63
15    -3.66     0.72
16     7.20    -8.12
-------------------------------
```

The difference between the two sets of lineaments is interesting. The TM lineaments have a NW-SE bias showing lineaments at a high angle to the NE-SW trend of the quartzite ridges. The air photos have a bias to N-S trending structures. Because the area is located near 3 degrees south of the equator an illumination bias is not considered likely. The stereo effect of the air photos is sus-

pected as being the main source of bias. Whatever the reason, both trends are gold-bearing and the importance of well rectified imagery from more than one source of data is to be noted.

GEOPHYSICS

Geophysical data were converted from magnetic tape to floppy disk and contoured within AutoCAD using another CAD compatible add on called Quicksurf (tm). An example is shown in Figure 10 where a magnetic contour map with positive magnetic anomalies over the northern end of the Butihinda Block is shown. The anomalies are generally N-S trending and cut the NE-SW trending ridge at a high angle. Some anomalies are directly observed in the field as corresponding with N-S trending argillites. However, argillite-bearing outcrops often have quite complex trends in detail because they are structurally incompetent. The overall geophysical trends often correspond with subtle topographic inflections in the quartzite ridges suggesting an overall trend for the argillite.

Although flight line spacing may bias azimuthal trends to some degree, various contouring tests of the data consistently show the northerly trend to be real. On a smaller scale the northerly trend is apparent on a subsampled data set (Figure 11). On the subsampled data set the sampling on the flight line is more equivalent to the spacing between flight lines. The subsampling minimizes directional bias in the contouring. Although the decreased resolution does not distinguish individual argillite sheets along the flight lines, the regional structure is apparent.

Radiometric data have also been contoured, but the combined K, U, Th are best viewed together as a false colour composite. One image has been produced of NE Burundi showing K, U, Th as blue, green, red (Figure 12). A 512x512 image was constructed by gridding the radiometric data (minimum curvature with cubic spline interpolation) for each element.

Another image was produced using TM band 5 as a background. The 512x512 images (one for each element) were geometrically resampled to the 2048x2048 TM band 5 image. For each element a new 2048x2048 image was constructed by linear combination:

(TM 5 + K)/2;
(TM 5 + U)/2;
(TM 5 + Th)/2

After interactive stretching of the individual images a false colour composite was produced:

Figure 10: Magnetic contour map of the
northern end of the Butihinda Block. The
quartzite is dashed and the roads are a
thick solid line. The thin solid lines
show aeromagnetics contoured at 5 nT intervals (BGR, 1981). The high contour areas all represent magnetic highs. On the
quartzite ridge the magnetic highs correspond with argillites. Notice how the
magnetic highs, or argillite intercalations, suggest some N-S trends. This can
be verified in outcrop exposures.

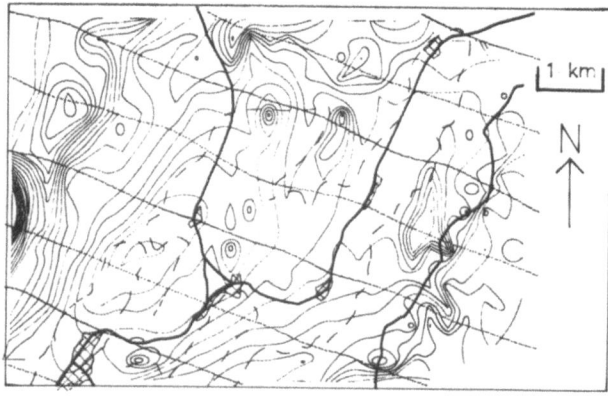

Figure 11: Regional magnetic map produced
from a subsampled data set (every tenth
point along a flight line). Notice the
number of N-S lineaments one can find in
the geophysics. This is area 4 from the
BGR (1981) report (Figure 13).

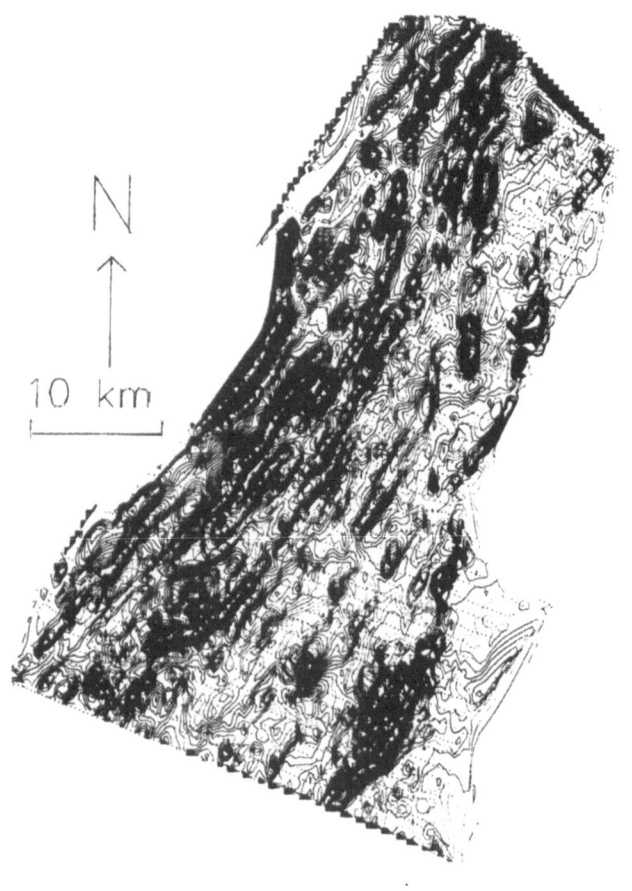

Figure 13: Location map of Area IV from
the BGR (1981) report.

Red = Th (+ TM);
Green = U (+ TM);
Blue = K (+ TM)

The image in Figure 12 does not have
the TM band 5 background as the colours
were a bit pastel for publication. This
image however can be located on the key
(Figure 13) and compared with Figures
3-5.

Figure 12 shows a great deal of detail
in the argillaceous sequences which are
difficult to map as they erode out more
easily than the quartzites. High K values
may distinguish Rhyodacitic tuffs, although more field work are needed to
verify the consistency of this association throughout the belt. The quartzite ridges are displayed as dark radiometric lows and verify the TM and
airphoto interpretations. Of particular
importance is the lack of radiometric
signatures around the areas of known gold
mineralization, verifying that remote
sensing interpretation and field work
have not missed any exposed granite intrusions.

GEOLOGICAL MODEL

The data integration is revealing a great
deal of information on regional structures and lithology. A broad kinematic
model has already been proposed by
Theunissen (1989). Theunissen proposed a
model for the overall shear of the
assymetric package of Burundian sediments
in NE Burundi and neighboring Tanzania.
The structures found in this study may
indicate second order structures applicable to Theunissens model. Figure 14 is

184

a schematic model of the assymetrical sedimentary package. In the overall left lateral stress system a number of second order stress systems will be operating, producing not only NW-SE compression, but also left lateral shear along a NE-SW trend. Given the NE trend, competent quartzites would tend to be subjected to thrusting as well as a left lateral shear parallel to their contact. Evidence for the shear is particularly evident on the TM imagery in the Cankuzo region where drag folds and fracture in the quartzites allow determination of the left lateral shear (Figure 15). In addition, incompetent argillites would tend to be refracted to a northerly direction as a second order structure. It is possible that the observed thrusting, left lateral shear, and counterclockwise refraction of the incompetent argillites may occur throughout the belt under a single regional stress system. The kinematic features found at the regional scale and individual prospect scale are presented in more detail elsewhere (Bedell et al. in press)

To summarize, regional geology has been outlined by the TM imagery and verified in many places by field investigations and ancillary data, including stereo air photos, and geophysical data. The structures outlined are interpreted as being dominated by thrust tectonics during the Kibaran Orogeny. All the early structures observed in the external zone during this study can be explained under one kinematic model involving simultaneous wrench and thrust tectonics. Later structures are dominated by block faulting, presumably related to the East African rift system. We will now describe the primary gold mineralization and attempt to relate what is presently known about the mineralization to the regional geology.

Figure 14: In addition to the compressive stresses described by Theunissen it may also be possible to generate left lateral strike-slip movement within the sediments. Competent quartzite contact zones will be particularly susceptible to strain and the argillites will be quite sensitive strain indicators. Many of the argillites demonstrate a refraction from the dominant NE-NNE trend quartzite trend. North is approximately vertical in this figure.

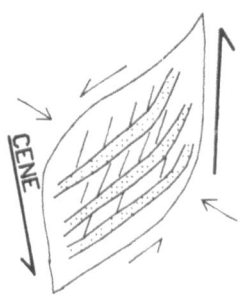

Figure 15: The Archean age granite NE of Cankuzo shows some magnificent decollement structures. Mylonites can be found along the decollement in contact with the granite. There is also much evidence in drag folds and faulting of the left lateral strike-slip faulting.

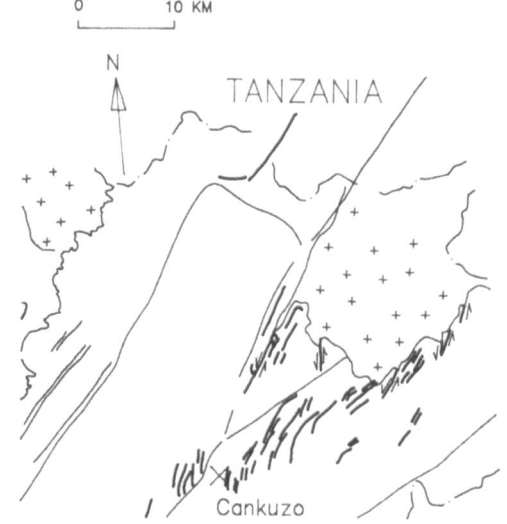

PRIMARY MINERALIZATION

The gold mineralization in the NE of Burundi is relatively simple. All the gold is hosted by quartz veins largely confined to the mechanically competent quartzites. The gold is largely free gold with little associated pyrite. Associations are sporadic in both the hand specimen mineralogy and regional geochemistry. In some areas As or W shows an association, but not in others. There is no obvious spatial association between igneous rocks of any kind and the mineralization. It seems likely that the gold was in the basin and then structurally emplaced sometime during or after the Kibaran Orogeny. Coarser quartzites showing obvious shallow water features are predominant to the east suggesting an easterly source region. It has been hypothesized (Bedell et al. 1988) that the source of the gold may have ultimately been the stable Archean craton to the east in Tanzania including granite-greenstone terrain with abundant gold mineralization. Basins forming to the west of the greenstone belts would be likely recipients of gold-bearing detritus. A similar model has also been suggested as a source for gold in a Mid Proterozoic basin in Zambia (Andrews-Speed, 1989).

An alternative model, not in conflict with the present status of knowledge, is an association with volcanic activity contemporaneous with basin formation in the Mid Proterozoic. Volcaniclastic material has been found in the vicinity of some gold mineralization, although it is very difficult to distinguish from other argillaceous material due to weathering. To prove such a model, detailed lithological mapping of the argillaceous sequences must be undertaken to demonstrate a spatial association with the gold mineralization. Field work in conjunction with the radiometric data may be particularly useful. Detailed geochemistry is also necessary to establish what, if any, genetic link may exist between volcanism and the gold.

Regardless of the source of the gold, its final em- placement is structurally controlled and is of paramount importance to exploration. From the data collected thus far, geochemical path finders do not exist. Even visible alteration haloes are not apparent, because the silica rich fluids were emplaced into a quartz dominant rock. The only reasonable exploration method is to understand the structural emplacement of the mineralization and extrapolate to the regional scale.

The gold-bearing veins form simple dilational structures (often as conjugate sets), as well as complex stockwork systems. Dark quartz veins are found in many areas, and when present are always found to be early and sometimes have a curvilinear form. White quartz veins are always found to be cross-cutting the black veins and inevitably have planar boundaries. Gold is associated with both dark and white veins.

Chartry (1989) published a paper on some of the prospects in the northern Butihinda Block. He used the 1986 maps as a basis for his interpretations and related the veins he measured to a conjugate shear set generated during folding. The same conjugate set would be produced by thrusting during NW-SE compression. More detailed work has demonstrated that structural details are actually more complex but the conjugate shear set described by Chartry is statistically significant. Details of the stereonet analysis are reported elsewhere (Bedell et al. in press), but include both coaxial and noncoaxial strain.

Reconnaissance work on fluid inclusions has been carried out on the dark veins. The inclusions found are associated with a secondary resilicification with temperatures clustering around 330 degrees, low salinity, and CO_2 bearing. This suggests episodic silicification by relatively hot fluids.

Timing of the Mineralization
The timing of mineralization is still equivocal, but geological data suggest the quartz veins were emplaced as syn to late Kibaran. Locally quartz fibre growths may be found. Fibre growths are typically perpendicular to the vein margin, however, angular and sigmoidal growths are not uncommon in both the white and the dark quartz veins. Most of the fibres are classified as "stretched fibres" using the classification of Ramsay and Huber , 1983 (figure 13.9d).

Figure 16 describes stretched quartz fibre growths found in the Muyinga Quartzite associated with the apparent SE movement of a small block of quartzite. This structure is found on a regional scale where larger blocks with differential SE directed movement form a flexure in the quartzite just north of Muyinga (Figure 16). Note that Figure 16 was drawn largely from 1:25,000 scale TM imagery and verified in the field. Faults and relative offsets in the ridge are clearly discernible on the TM imagery.

In addition to syntectonic quartz growths, slickensides have been found on quartz veins. One outcrop at Masaka 2 is particularly obvious showing a reverse sense of movement to the SE. Clearly, some quartz veins must have been emplaced prior to some reverse faulting.

Mineralization is therefore bracketed as being emplaced along dilational structures produced during Kibaran thrusting. Additional vein trends occur along conjugate shear fractures. Outcrops demonstrate that quartz fibre growth occurred during SE directed movement and quartz veins were deformed by SE directed thrusting. Significant numbers of quartz veins are also found along the Late Kibaran N-S trend.

Figure 16: Cross faulting in the Muyinga Quartzite. Some quartz veins must have precipitated during the cross faulting stage as evidenced by the quartz vein fibre growths found in the Muyinga Quartzite. The same structures are found on the regional scale as shown by the Muyinga quartzite offsets determined by TM interpretation and verified by field mapping.

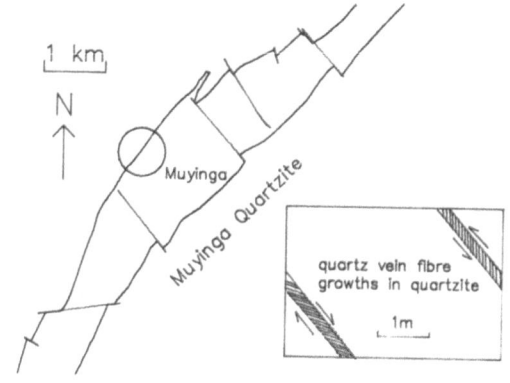

DISCUSSION

Although a granite source cannot be entirely ruled out, there is no compelling evidence to suggest that the auriferous fluids were not generated entirely within the sedimentary basin during thrusting. The lack of any known spatial association and consistent mineral association with granitic products is important. It is also important to highlight the observation that plenty of granites are found in NE Burundi and no spatial association with gold mineralization is found. Most of these granites are deformed and do not indicate partial melting. Later granites are also found throughout the Kibaran with associated tin mineralization (e.g. Pohl, 1987).

Reconnaissance fluid inclusion work suggests the presence of at least one generation of relatively hot (up to 400 degrees C) low salinity, CO_2 bearing inclusions. Firstly, the low salinity and CO_2 bearing inclusions do not correspond with classic granitic fluids. Secondly, at these relatively high temperatures one might expect to find some other granite related minerals still in solution.

Although there are strong arguments against a granite source this cannot be entirely ruled out. It is important to note that these quartzite hosted vein systems are similar to the Telfer gold deposit in Western Australia (e.g. Goellnicht et al 1988). At Telfer they eventually found a related granite at depth. However, fluid inclusions at Telfer have high salinities (21-54%) (Goellnicht et al. 1988) as opposed to the very low (no greater than a few percent) salinities encountered in NE Burundi thus far.

In summary, all the data taken together suggests that there is no related granitic source from:

1) field mapping
2) geophysics
3) satellite imagery
4) aerial photos
5) regional geochemistry
6) fluid inclusions
7) vein mineralogy.

All the evidence to date is compatible with veins emplaced during thrusting by fluids generated within the basin. A working model has therefore been put forward to explain the link between thrusting and auriferous fluid emplacement. Figure 8 describes the general quartzite distribution on the northern end of the Butihinda Block and suggests that this is some kind of NE plunging complex antiformal stack. Given the timing constraints of quartz vein emplacement it is likely that the siliceous and auriferous fluids were generated sometime during thrusting. It is well established that a periodicity in thrust development

occurs whereby thrusts advance typically from the internal zone outward to the foreland. Fluids liberated during high and sudden overpressures created during thrusting have been put foward as a mechanism to perpetuate the growth of thrust domains. Rubey and Hubert (1959) demonstrated the role of fluid pressure in the mechanics of overthrust faulting. Their ideas were advanced by Gretner (1972) who described the periodicity of overthrust faulting with the development of high fluid pressures using both geological observations and experimental data. Fluid pressures build up in the incompetent layer until the stress is transferred to the competent layer whereby fracturing is initiated, a step then forms, collapse occurs, and the load is once again transferred to the incompetent layer. This is in effect a pumping mechanism which is not unlike the seismic pumping mechanism proposed for many epigenetic mineral deposits throughout the world (Sibson et al. 1975).

Figure 17 is a sketch of the mechanism envisaged and can help to explain the concentration of siliceous veins in competent and structurally complex hanging wall horsts such as the Butihinda Block. The volume of quartz veins in this case may represent incremental generations of step development and fluid expulsion. Certainly there is evidence from the differences in the dark and light quartz veins that incremental strain patterns exist (Bedell et al. in press). Such a model may be very useful in explaining the complex incremental strain patterns found in other stockwork systems.

Figure 17: The link between fluid migration and episodic thrusting is suggested in this figure Gretener (1972). High fluid pressures build up in the argillaceous rocks under areas of maximum loading. Once fluid pressure exceeds lithostatic load the competent quartzites fracture. The fracturing may produce a step and the whole sequence rides on a lower thrust surface greased by the high fluid pressures. The resulting structure is what is envisaged for the north end of the Butihinda Block with a NE plunging antiformal stack. Compare Figure 8.

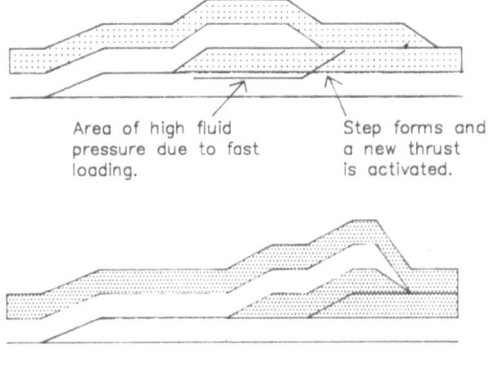

Area of high fluid pressure due to fast loading.

Step forms and a new thrust is activated.

The old thrust might be deactivated and carried "piggyback"

CONCLUSIONS

The methods employed in manipulating spatial data in Burundi have been described and allow for cost effective and adaptable data input - output necessary for any exploration program. The use of remote sensing data in particular has been very helpful in redefining the regional geology. Many of the large thrust structures are obvious on the TM imagery. Air photos, enlarged TM imagery and geophysics have been particularly useful in outlining the complex geology of the Butihinda Block and other areas. Detailed analysis of these data around known prospects allows further use of the regional spatial data to extrapolate for regional exploration.

The association between known gold occurrences and their emplacement into thrust related structures gives regional exploration some prospecting tools. Quartz veins are most likely to be found:

1) In the competent quartzites.

2) In areas which exhibit anomalous tectonic thickening (e.g. antiformal stacks).

3) Along major cross fractures to the quartzite ridges (e.g. ramps or pure dilational structures).

4) Along shear fractures that might provide dilatant splays (e.g. Bedell et al. in press).

In addition, a genetic model relating fluid pumping to the mechanism of thrust tectonics and gold mineralization is put foward. This model which can generate hot episodic fluids may be useful in exploration for other turbidite hosted deposits in thrusted terrains.

Finally, recent exploration has located three more gold occurrences using the TM imagery and airphotos near the Butihinda Block. These areas exhibit cross faulting or tectonic thickening (possible breached antiformal stack). In each of the three sites the main structure hosting the gold was previously interpreted on the TM imagery. Exploration has only recently verified the presence of gold in quartz veins by qualitative crushing and panning, but will follow up with quantitative sampling over the coming year.

ACKNOWLEDGEMENTS

We would like to thank the government of Burundi and John Cole-Baker the Managing Director of BUMINCO for their assistance and permission to publish. This work has benefited from discussions with many people including Dr. Audace Ntungicimpaye Director of Geology and Mines, Professor Luc Tack, Professor Gerard Nimpagaritse, Professor Egide Nzojibwami, and Professor Louis Nahimana from the University of Burundi, and Professor Francois Dimanche from the University of Liege. We would also like to thank everyone at the Royal Museum for Central Africa for their assistance, and particularly Dr. Carl Theunissen for discussions on regional tectonics. Dr. Norman Jackson at Kingston Polytechnic in the UK was of great assistance on the reconnaissance fluid inclusion work. Chris Legg and Tony Harding at the National Remote Sensing Centre in the UK were helpful with discussions on image processing. Lastly, but most importantly we would like to thank our colleagues at BUMINCO who have assisted us including: Zenon Nimpagaritse on detailed prospect geology, Alphonse Rubayika and Gerard Munongo on computer aided drafting, and Assumpta Mpitabakana, Marina Umurungi, and Concilie Sirabahenda on secretarial support.

REFERENCES

Andrews-Speed, C.P. 1989. The Mid-Proterozoic Mporokoso Basin, Northern Zambia: Sequence Stratigraphy, Tectonic Setting and Potential for Gold and Uranium Mineralisation; Precambrian Research, Vol.44, Elsevier Science Publishers B.V.,1-17.

Baudet, D., Hanon, M., Lemonne, E. and Theunissen, K. 1988. Lithosgraphie du Domaine sedimentaire de la Chaine Kibarienne au Rwanda; Annales de la Societe Geologique de Belgique, Vol.112(fasc.1), 225-246.

Bedell R.L., Bowyer, G., Neal, T., Mageed, A. 1988. Regional gold exploration in Burundi; Abstract presented at the Mineral Deposits Studies Group meeting in December 1988, Royal Holloway and Bedford New College, Egham, UK.

Bedell R.L. 1989. Solving problems with digital maps; Computer Oriented Geological Society (COGS) Newsletter November 1989 pp. 3-5.

Bedell R.L., Odinga, M., Niyondezo, S. (in press) Kibaran Thrust Tectonics and its relation to gold mineralization in Burundi, Central Africa; Proceedings of the international meeting on Thrust Tectonics, Royal Holloway and Bedford New College, Egham, UK April 2-5, 1990.

BGR 1983 Bundesanstalt fur Geowissenschaften und Rohstoffe; Leve Aerogeophysique au Burundi 1981, Rapport d'interpretation.

BRGM 1981 Bureau de Recherches Geologiques et Minieres Recherche Miniere dans le Nord-Est du Burundi Campagne 1981, Rapport du BRGM 82 RDM 058 AF.

Chartry, G. 1989. Les Mineralisations Auriferes Primaires de la region de Muyinga (Burundi); IGCP n°255 Newletters/Bulletin, Vol.2, 9-14.

Drury, S. A. 1986. Remote Sensing of geological structure in temperate agricultural terrains; Geol. Mag. vol. 123, pp.113-121.

Fernandez-Alonso, M., Lavreau, J. and Klerkx, J. 1986. Geochemistry and Geochronology of the Kibaran Granites in Burundi, Central Africa: Implications for the Kibaran Orogeny; Chemical Geology, Vol.57, Elsevier Science Plublishers B.V., 217-234.

Goellnicht, N.M., Dimo, G., Groves, D.I., McNaughton, N.J. 1988. An epigenetic origin for the Telfer gold deposit; Proceedings of the Bicentennial gold 88, Melbourne, May, 1988, pp.79-84.

Gretener, P.E. 1972. Thoughts on overthrust faulting in a layered sequence; Bull. Can. Petrol. Geol., vol. 20, no.3, pp. 583-607.

Hunting Geology and Geophysics. 1975. Mineral Survey of a selected area of Burundi, airborne geophysical survey; United Nations Development Program.

Klerkx, J., Liegeois, J.P., Lavreau, J.and Claessens W. 1987. Crustal Evolution of the Northern Kibaran belt, Eastern and Central Africa; Geodynamics Series, Vol.17, American Geophysical Union, 217-233.

Lavreau, J., Tack, L. and Theunissen, K. 1989. The N-S accident of Burundi. Remarks about a paper by J. Chorowicz, T. Nkanira and G. Tamain; Tectonics, C.R.1-C.R.6.

Pohl, W. 1987. Metallogeny of the northeastern Kibaran belt, Central Africa; Geological Journal, Vol.22, John Wiley & Sons, Ltd, 103-119.

Ramsay, J.G., and Huber, M.I. 1983 in: The Techniques of Modern Structural Geology, Volume 1: Strain Analysis, Session 13 Measurement of Progressive Deformation, 1. Extensional Veins; Academic Press.

Rubey, W.W. and Hubbert, M.K. 1959. Role of fluid pressure in mechanics of overthrust faulting; Geol. Soc. Am. Bull., vol. 70, no. 2, pp.167-205.

Sibson, R.H., Moore, J.M., and Rankin, A.H. 1975 Siesmic Pumping-a hydrothermal fluid transport mechanism; J. Geol. Soc. London, Vol. 131, pp.653-659.

Tack, L., De Paepe, P., Liegeois, J.P., Nimpagaritse, G., Ntungicimpaye, A., Midende, G. (in press) Late Kibaran Magmatism in Burundi; Journal of African Earth Sciences, Vol. 10, No. 4.

Theunissen, K. 1988. Kibaran Thrust Fold Belt (D1-2) and Shear Belt (D2); IGCP n°255 Newsletters/Bulletin, Vol.1, 55-64.

Theunissen, K. 1989. On the Rusizian Basement Rise in the Kibara Belt of northern Lake Tanganyika Collision Belt Geometry or Restraining Bend emplaced in the late Kibaran Strike-Slip Environment; IGCP n°255 Newsletter/Bulletin, Vol.2, 85-92.

Interpretation of remotely sensed data over the South Armorican shear zone, Brittany, France: contribution to exploration for gold mineralization

C. Braux
G. Delpont
D. Bonnefoy
D. Cassard
M. Bonnemaison
BRGM, Orléans, France

ABSTRACT

A methodological investigation to define favourable areas for the development of gold-bearing
shear zones within regional tectonic belts was
centred on the South Armorican shear zone (Brittany, France), which is the largest shear in the
European Hercynian basement. Structural analysis
(carried out over the last few years, mainly on
the gold deposits of the French Massif Central)
has shown a relationship between phases
of deformation and polyphase emplacement of
mineralization. The hydrothermal processes
responsible for the economic concentrations of
gold are related to Neo-Variscan (350 to 290 Ma)
transcurrent faults contemporaneous with the
South Armorican shears which they locally extend
to the east. The South Armorican shear zone,
therefore, has great potential for gold
mineralization.

The present study relied on Landsat Thematic
Mapper imagery. The observed discontinuities were
interpreted according to a reference model established from the main gold deposits of France,
and the investigation consisted of looking for
areas equivalent to this model. By then integrating data from numerous other sources
(geological, metallogeny, structural, geophysical, mineralogical, geochemical, petrographic)
using a software programme (SYNERGIS) developed
by BRGM, it is possible to propose a hierarchy
for the target areas.

INTRODUCTION

The use of satellite photo-imagery can be important in the exploration for gold mineralization
controlled by shear faults. On a local scale
(10-100 km^2), satellite imagery helps structural
studies by giving a reliable cartographic plot of
the prospected shear zones. Multicriterial
analysis, using other techniques such as
metallogeny, structural geology and
geomorphology, enable a selection, from among the
discontinuities interpreted as gold-bearing shear
zones, of those that correspond to priority
sections for detailed prospecting (Bouchot,
1989).

On a province scale (n x 100 km^2), the prime
concern, and the major problem, is delimitation
of the a priori most favourable zones for gold
mineralization within a regional tectonic corridor. The present research was therefore centred
on establishing a tool for the very early selection of such potentially auriferous zones within
a province where outcrop conditions are moreover
very poor.

The research was based on Landsat Thematic Mapper
imagery, with the selection of favourable zones,
computer filtering of spectral discontinuities,
integration of multi-source data (geophysics,
regional geology, structural studies, heavy
concentrate mineralogy, local stream-sediment
geochemistry) and a renewed knowledge of the gold
deposit types being looked for.

The strategy consisted of the following stages:

1 - Interpretation of Landsat Thematic Mapper satellite photo-imagery.

2 - Compilation of a computerised data base including, in addition to the above interpretation, any other available information on the study area.

3 - Acquisition of additional structural data and of information on the geology and deposit types.

4 - Establishing a reference from geometric model(s) controlling the type of gold deposit being looked for.

5 - Analogue search of the entire area for sectors corresponding to the reference model(s).

THE STUDY AREA

Choice of area

The studies covered 20,000 km² of the western part of the South Armorican shear zone (Fig. 1),

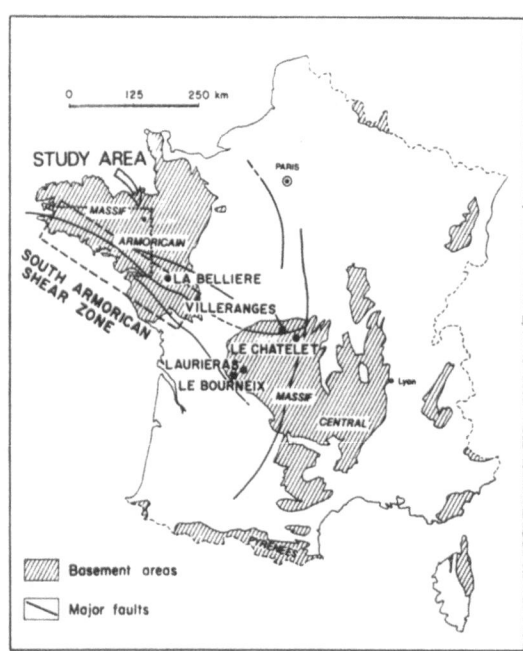

Fig. N°1: Location of study area

which forms the largest shear in the European Hercynian basement. Easterly extensions of this shear fault controlled discordant gold mineralization, such as the deposits of Belliere (10 t of gold at 11 g/t) in Vendée, Châtelet (11 t of gold at 24 g/t) and Villeranges (at present being investigated) in the north of the French Massif Central). Gitological and structural investigations carried out over the last few years on these deposits, and on the discordant Laurieras and Bourneix deposits in the Limousin (western Massif Central) show a relationship between the phases of deformation and the polyphase emplacement of the mineralization (Bonnemaison, Marcoux 1987; Bonnemaison, 1988; Bouchot, 1989). The hydrothermal processes responsible for the economic concentrations are associated with Neo-Variscan (350 to 290 Ma) strike-slip faults contemporaneous with the South Armorican shear zones.

Thus, this part of the South Armorican fault zone, until now hardly explored for this type of gold mineralization, is considered to be an important objective.

Geological and structural setting

The Armorican Massif can be divided into three large domains separated by the major North Armorican and South Armorican shear zones (Cogné, 1960; Chauris, 1969; Cassard, Chantraine, 1989; Fig. 2):

Fig. N°2 Geological skech map

192

1 - **The Cadomian domain** in the north and northeast which behaved as a craton during the Variscan cycle. This domain lies outside of the area studied.

2 - **The intermediate Central Armorican domain** which behaved as a passive margin during the Variscan cycle. It corresponds to the Cadomian hinterland separated from the the Cadomian domain _sensu_ _stricto_ by the North Armorican fault and the Central Variscan through (late central pull-apart basin). It can be divided into two subdomains:

a. The "Central Brittany" subdomain occupied by post-schistose Brioverian (Upper Proterozoic) formations unconformably overlain by Paleozoic detrital formations of Ordovician to Devonian age. These formations were all affected by a single Hercynian phase of deformation (Carboniferous) characterised by folds, varying from the vertical to overturned to the south (Le Corre, 1978), associated with right-lateral movement of the North and South Armorican shear zones (Gapais and Le Corre, 1980).

b. The "West Breton basin" subdomain which is an active marginal basin on a thinned crust. The Brioverian subbasement is overlain by a Paleozoic series, of Lower Ordovician to Devonian age, with abundant extension volcanism (MORB) having taken place during the Silurian-Devonian. The series is intruded by early hybrid granites (fusion of the lower crust) that were orthogneissified during the Variscan evolution.

3 - The Variscan domain, in the south and southeast, which behaved as an active margin during the Variscan cycle and underwent late shearing by the different branches of the South Armorican shear zone. It comprises a pile and/or a juxtaposition of complex lithostructural units of Paleozoic age, and contains several subdomains:

a. The "Landes de Lanvaux" subdomain comprising (i) the Lower Paleozoic Saint-Georges-sur-Loire unit whose volcanic character is marked by the accumulation of pyroclastites and epiclastites with intercalations of basic sills and flows, and (ii) the Lanvaux unit composed of several Ordovician-Silurian wedges imbricated beneath the crustal nappe and underthrusting the fore land as far as the Landes-de-Lanvaux axis centred on a blade of granite.

b. The "Ligero-Vendeen" subdomain which, along with the Cadomian craton and the Chatonnay basin, consists of crustal nappes formed by tectonic melanges that were mainly orthoderived and are characterised by intense tectonism and a meso- to catazonal metamorphism with relicts of rocks that had undergone previous high-pressure metamorphism (Marchand, 1981). These leptiteamphibolite complexes were emplaced in the main intracrustal shears of the orogeny that are marked by outliers of serpentinised peridotite.

c. The "South Brittany and "West Vendeen" subdomain consists of underthrust wedges of orthoderived material. The orthogneisses originated from granites emplaced during Ordovician intracontinental extension; they are anatectic in the Central Breton part where autochthonous to parautochthonous anatexis granites dated 370-380 Ma (Vidal, 1973) were succeeded by diapiric granites (330 Ma; Peucat, 1973). The subdomain also contains wedges of volcanogenic formations of the active margin type associated with thick silicic outpourings, and wedges of Paleozoic terrigenous formations.

The Hercynian evolution was characterised by two main periods (Audren, 1987):

1 - A collisional Eo-Hercynian period (Silurian to Middle Devonian) where the mechanisms of deformation were diapirism and leftlateral transcurrent movement with the emplacement of anatexis granites (see above). The direction of shear is N80°W (northern branch of the South Armorican shear zone). Outside of the diapiric zones, the deformation was essentially tangential.

2 - The post-collisional Hercynian period in the strict sense (Carboniferous), where the mechanism of deformation was a right-lateral transcurrent shear associated with the emplacement of leucogranites. This emplacement was either associated with deep intra-crustal faulting (alumino-potassic granites of 350-330 Ma, and hyperaluminous granites of 330-315 Ma) or with flat thrusts (granites of 300 Ma). The southern branch of the South Armorican shear zone was formed during this period.

Geomorphological setting

The area studied corresponds to an erosional paleosurface in which the older formations crop out scarcely, most commonly covered by alterites. The resultant morphology, with little contrast, is further softened by the common presence of Quaternary formations (loess). Locally, especially near the coast, recent regressive fluviatile erosion has revived the landscape. The surficial formations, wheathered formations and loess, are a great hindrance to mineral exploration both in the simple investigation of outcrops and in the use of more complex techniques such as geochemistry. Agricultural use of the ground aggravates this situation since the innumerable pastures and forests further conceal indications of the subsoil. Such agro-pedo--geological conditions enhance the interest of remote sensing.

Interpretation of geophysical data

Systematic gravimetric data, at a scale of 1:80,000 interpolated on a 1 km grid, are available for the South Armorican fault zone. The Bouguer anomaly and vertical gradient maps show lows related to outcropping leucogranites or to granites at shallow depths (North Redon, North-west Pontivy). The highs demarcate the West basin which contains basic volcanites. Structural analysis of the gravimetric data was carried out by automatic methods; other than the large N60°-80°W-striking South Armorican faults, the structural map shows N50°-70°E-striking structures in the Quimper sector (Fig. 3).

LINEAMENT ANALYSIS

Use of Remote Sensing

Images

Landsat Thematic Mapper imagery was chosen for its spectral bands - Band 5 (1.57-1.78 Mm) particularly appears to give fine discrimination at bare ground level depending on the state of humidity. The combination of channels finally adopted for the study was Band 4 (red), Band 5 (green), and Band 2 (blue) - these bands proved the least connected with their composition being the richest in original information on the state of the vegetation and bare ground.

Processing

The study area is covered by four quarters of different scenes. Due to the fact that different images were taken on different dates, a preliminary radiometric adjustment was required to partly homogenise the spectral responses. The images were geometrically corrected and assembled as a mosaic, then presented as spacemaps fitting the established 1:50,000-scale map grid of France as well as the same cartographic reference system as the rest of the mining and geological documents used for the national inventory. This enables direct comparisons, either visually or using a Geographic Information System, the precision of the orthophotomaps, with a pixel size of 30 m, being completely compatible with these documents. The corrected image was enhanced by stretching in order to optimize the information contained in each spacemap.

Interpretation

The geological target is a tectonic structure (fault, shear, etc.) that could contain quartz veins, of variable size, and in a structural shear zone setting. The preferential remote-sensing pattern is thus the linear image--discontinuity likely to represent a structural feature. In the geomorphological and agro-pedological context of the area such discontinuities are of two main types:

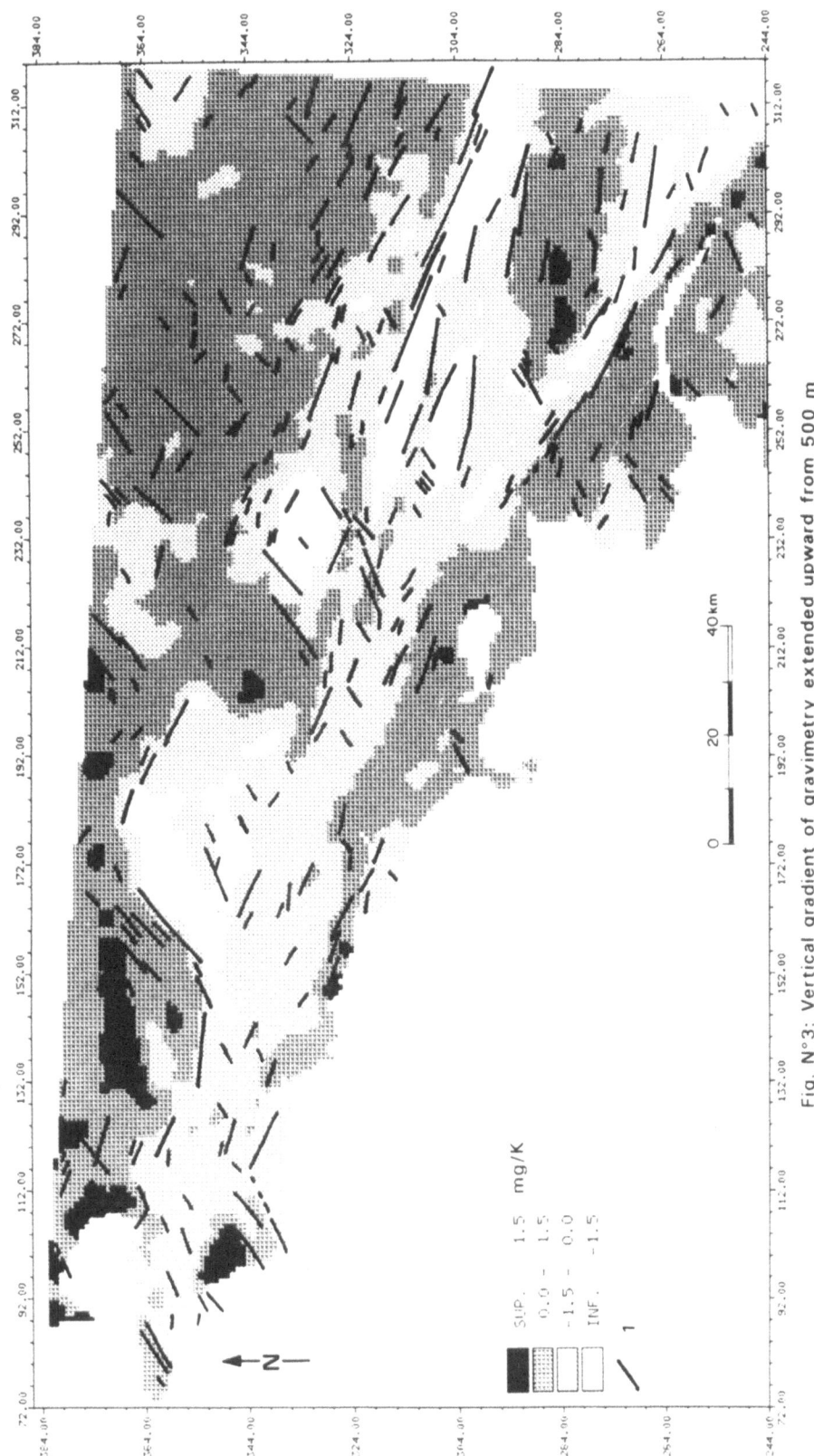

Fig. N°3: Vertical gradient of gravimetry extended upward from 500 m

1: structural analysis

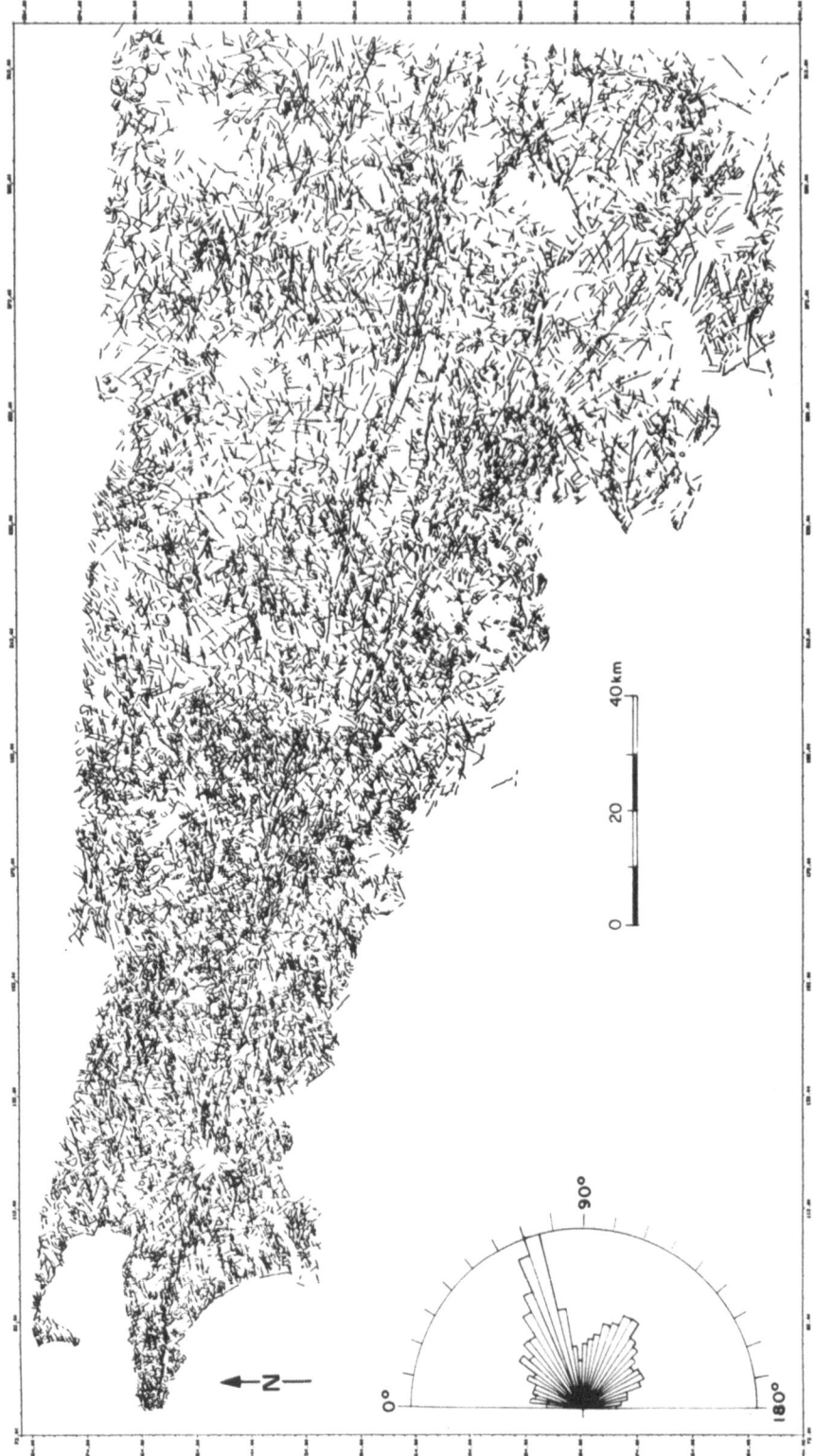

Fig. N°4: Discontinuities from Landsat imagery in south armorican zone

196

1 - Morphological type - consisting of recti-
linear discontinuites caused by the effect of
shadow on the morphology such as deep
valleys, talus slopes or cliffs. These
features are well observed when they are
orthogonal, or at least oblique, to the
direction of illumination (SE-NW); they can
be totally obliterated if they are parallel
to this direction, even if they are large
features discontinuity.

2 - Radiometric type - corresponding to abrupt
colour changes with bound patches of different colour, or to isolated lines situated
within these same patches. The former can,
for example, represent changes in vegetation
population and the latter could be alignments
of vegetation intersecting the axis of a
small stream. Most commonly, however, these
lines indicate a change in the parcel arran-
gement of the land, which can locally reflect
complex relationships between geology and
land use.

The boundaries of the above described tonal
zones over bare ground were locally interpreted, and superpose quite well to known
geological boundaries and allow the interpreter to draw the discontinuities with their
horizontal movement.

Method of interpretation:

The methodology used during interpretation was
considered from an economic exploration point of
view in which financial constraints are a
priority. Four hours only were spent studying
each spacemap (approximately 550 km^2); this was
done by visual interpretation of the tones and
the morphology. This very short interpretation
time allowed only the major or more obvious
structures to be drawn ; but this interpretation
is however representative of the structural
pattern of the area. This became obvious once the
interpretations of the 40 spacemaps were, after
digitization, assembled onto a single
1:250,000-scale document (Fig. 4).

Examination of the discontinuity map

The discontinuity map, which covers the whole of
the study area appears relatively complex with

17700 discontinuities. However, the spatial and
directional pattern shows three main families
(rose diagram of Fig. 4):

1 - Groups striking N60°-80°W, which superpose
perfectly on the known large South Armorican
shears.

2 - Numerous discontinuities striking N55°E and
N70°E confirming the structures of the
gravimetric data.
Ductile N50°-70°E-striking left-lateral
faults are reported at the edge of certain
granite massifs which mark the northern
branch of the regional shear zone (Vigneresse
and Lefort, 1976; Jegouzo, 1980); these
authors interpret them as secondary conjugate
shears to the main right-lateral South
Armorican shear striking N60°-80°W. In the
Quimper sector, these discontinuities are
superposed commonly on ductile-brittle to
brittle left-lateral strikeslip faults.

3 - Discontinuities striking N0°-30°W, which are
less numerous than in the other two families.
Groups of right-lateral fractures, marked by
cataclasites and of Late Hercynian age, are
known all across Bretagne.

THE GEOMETRIC REFERENCE MODEL

Satellite imagery interpretations, identical to
those carried out on the South Armorican zone,
were carried out on the environment of the
Châtelet, Villeranges (north Massif Central) and
La Belliere (Vendée) deposits in order to define
a (or several) reference model(s).

The Châtelet and Villeranges deposits are
contained in a migmatitic and granitic domain
affected by large, N50°-80°W-striking regional
faults (Marche and Chambon dislocations). The
mineralization at Châtelet (Braux et al., 1989)
was controlled by anastomosing structures with a
southerly strike close to the Chambon dislocation, whereas at Villeranges the mineralization
occurs in formations affected by the Chambon
dislocation. Both deposits are characterised by
the absence of free gold; the gold is integrated
in the arsenopyrites lattice (Marcoux et al.,
1989) and could have been missed by traditional
exploration methods (heavy mineral concentrates).

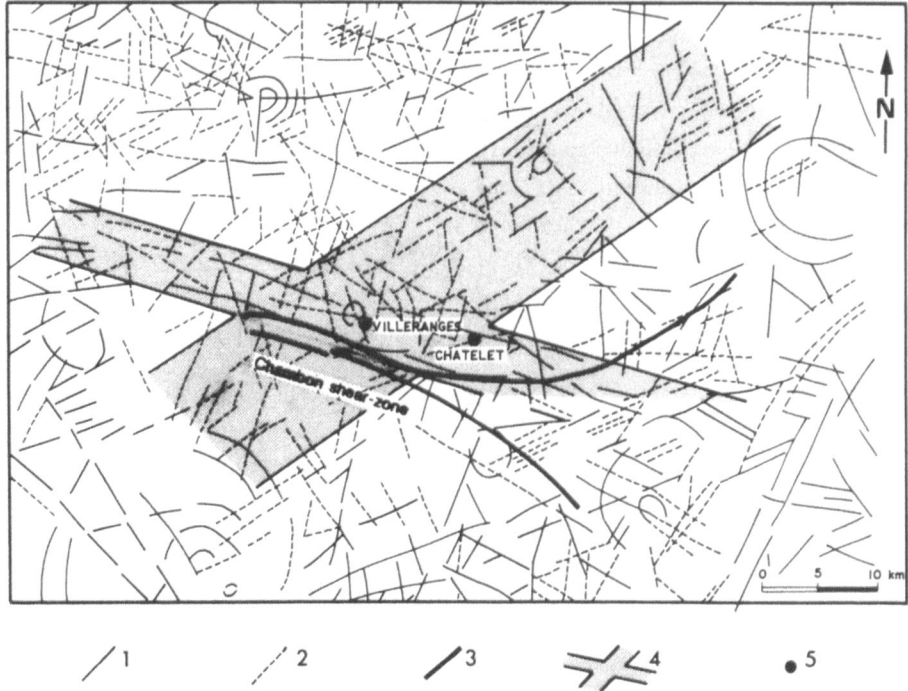

1: morphological image discontinuities , 2: radiometric image discontinuities
3: Marche-Chambon shear zone , 4: discontinuity groups , 5: gold deposits

Fig N°5 : DISCONTINUITIES FROM LANDSAT IMAGERY IN THE VINCINITY OF LE
CHATELET AND VILLERANGES GOLD DEPOSITS

On the satellite photo-image (Fig. 5), the two deposits lie in the intersection zone between a group of N70°-80°W-striking discontinuities, which correspond to the Chambon dislocation and a group of N50°-70°E striking discontinuities.

The Bellière deposit, the closest to the study area, occurs in Brioverian formations (Upper Proterozoic) where interbedded pelites and graywackes are affected by N70°-90°W-striking left-lateral shears. The mineralization is hosted by N60°E-striking veins along brittle to ductile-brittle structures.

Unlike Châtelet and Villeranges, the La Bellière mineralization is characterised by the presence of free gold. On the satellite imagery (Berthiaux et al., 1987), the deposit appears at the intersection between a group of N50°-70°E-striking discontinuities which superpose on the mineralized veins, and a group striking N80°-90°W which corresponds to the shear direction.

Thus the reference model retained is defined as the intersection of two preferential directions of image discontinuities striking N50°-70°E and N60°-90°W But the envelopes of certain types of Variscan granite are of considerable importance.

It would appear that this geometric model could also be applied to the Saint-Yrieix gold district (west Massif Central) which groups the Laurieras and Bourneix deposits, presently under mining activity. Here, the mineralization is controlled by N60°E-striking shear structures which are probably related to an east-northeast-striking ductile structure that would have controlled the emplacement of a hidden granite ridge (Bouchot, 1989). These N60°E-striking strike-slip faults are marked by groups of image discontinuities located next to poorly marked N80°W-striking discontinuities.

The reference model is a geometric model for which the limits of validity are known, and it is in the process of being refined through structural studies (chronology of the deformations) and metallogenic studies of the deposits.

INTERPRETATION

Computer tool

Data processing was carried out using a Geographic Information System (SYNERGIS, developed by the BRGM) that enables data of various origins (exploration geochemistry, geophysics, remote

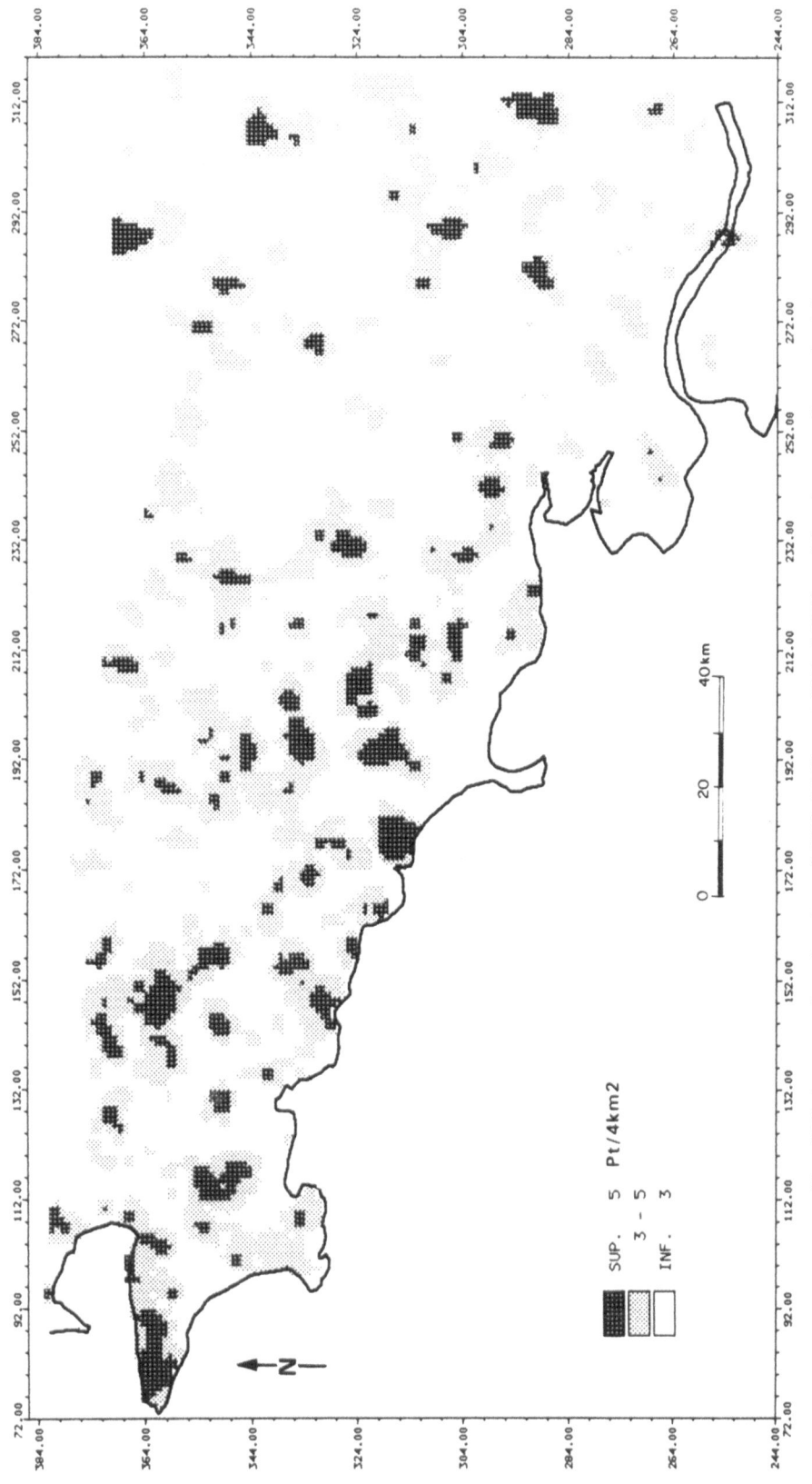

Fig. N°6: Intersection density of N50-70°E and N60-90°W striking image discontinuities

SUP. 5 Pt/4km2

3 - 5

INF. 3

199

sensing, gitology, geology, etc.) to be processed together (Bonnefoy, Guillen, 1988). Data are represented in:

1 - "Grid" mode in which each grid carries a value describing the feature, and

2 - "Vector" mode which completes the grid presentation for each geographic component (point, line, area) and gives it, in addition to the attributes of visualisation, a suite of intrinsic values such as length, orientation, area.

The grid images can be automatically processed through statistical means, signal functions (filtering, shadow, convolution), graphic display and a combination module. Processing the vector images involves grouping study functions with geometric relationships between graphic objects (distance, inclusion, intersection). Specific functions allow communication between the two modes.

SYNERGIS runs with UNIX or VDS operating systems and uses standard languages (Fortran 77,C) and standard graphic libraries (GK, X-WINDOWS version 11).

Location of zones identical to the model and hierarchical selection of targets

Locating targets equivalent to the proposed model is carried out in two stages. The first consists of looking for intersections between the two described families of directional discontinuities. The second consists of calculating the density of intersection points and showing them as a coloured equidensity image (Fig. 6: part of the study area located on Figs 3 and 4).

The hierarchical selection of targets according to their mineral potential is done by considering available lithologic, geophysical and structural data corresponding to the Châtelet, Villeranges and La Bellière gold deposits. The hierarchical selection was also inspired by the Laurieras and Bourneix deposits (Limousin, west Massif Central).

The main discordant gold deposits of the French Hercynian (like the Archaean mesothermal deposits; Colvine et al., 1984, 1988; Eisenlohr et al., 1989) are located along brittle or ductile-brittle structures related to regional ductile faults. Moreover, most of these French gold deposits are hosted by rocks attributed to the Upper Proterozoic to infra cambrian, it is possible that either the gold originated from the rock series, or that the formations acted as a physico-chemical barrier.

The studies carried out on the Saint-Yrieix district (Bouchot, 1989) show that the geometry of the brittle to ductile-brittle structures is influenced by the presence of dykes or small leucogranite bodies that are the surface expression of a hidden leucogranite ridge. It can be inferred, therefore, that the occurrence of auriferous structures would be conditioned both by the presence of regional ductile tectonic corridors and by the presence of hidden or subcropping granite.

With the Villeranges, Châtelet and Bellière deposits the relationship between mineralization and structures (brittle to ductile-brittle and ductile) is obvious (see earlier). Moreover, the presence of specific geochemical Li ± W signatures, the local presence of hightemperature minerals and the presence of bismuth minerals indicate the existence of nearby granitic bodies.

The map (Fig. 7) shows the density of image-discontinuity intersections combined with geophysical (negative anomalies) and geological (main Hercynian granites) data.

Sectors 1 and 2 (East Quimper) are located near the South Armorican ductile zone within Proterozoic to infra cambrian metamorphic formations intruded by leucogranites. The granite massifs, however, are not homogeneous units (Marcoux, 1982); they show granodioritic trends as well as the classic leucogranites one of which very probably corresponds to a very evolved granite cupola containing concentrations of the volatile elements (Sn, W, Li, Be). With such a setting, these are very favourable sectors, which is confirmed by the presence of anomalous alluvial gold and streamsediment geochemical As, Sb, Pb, Ag anomalies (an association characteristic of auriferous shear zones of the French Hercynian basement).

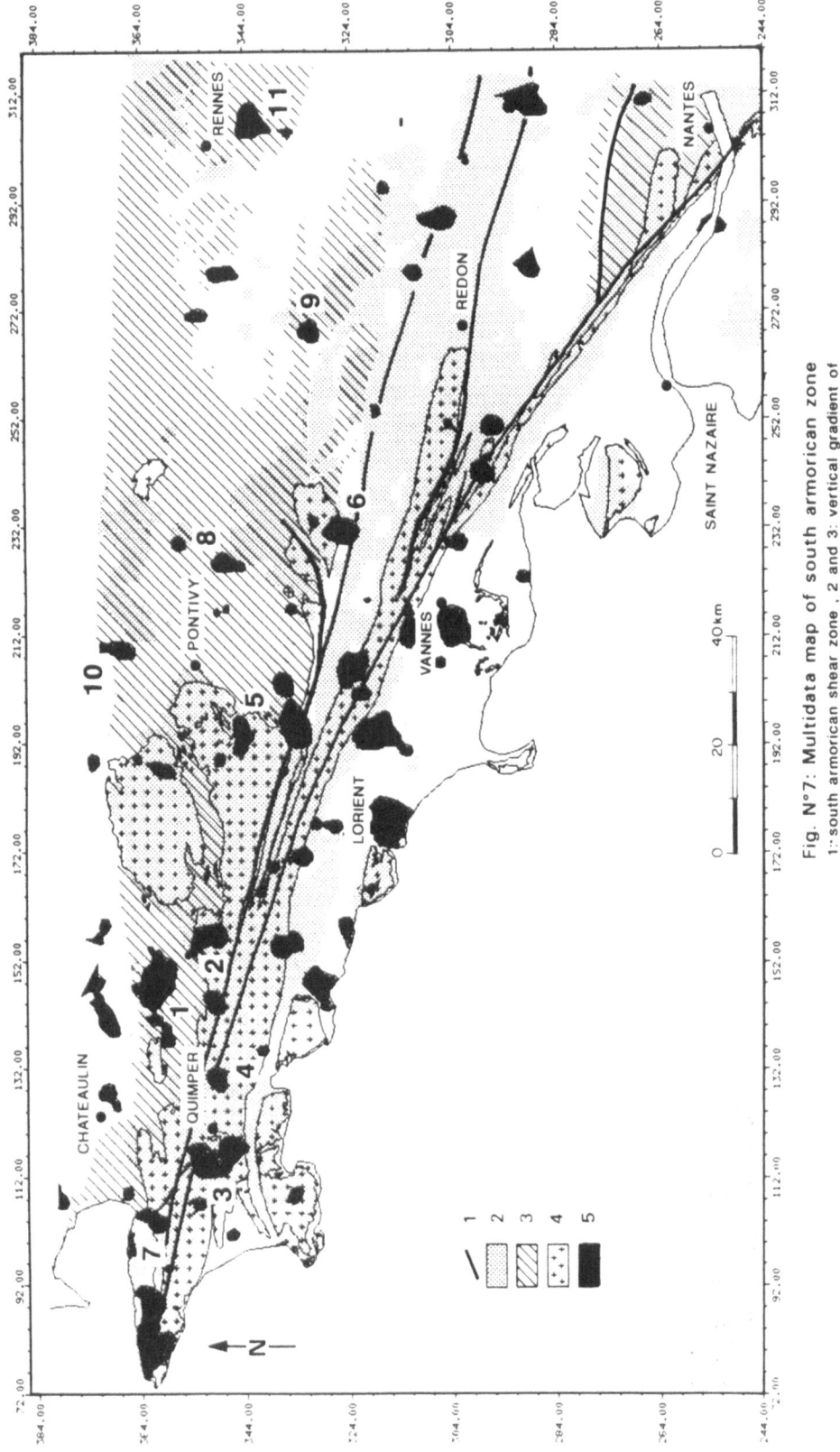

Fig. N°7: **Multidata map of south armorican zone**
1: south armorican shear zone , 2 and 3: vertical gradient of
gravimetry (2=0 to -1.5 mg/km2 , 3= <-1.5 mg/km2).
4: Proterozoic . 5: Variscan granites . 6: density of image
discontinuity intersections (> 5/4km2)

Sectors 3 and 4 are located in a granitic environment affected by the N80°W-striking South Armorican ductile shear. They also show a very clear N50°-70°E-striking structure that corresponds to a ductilebrittle to brittle deformation. These sectors contain As, Pb, Sb anomalies and a differentiated leucogranite with occurrences of tungsten and lithium.

Sectors 5, 6 and 7, which have a comparable lithological context, should also be considered as favourable, in spite of the lack of geochemical data.

Sectors 8, 9, 10 and 11 although of a lower level than the others, should also be retained. They are situated in Proterozoic to infra cambrian formations contained locally within a gravity low which probably indicate an underlying granite. These sectors also contain geochemical As, Pb anomalies, alluvial gold anomalies and signs of mineralization.

CONCLUSIONS

The studies carried out on the South Armorican shear zone have established a procedure for evaluation satellite imagery discontinuities. It is a tool that allows the user to select favourable sectors of several square kilometers (permit size) for exploration of "auriferous shear-zone" type gold deposits, within a regional tectonic corridor of several thousands of square kilometers. This remote sensing approach gives results that correlate well with geochemical data and existing occurrence maps and demonstrate that the method is reliable.

By using this multidata approach with the flexibility of the computer tool SYNERGIS developed by the BRGM, it should be possible to define several reference models and to edit the corresponding maps. As a second phase, more detailed studies can be conducted on the indicated targets; interpretation of satellite imagery discontinuities can give a reliable plot of the shear zones to be used as a guide for more classical exploration methods such as field studies, geochemistry, structural analysis and geophysics.

ACKNOWLEDGEMENTS

The authors would like to thank the scientific committee of the BRGM for supporting this work as part of Research Project RM13 (Gitology and exploration of auriferous shear zones). They would also like to thank J.P. Milési, G. Kozminski for numerous discussions and for reviewing the article. The English translation was done by Sir Patrick Skipwith, Bt.

REFERENCES CITED

Audren C. - Evolution structurale de la Bretagne méridionale au Paléozoïque. Mém. Soc. géol. minéral. Bretagne, n° 31, 1987, 365 p.

Bertiaux A., Clement J.P., Minoux L. and Wynns R. - La région des Mauges; Analyse et comparaison des données spatiales spot et thématic mapper. Corrélation avec les connaissances géologiques acquises récemment. Rapport inédit BRGM N° 87 DT 043 TED, 1987.

Bonnemaison M. and Marcoux E. - Les zones de cisaillement aurifères du socle hercynien français. Chron. rech. min., n° 488, 1987, p. 29-42.

Bonnemaison M. - Les concentrations aurifères dans les zones de cisaillement. Métallogénie et prospection. Thèse de doctorat d'Etat, Toulouse, 1987.

Bonnefoy D. and Guillen A. - Mappable data integration techniques in mineral exploration. Symposium "computer application in resource exploration" Cogeodata, Espoo (Helsinki) Finland, 1988.

Bouchot V. - Géologie et tectonique du district aurifère de Saint-Yrieix. Thèse 3ème cycle, en cours. Université d'Orléans, 1989.

Braux, C., Prévot J.C. Bonnemaison M., Maurin G. and ZEEGERS H. - Le gisement d'or du Châtelet ; gitologie et prospection. Congrès "Gold 89 in Europe". Toulouse, France, 1989.

Cassard D. and Chantraine J. - Commentaires de la carte de synthèse à 1/320 000 de la partie méridionale du Massif armoricain. Rapport BRGM inédit R 30006 GEO SGN 89, 1989.

Chauris L. - Sur un important accident structural dans le nord-ouest de l'Armorique. C.R. Acad. Sci., Fr., t. 268, 1969, p. 2859-2861.

Cogné J. - Schistes cristallins et granites en Bretagne méridionale : le domaine de l'anticlinal de Cornouaille. Mém. Serv. Carte géol. Fr., 1960, 382 p.

Colvine A.C., Andrews A.J, Cherry M.E., Durocher M.E., Fyon A.J., Lavigne M.J., Macdonald A.J., Marmont S., Poulsen K.H., Springer J.S. and Troop D.G. - An integrated model for the origin of Archean lode golg deposit. Ontario Geol. Survey, O.F.R. 5524, 1984, 99 p.

Colvine A.C., Fyon J.A., Heather K.B., Soussan Marmont, Smith P.M. and Troop D.G.- Archean lode gold deposit model in Ontario, Ontario Geol. Survey, Miscellaneous, Paper 139, 1988, 136 p.

Eisenlohr B.N., Groves D. and Partington G.A. - Crustalscale shear zones and their significance to Archaean gold mineralization in Western Australia. Mineral. Deposita, vol. 24, 1989, pp. 1-8.

Gapais D. and Le Corre Cl. - Is the Hercynian belt of brittany a major shear-zone ? Nature, vol. 288, n° 5791, 1980, p. 574-576

Jegouzo P. - The South Armorican shear zone. Journal of structural Geology, vol. 2, n° 1/2, 1980, p. 39-47.

Le Corre Cl. - Approche quantitative des pro cessus synschisteux : l'exemple du segment hercynien de Bretagne centrale. Thèse d'Etat, Rennes, 1978, 381 p.

Marchand J. - Ecaillage d'un "mélange tectonique" profond : le complexe cristallophyllien de Champtoceaux (Bretagne méridionale). C.R. Acad. Sci., Fr., t. 293, série II, 1981, p. 223-228.

Marcoux, E. - Le massif de Pontivy (Massif armoricain, France) : une association géographique de trois unités leucogranitiques. C.R. Acad. Sci., Fr., t. 294, série II, 1982, p. 1095-1098.

Marcoux E., Bonnemaison M., Braux C. and Johan Z. - Distribution de Au, Sb, As et Fe dans l'arsénopyrite aurifère du Châtelet et de Villeranges (Creuse, Massif central français). C. R. Acad. Sci., Fr., t. 308, série II, 1989 p. 293-300.

Peucat J.J. - Les schistes cristallins de la baie d'Audierne (Massif armoricain, France). Etude pétrographique et structurale. Thèse 3è cycle, Rennes, 1973, 114 p.

Vidal Ph. - Premières données géochronologiques sur les granites hercyniens du sud du Massif armoricain. Bull. Soc. géol. Fr., (7), t. XV, 1973, p. 239-245.

Vigneresse J.L. and Lefort J.P. - Les mouvements tardihercyniens au niveau de la zone broyée sud-armoricaine. Apports des données géophysiques. 4è RAST, Paris, 1976, p. 392.

Use of SPOT and airborne radar imagery for gold exploration in Qiaben, northern Xinjiang, China

Guo Huadong
Lin Shudao
Li Lin
Lin Qizhong
Li Naihuang
Shao Yun
Institute of Remote Sensing Applications, Academia Sinica, Beijing, China

ABSTRACT

The Qiaben study area lies within the Altai mountain range, close to the border between Xinjiang Uygur Autonomous Region and Russia. The study area is located at the junction of three major faults, and is underlain by Palaeozoic granodiorites, gabbrodiorites and metavolcanic rocks. During 1987 and 1988, SPOT and airborne X-band SLR (Side-Looking Radar) imagery were used for gold exploration. In order to enhance fine distinctions of spectra as well as gold-bearing geological bodies, a variety of image processing techniques were used including the IHS transformation The gold host-structure was identified using high resolution radar imagery. One hundred gold-bearing quartz veins and one gold-bearing altered fracture zone 4 km in length have been found by combining image interpretation with detailed fieldwork. Tens of tons of gold reserves have been assessed by geochemical, geophysical and geological means.

1. INTRODUCTION

Spaceborne SPOT HRV imagery contains a wealth of multispectral information and has a significantly higher resolution than other satellites, which is very usful in detecting mineralized geological bodies. Airborne synthetic aperture radar imagery has strong stereoscopic capabilities which show relief clearly and its prevailing ability is to identify structures which control mineralization.

As part of a national project entitled 'Remote Sensing for Mineral Exploration , a remote sensing investigation for gold deposits using SPOT HRV and SAR data has been carried out in the Qiaben area in the last three years. The Qiaben area is located in the farthest northwestern part of the Xinjiang Uygur Autonomous Region of China. It lies in piedmont on the southern side of Altai Mountains. In Mongolia Altai Mountains means ' Gold Mountains ' People have panned gold in this for hundreds years, but the source gold d eposits had never been found.

The purpose of this study was to discover the original gold deposits. SPOT and SAR imagery were choosen as the main data sources as they have significant advantadges for lithological characters discrimination and structure interpretation. In order to efficiently identify gold-bearing geological bodies such as quartz veins the IHS transformation of SPOT data was utilised and both SPOT and SAR data were analysed comprehensively. On the basis of the discovery of a gold mineralized anomaly, .geophysical and geochemical methods were applied to this area. Subsequently, the study area was tested by means of geological engineering methods. Finally, it was proved that the Qiaben gold deposit exists in the study area.

2. GEOLOGICAL BACKGROUND

The study area lies on the eastern side of Zashang depression and southern side of the Altai Fold belt. Its tectonic position is controlled by the major faults of the Arsele-Yanhu fault system and the Salesuke fault system, striking in a NW direction and it lies at the conjunction of these two faults. The minor faults and joins are well developed and cross each other. They meet the regional major faults with small angles and mainly striking in a NW and NNW direction.

Magmatic rocks exposed in this area belong to the Habahe complex, mainly are Hercynian plagioclase granites, gabbro-diorites, diorites and quartz-diorites etc. All of the magmatic rocks are strongly metamorphosed by late granite intrusions and post magmatic pneumatolytic-hydrothermal metamorphism. The chloritization, epidotization, silicification, pyritisation and association of pyritization-sericitization-silicification have been found to be well developed and widely distributed in the Qiaben area by field and microscope observation. The country rocks are Devonian and Carboniferous sandstones, shales and conglomeratic rocks in a variety of metamorphic grades.

Mafic complexes are usually strongly metamorphosed

They appear as flat, kidney-shaped rolling hills
on the imagery due to weathering. In the field,
most of these rocks are greyish green-light green
in colour, of massive texture and with visibal
schistosity. It has been found under microscope
observation that almost all of the original
minerals have disappeared and been replaced by
late alteration minerals, only a few of their
pseudomorphism have been left. According to observation and analysis of the alteration features, it
can be concluded that the original rocks were
strongly metamorphosed by late stage alteration
and highly compressed to generate schistosity and
strongly oriented minerals. Generally speaking, the
original minerals are relatively simple, mainly
mafic plagioclases and mafelsic minerals. The
initial alteration was by selective metasomatism,
metasomatic zoisite retained the original
rock texture very well. As the hydrothermal
alteration became stronger, the chlorites meta-
somatically started to replace the actinolite and
zoisite, until it destroyed the original texture.

The medium and coarse grained granites are not
strongly metamorphosed. In the field, they are light
flesh pink in colour, medium-coarse grain. They
appear as tumour-shape on airborne infrared colour
photography. The major minerals are plagioclases,
quartz and a few biotites.

The contact zones of the different rock units
within the complex are regarded as favourable
places for mineral deposits in accordance with
regional gold metallogeny, while altered belts
within the contact zones should be regarded as a
focus for mineral exploration.

3. INFORMATION EXTRACTION ON GOLD-BEARING GEOLOGICAL BODIES FROM SPOT IMAGERY

Several bright stripes against the dark background
are intrusions that have been discovered by interpreting the Spot image. After detailed analyses,
they were identified as quartz veins strongly
associated with gold deposits. Many quartz veins
cohabit with altered hornblende gabbros.
Therefore, extracting altered hornblende gabbros
and quartz veins from the imagery has become one
of the key steps of remote sensing for gold
exploration.

3.1 IHS Transformation for Extracting information of Altered Rocks

The colour composite SPOT imagery shows hornblende
gabbros clearly, but shows very few differences
between hornblende gabbros and altered hornblende
gabbros. There are two reasons: one is that

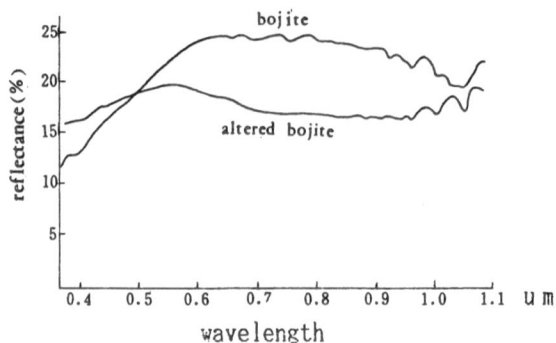

Figure 1. Spectral Curve of Altered and Unaltered
 Bojites

spectral reflectance of altered hornblende gabbros
(Fig. 1) is lower than unaltered bojites;
another important reason is that the distribution
of altered hornblende gabbros is smaller and more
scattered; the more widely distributed unaltered
hornblende gabbros predominate over the smaller
and scattered altered hornblende gabbros.

Colour saturation and hue information are not
well shown the SPOT imagery due to a high
correlation amonge the three bands. The IHS colour
space transformation overcomes the limitations of
SPOT imagery and exaggerates the colour saturation
and hue independently of the intensity, which has
been proven to be very useful for extracting geological information.

3.1.1. Procedures of IHS Transformation

Considering the spectral characteristics of SPOT
imagery; the correlation co-efficient of band 1
with band 2 is large, and the brightness value is
similar; while the correlation co-efficient of band
3 with band 1 and band 2 is smaller and brightness
is different. The diffeence between band 2 and
band 1 is used to control the hue; band 3 minus
the arithmetic mean of band 1 and band 2 controls
the colour saturation. If B1, B2 and B3 represent
the grey levels of band 1, band 2 and band 3 of
SPOT imagery, the transformation function can be
demonstrated as follows:

positive transformation:

$$I = 1/3 (B1 + B2 + B3)$$
$$H = B2 - B1$$
$$S = B3 - 1/2 (B2 + B1)$$

negative transformation:

$$B1 = I - S/3 - H/2$$
$$B2 = I - S/3 + H/2$$
$$B3 = I + 2/3 S$$

The above functions have been performed as an image processing program. The procedures of transformation are shown in Fig. 2.

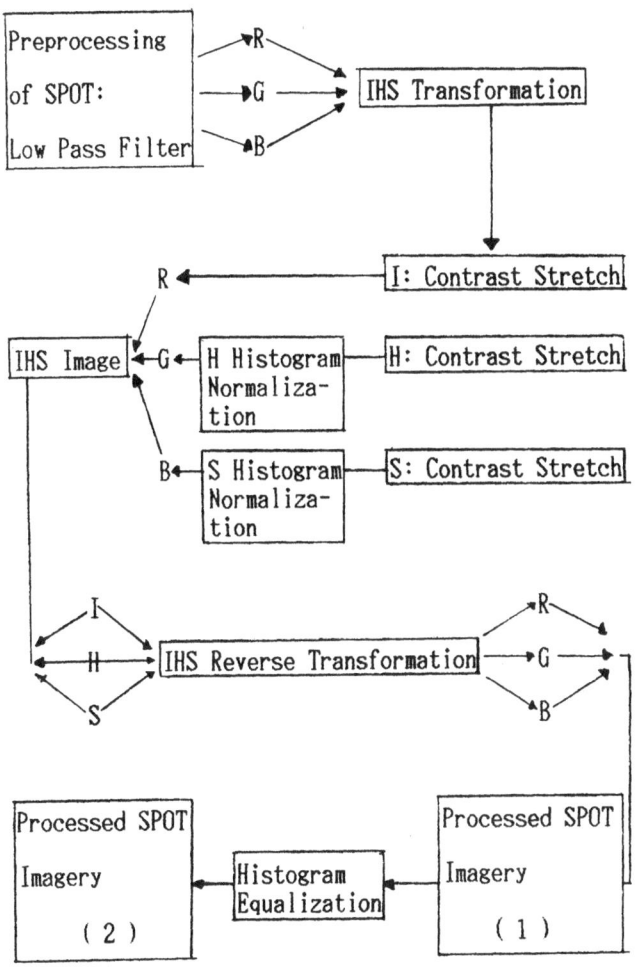

Figure 2.   Procedure of IHS Transformation

3.1.2. The Results of IHS Transformation

The imagery processed by IHS transformation contains more information, has larger range of colours and makes details more distinguishable.

On the image, the altered gornblende gabbros appear in green, whereas the unaltered hornblende gabbros are blue. The boundaries between them are easy to identify. The processed imagery shows altered hornblende gabbros better than original SPOT imagery.

The extraction of altered hornblende gabbros plays an important role in gold exploration in the Qiaben area. Quartz veins developed within altered hornblende gabbros are probably gold-bearing. The altered hornblende gabbros can be indicators of gold-bearing quartz veins and targets of sampling in the field. Altered hornblende gabbros can also

themselves have significant gold contents, as well as hosting gold-bearing quartz veins. One hornblende gabbro sample reached 11.24 g/t Au, which is a promising lead in the search for alteration type of gold deposits in the study area.

3.2 IHS Transformation for Extracting Information on Quartz Veins

Some quartz veins have been interpreted from the original SPOT imagery. If the quartz veins appear as a bunch or a group against a complicated background such as that of Habahe complex they are more easily identified than single quartz vein. Some of the small veins, however, are hidden in a complex background. The surrounding rocks gloss over the quartz veins, so that the latter are not obvious on the original SPOT imagery. In order to enhance the quartz veins, the Munsell coordinate IHS transformation was used.

3.2.1. Procedures of Transformation

B1, B2 and B3 represent band 1, band 2 and band 3 of SPOT imagery. According to the spectral characteristics of the three bands of SPOT imagery, the transformation functions are shown as follows:

positive transformation:

$$Xi = 2 (B2 - B1)$$
$$Bi = \sqrt{6}/3 \, B3 - \sqrt{6}/6 (B2 + B1)$$
$$I = \sqrt{3}/3 (B1 + B2 + B3)$$
$$H = arctg \, Xi/Bi$$
$$S = (Xi2 + Bi2) 1/2$$

negative transformation

$$Xi = S * Sin \, H$$
$$Bi = S * Cos \, H$$
$$B1 = \sqrt{6}/3 \, Bi + \sqrt{3}/3 \, I$$
$$B2 = \sqrt{2}/2 \, Xi + \sqrt{3}/3 \, I - \sqrt{6}/6 \, Bi$$
$$b3 = \sqrt{3}/3 \, I - \sqrt{2}/2 \, Xi - \sqrt{6}/6 \, Bi$$

After positive transformation, contrast stretches have been applied to the I and S component respectively, and histogram normalization has been applied to the S component. The H component have been divided by a constant. According to results of colour coding, 60 was choosen as the constant to eliminate the value of H component. Then, using I, S, H component as red, green and blue components of a colour composite image, a negative transformation has applied to get final processed imagery.

3.2.2. Results of Munsell Transformation

The imagery processed by Munsell coordinate IHS transformation shows quartz veins very obviously. The quartz veins
appear as dark orange, red dots or mixed coloured dots against the dark background of hornblende gabbros. The distinguishable quartz veins are easier to identify than on the original SPOT image.

The processed SPOT imagery is capable of determining the quantity of quartz veins and their distribution regularity, and allows the direct analysis of the relationship between quartz veins and structures in the study area. Most of the orange colour anomalies correspond to quartz veins, which is very useful for position and orientation in the field. In the study area, the quartz veins usually occur in clusters, consequently, the gold-bearing quartz veins concentrate in a relatively small area.

The results of geological interpretation of SPOT imagery are shown in Fig. 3. More than one hundred quartz veinshave been determined in the study area. One quarter of them were determined as gold-bearing quartz veins which reach industrial grades in accordance with the results of sample chemical analysis. The altered hornblende gabbro is an important indication of gold deposits as well.

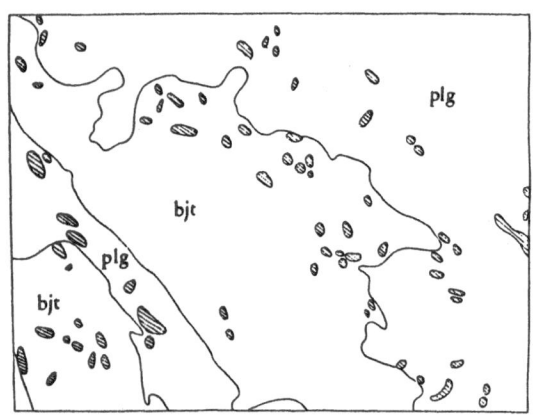

LEGEND   plg : plagioclase granites
            bjt : bojites
      | ❧   quartz veins

Figure 3. Quartz Vein Interpretation Map

4. IDENTIFICATION OF GOLD-BEARING COMPRESSIVE ZONE USING SAR IMAGERY

The SPOT imagery shows the distribution and regularity of quartz veins in the study area, and indicates that the quartz vein type of gold deposit exists in the Qiaben area. However,

considering faint image features on SPOT imagery and metallogenetic regularity, it is possible that another type of gold deposit exists in Qiaben area, ie, altered fracture zone type of gold deposit. Identification of this type of gold deposit needs a good understanding of the structure and tectonics. Radar imagery is an efficient tool for structureal interpretation. Therefore, an airborne synthetic aperture radar survey was flown over the study area.

4.1 Acquisition of Radar Data

The radar remote sensing study has been carried out in parallel with SPOT imagery. The system is airborne, X-band, synthetic aperture side-looking radar with HH polarization and high range and azimuth resolution (Table 1) belonging to National Remote Sensing Centre in China.

Parameter	Value
Wavelength	3 cm
polarization	HH
Depression Angle	23'
Range Resolution	3 m
Azimuth Resolution	3 m
Data Record	Optical

Table 1   Characteritics of SAR System

In order to acquire optimum information on geological bodies the radar was flown in north- south direction and the pulse beam is therefore from east to west, which is different to the solar illumination direction on the SPOT imagery,so that it is able to show different geological features.

4.2. Interpretation of Radar Imagery

The radar imagery is a kind of visual image recording the microwave scattering characteristics of targets and their geometric distribution. The imaging mechanism of the radar system is totally different from the visible and infrared imagery for various geological bodies. It provides information which the visible and infrared image cannot acquire. Relief features are extremely distinct on radar imagery. Mafic veins distributed within plagioclase granites can be easily interpreted due to outcropping veins over granites acting as corner reflector .

Figure 4. Interpretation Map of SAR Image

Four major faults have been interpreted from the SAR imagery (Fig. 4). The northernmost and southernmost are straight linear faults, ie, F1 : Salesuke fault; F2 : Arsele-Yanhu fault. Both the centrally located and east-west striking faults are arcuate faults which are FG1 and FG2. FG1 is a straight linear fault on SPOT imagery, whereas it is arcuate on SAR imagery, which is called Axile faulting fracture zone. According to regional metallogeny, this zone of faulting fracturing has favourable conditions for developing altered fracture zone type of gold deposits. In the field, geological and geochemical investigation have been undertaken. The geochemical analysis shows that not only do the quartz veins have high a gold content, but altered granite-diorites and diorite-gabbros etc. are also enriched in gold. The trench results show that a gold anomaly is obvious with a several meter wide mylonitic belt,which promises that the altered fracture zone type of gold deposit probably exists in this belt.

5. FOLLOW UP USING GEOPHYSICAL AND GEOCHEMICAL TECHNIQUES

In order to confirm the results of information extraction and image interpretation of SPOT and SAR images that geophysical,geochemical geological field investigations have been undertaken in the Qiaben area.

5.1. Gold-bearing Quartz Veins Interpreted from SPOT Imagery and Proven by Geophysical Data

One large quartz vein chosen from numerous quartz veins which were interpreted from
SPOT imagery and appeared to have good mineralization potential is coded as Q15 and has been proven by means of geo-electrical sounding and electrical extraction. In the study area, quartz veins exist in hornblende gabbros and plagioclase granites and have different dielectric properties from the country rocks.Most of gold-bearing quartz veins have pyritization in this area. This kind of quartz vein contains impregnations disseminated pyrite and their resistivity and apparent polarizability are significantly dissimilar to the surrounding country rocks.It is therefore possible to measure the downward extent of the gold-bearing quartz veins from the changes of apparent resistivity and polarizability at different depths. The measurements show that the mineralized part of Q15 quartz vein extents to a depth of 60 meters. It is like a chain of pearls, so probably Q15 is getting smaller below 60 meters, forms a dead mineralized part, then, getting larger, wider and more mineralized.The gold-bearing level of Q15 has been measured by means of electric extraction method, in which a series of measurement points are arranged at intervals along the profile line,

a graphite pole is buried at each measurement points and a current is applied for 24 hours. The gold ions will be attracted to the pole.Therefore, the quantity of gold ions absorbed around the pole can be measured and analysed, which will indicate the diffusion halo of gold in the area surrounding the pole. For Q15 quartz vein, 8 profiles and 83 point measurements have been taken. A selection of the measurements are illustrated in Fig.5.It is

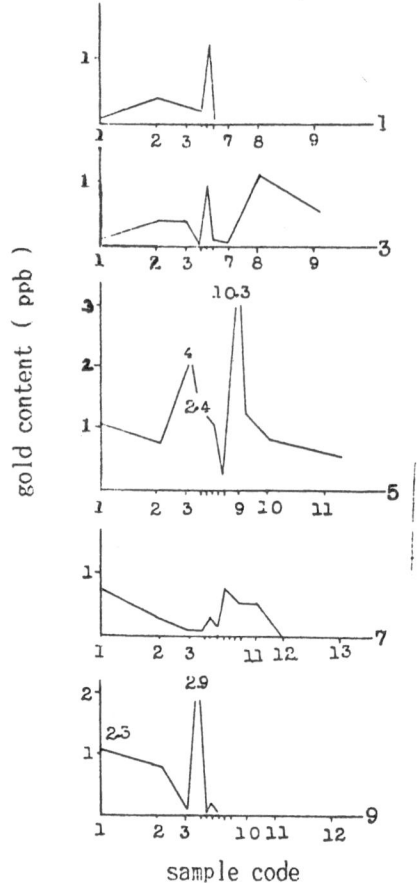

Figure 5. Electrical Extraction Gold Content of Quartz Vein Q15

inferred from Fig. 5 that the electric extraction measurements show many points having higher levels than background. In order todetermine the practical significance of those higher level a set of corresponding rock samples have been chemically analysed. The gold content of quartz vein taken at eletric extraction level 10.3 is 107.5 g/t. It can be concluded that the anomalous level o electric extraction shows higher gold content in some parts of quartz vein which reaches economic grades.

A lot of quartz veins have similar apperances to Q15, some of them are even larger and better mineralized. As a result, the geophysical measurement of Q15 is a significant reference for assessment of gold deposits over the whole study area and a powerful proof of success of gold-bearing quartz vein interpretation from SPOT imagery.

210

5.2. Interpretation of SAR Imagery Proven by Geochemical Data

The Axile altered fracture zone interpreted from SAR imagery has been examined by means of geochemical measurements. 56 soil geochemical profile cross the major part of the fracture zone.The line distance of each profile is 300 - 500 meters and point distance is 100 meters. 223 soil samples have been taken. The analytical results of soil samples show that most samples gold contents 10 times higher than gold Clark value. The highest one is 900 ppb, which suggests that the Axile altered fracture zone is a strongly anomalous zone of a gold diffusion halo. The strong anomaly concentrates in a section, 6 km to the southeastern end of the fracture zone. 55 samples have been taken from 11 profiles over this section, 18.2% of samples are over 100 ppb. The

results are demonstrated in Fig. 6. The soil geochemical profiles show that peak values appear regularly. The altered fracture zone lies on a slope which dips to northeast and the southwest side is higher, but the relief does not confine the distribution of peak values. It is located in central part of fracture zone. As a result, it is believed that peak values are the result of gold-bearing hydrothermal migration along the fracture zone rather than the transportation of the gold-bearing sedentary products.

5.3 Results of Geological Engineering Test

1)More than 100 quartz veins have been interpreted from SPOT imagery in the study area. Fieldwork has proved their existence. A quarter of the quartz veins are gold-bearing quartz veins proven by means of sampling and analysis. Geological engineering methods have been used to study some of quartz veins, e.g. 5 trenches over Q15. The highest content of trench sampling in Q15 reaches 16.9 g/t.The thickness of industrial ore bodies is more than 11 meters. The acheivements of a large amount of fieldwork show that Qiaben is a promising area of gold deposits.
2)The Axile fracture zone which is identified on SAR imagery has also been proven by geochemical data. More detailed fieldwork and geological engineering tests have been carried out in the study area. The fieldwork determines that Axile is a altered fracture zone and formed during long geological history through many geological events. Strong cleavages, mylonitization, medium-low temperature hydrothermal alteration and quartz veins are well developed within the fracture zone. Main alteration types are silicification,sericitization,yritization, kaolinization, chloritization, epidotization and zoisitization etc.. The highest gold content of quartz veins in the fracture zone is more than 300 g/t. geological engineering tests have been carried out in a strongly geochemically anomalous section. All results show that Axile fracture zone has potential significance for gold exploration. The geological engineering tests has strongly confirmed the interpretation of remotely sensed imagery.

6. CONCLUSIONS

SPOT HRV data and SAR data have been sucessfully used in gold exploration,which has resulted in the discovery and determination of the Qiaben gold

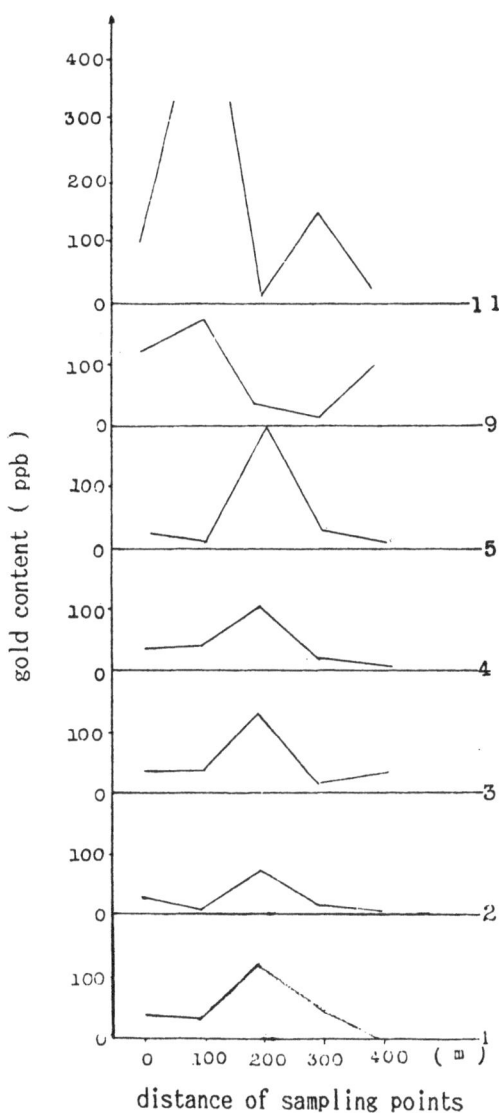

Figure 6. Soil Gold Content and Sampling Position

deposits. Current studies shows that Qiaben has the potential to be a gold mine. SPOT imagery is an a effctive tool for detecting gold-bearing geological bodies. Two kinds of IHS transformation methods have a positive effect in distinguishing altered hornblende gabbros from unaltered hornblende gabbros and quartz veins from country rocks, which are strongly associated with gold mineralization. SAR imagery is capable of identifying structures. The most important contribution of SAR data to this study is the discovery of Axile altered fracture zone.

Integratively using SAR and SPOT imagery helps us to find out and determine the distribution of Axile gold-bearing altered fracture zone. Fieldwork, geophysical and geochemical measurements and geological engineering tests are key parts for mineral exploration programme. In this study, all of above methods have been used step by step, which formed a complete methodology series of remote sensing, geophysical and geochemical, and geological engineering tests, so that fruitful successes have been achieved in remote sensing for gold exploration in the Qiaben area.

REFERENCES

BORENGASSER, Marcus X., BRANDSHAFT, Donald R. & TARANIK, James V. 1984. Geological application of enhanced Landsat 4 TM imagery of south central Nevada. Presented at the International Symposium on Remote Sensing of Environment Third Thematic Conference, Remote Sensing for Exploration Geology, Colorado Springs, Colorado.

HUNTINGTON, J. F., & GREEN, A. A. 1988. Recent advances and practical considerations in remote sensing applied to gold exploration. Bicentennial Gold, 1988, Melbourne, 246-258.

JASKOLLA, F. & HENKEL, J. 1988. Evaluation and digital processing of multispectral SPOT data. Int. J. Remote Sensing, 9, 1629-1637.

212

Exploration for bulk-minable precious metal deposits in the western Great Basin by integration of geologic, geophysical and remote sensing data

Thomas P. Lugaski B.S., M.S., Ph.D.
Harold F. Bonham Jr. B.S., M.S.
Nevada Bureau of Mines and Geology, Mackay School of Mines, University of Nevada, Reno, Nevada, U.S.A.
John C. Kepper B.S., M.S., Ph.D.
Boulder City, Nevada, U.S.A.
William H. Aymard B.S., M.S.
Mackay School of Mines, University of Nevada, Reno, Nevada, U.S.A.

SUMMARY

The use of enhanced remote sensing data has become increasingly important in the exploration for gold and other precious metal deposits in the western Great Basin. Systematic image analysis can determine both the major features and subtle geologic detail of regional geology. Remote sensing data can be used to map surface geology, structural features (lineaments, faults, etc.), geomorphology, geobotany, and hydrothermal alteration.

Landsat Thematic Mapper (TM) data, while it does not have the spatial resolution of other systems (e.g., SPOT panchromatic), does provide a variety of spectral signatures. These spectral signatures can be used individually, in various combinations, in ratios, or as principal components to produce panchromatic and false color composite images that can be successfully used by exploration geologists.

TM data can be combined with higher resolution SPOT panchromatic data to produce sharper images for structural geology analysis. TM and SPOT panchromatic data can be combined with geophysical (aeromagnetic, gravity, etc.) data, geochemical data, digital elevation models, radar images, and geographic information to produce useful mineral exploration guides.

Remotely sensed data has been shown to be useful in delineating the geologic characteristics of both volcanic-hosted and sediment-hosted disseminated precious metal deposits as well as lode-type deposits in the western Great Basin. The use of remotely sensed data by an experienced exploration geologist, especially in combination with other data sets, has potential for substantial economic impacts on exploration programs. These types of integrated analyses focus exploration efforts toward more productive targets and reduces the size of the areas that might otherwise require field examination.

Many of Nevada's precious metal deposits occur along linear zones or trends. These trends are associated with major crustal flaws which are the locus of faulting and of intrusive igneous activity of Mesozoic and/or Tertiary age. Parallel or cross-faults are important ore controls at individual deposits or within mining districts. Many of the mineralized trends in Nevada are coincident with or parallel to major structural features which are recognizable on regional aeromagnetic or gravity maps or Landsat TM imagery. Remote sensing in conjunction with regional geology, geochemistry and geophysics can be used to delineate these trends in Nevada and the ore deposits that occur along them.

INTRODUCTION

Historically, remote sensing techniques based on aerial photography and more recently satellite technology have been used in mineral exploration programs in the western Great Basin. Early Landsat multispectral scanner (MSS) images were originally used although they had major drawbacks. These drawbacks were coarse spatial resolution and limited and coarse spectral coverage.[1-3] The development of the Landsat Thematic Mapper (TM) system with its higher spatial resolution (30 m) and greater spectral coverage made that instrument of greater value for geologic remote sensing.[1,3-4] The development of the higher spatial resolution (10 m) panchromatic SPOT instrument is valuable in structural geology analysis. TM multispectral and SPOT panchromatic digital data has been used in recent years in combination with aeromagnetic and gravity geophysical data, geochemical data, digital elevation models, radar data, and geographic information to produce useful mineral exploration guides in the western Great Basin.[5-11]

TONOPAH, WEST-CENTRAL NEVADA AREA

The Tonopah Mining District, Nye County, west-central Nevada (Fig. 1) has been the site of major precious metal mining activity since the beginning of the 20th century.[12-13] In recent years this district and adjacent areas geologic and geochemical nature have been studied in detail by Bonham and Garside.[14-15] Kepper et al. were able to discriminate the various lithologic units, alteration patterns, and major structural blocks of the Tonopah area using TM data.[5] In addition the blind ore body at Tonopah was clearly detected using calibrated color ratio composite images. Using the calibrated TM data, a set of spectral curves for the various lithologic units, both altered and unaltered, were produced. Kepper et al. also digitized the geologic map of Bonham and Garside as an overlay for the geocoded and calibrated TM data.[5,12] A digital elevation model was also developed for the area. Kepper et al. further enhanced these studies by using digital aeromagnetic and gravity data (lineaments drawn on Fig. 2), regional structural data (Fig. 2) and regional geology.[6-7]

The Tonopah mining district and other near-by mining districts are part of the southern Nevada structural zone; Kepper et al. and are part of several special structural blocks (Fig. 3) in the newly defined Walker Lane Belt.[3,6-7,16] They have shown a relationship between structures as delineated in the geophysical, geological and TM data sets, regional overlap of Upper and Lower Tertiary volcanic units, and the precious metal ore production of the region (Fig. 4). The known precious metal producing areas in the Tonopah mining district occur in areas where major east-west faults and lineaments (geophysical or image) occur in conjunction with parallel or cross-faults and there is an over-lap of the Upper and Lower Tertiary volcanic rocks. These structures may be major crustal flaws that have created permeable zones that resulted in an extensive plumbing system to feed the volcanic and hydrothermal deposits.

Present research is designed to merge the existing geochemical data with Landsat TM, the geophysical data, geology and geologic structures.[14,15] It is anticipated this merging of these data sets will more adequately explain the relationships that exist between the geochemistry and the other data sets and in turn shed light on the formation of the ore deposits in this region.

ELY-MT. HAMILTON AREA, EAST-CENTRAL NEVADA

The Ely-Mt. Hamilton area of east-central Nevada (Fig. 1) which includes southern Eureka County and south-central White Pine County, Nevada has been

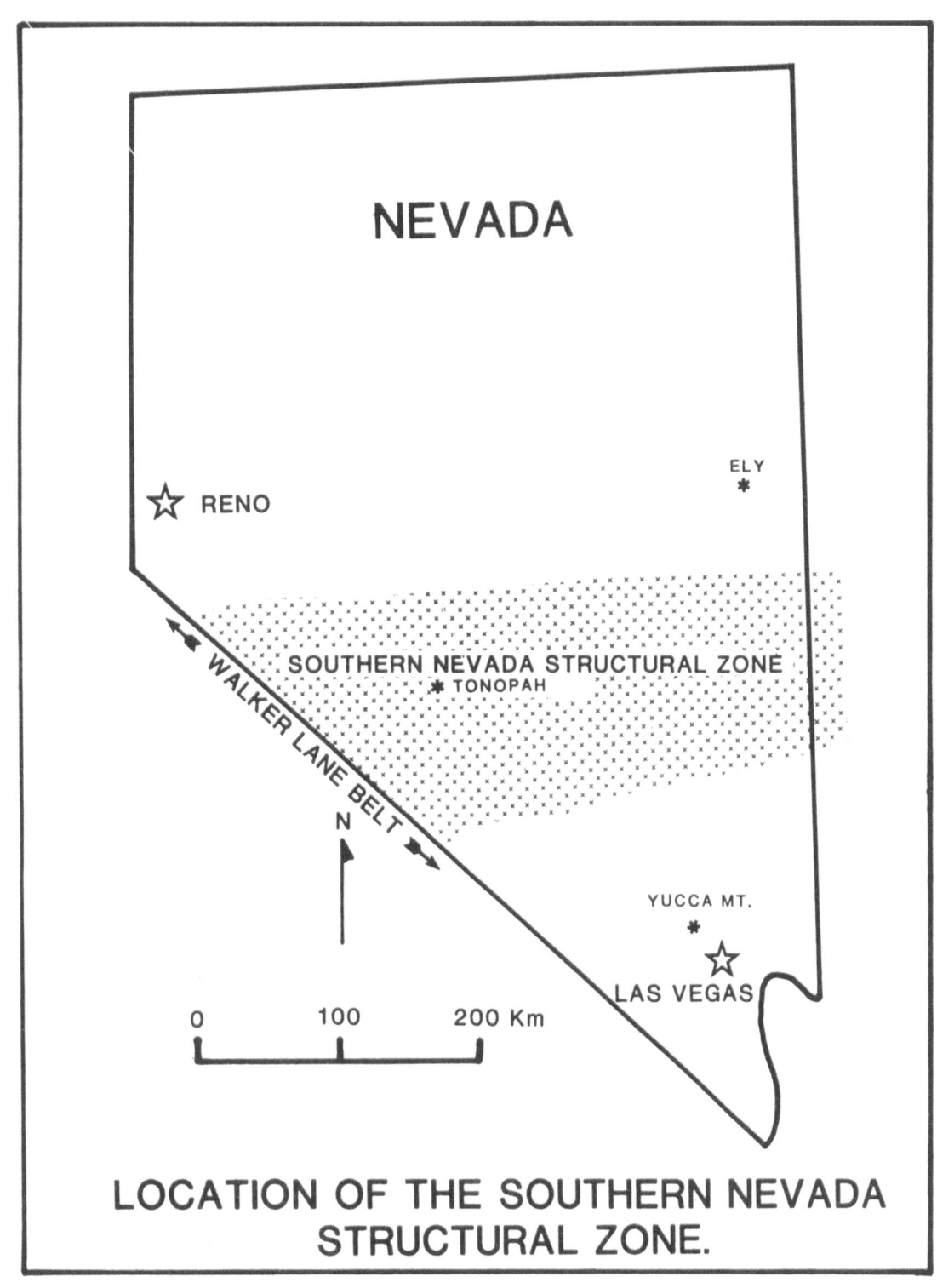

Fig. 1 Location of the study sites in Nevada, U.S.A.

preliminarily studied by Larson et al.[11] This area includes a number of historic mining districts. The White Pine mining district has copper-silver and lead-silver deposits found in association with granitic bodies. The Ward mining district includes lead-zinc-silver-copper ore deposits found in a variety of sedimentary rocks and associated with quartz monzonite dikes. The Taylor mining district where silver-lead and antimony ore deposits are found in a variety of sedimentary rocks and associated with rhyolite dikes,

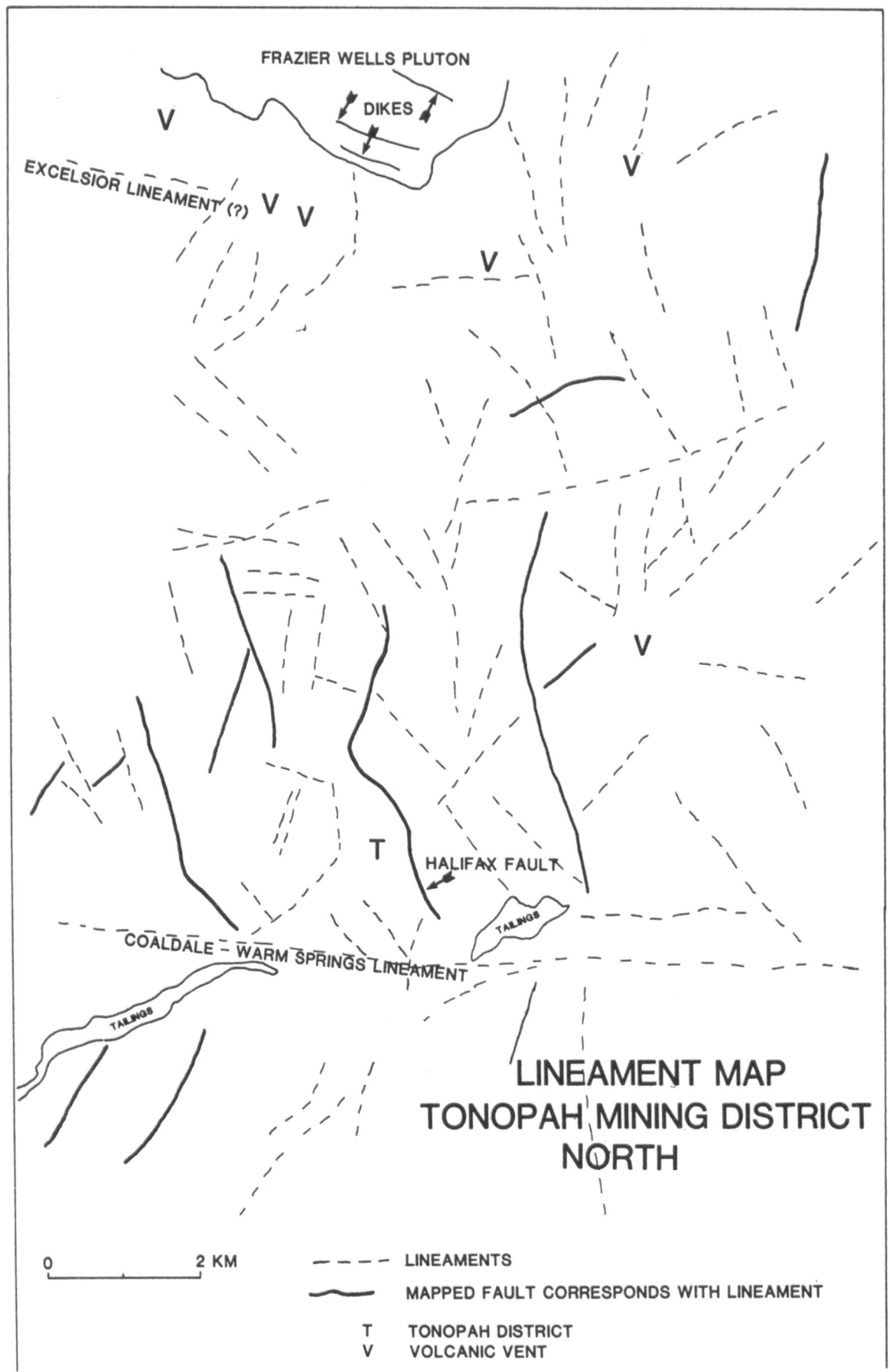

Fig. 2 A fault and lineament map of the Tonopah, Nevada area
compiled from Landsat image, aeromagnetic and gravity geophysical
data.

216

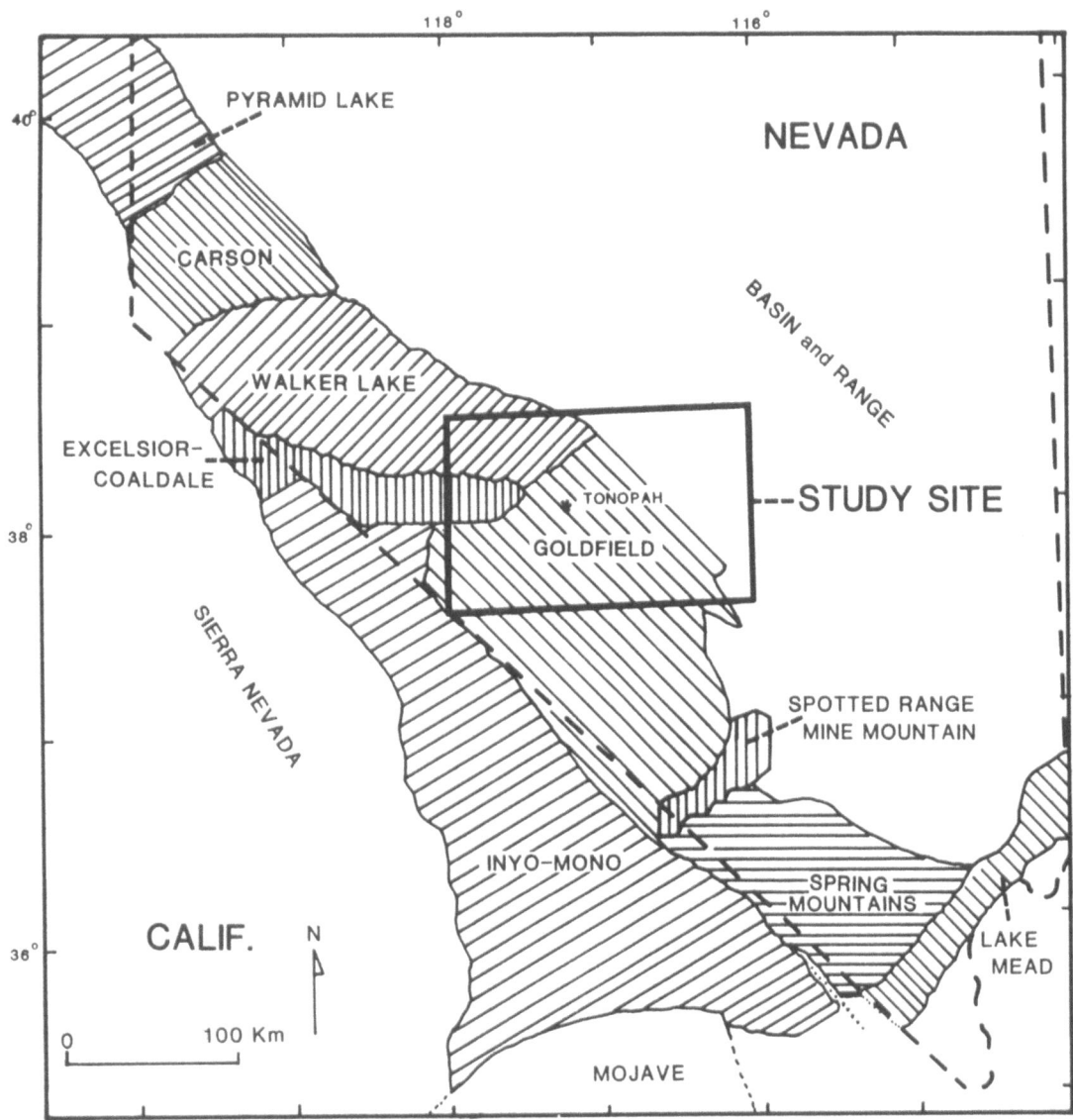

Fig 3 Map of the structural blocks in the Walker Lane Belt, modified after Stewart, 1988.

latite and dacite flows, and tuffs. The Robinson mining district contains a variety of deposits of porphyry copper, lead-silver replacement, manganiferous carbonates, disseminated and skarn-related low-grade gold and gold-bearing quartz veins and lead-silver veins.

The question arises as to whether these districts and perhaps other districts and/or prospects in the area are structurally and /or magmatically related to one another. In addition, do these deposits extend to the west of Mt. Hamilton and how is the Mt. Hamilton area and the areas to

its west related to the Battle Mountain-Eureka Mineral Belt Trends?

During the investigation of these areas existing geologic maps and literature were examined. TM digital data was examined and computer enhanced. Lineament analysis was undertaken and other structures were delineated as were areas of hydrothermal alteration. Digital aeromagnetic and gravity geophysical data was computer enhanced, reviewed and color images created.

The results of the

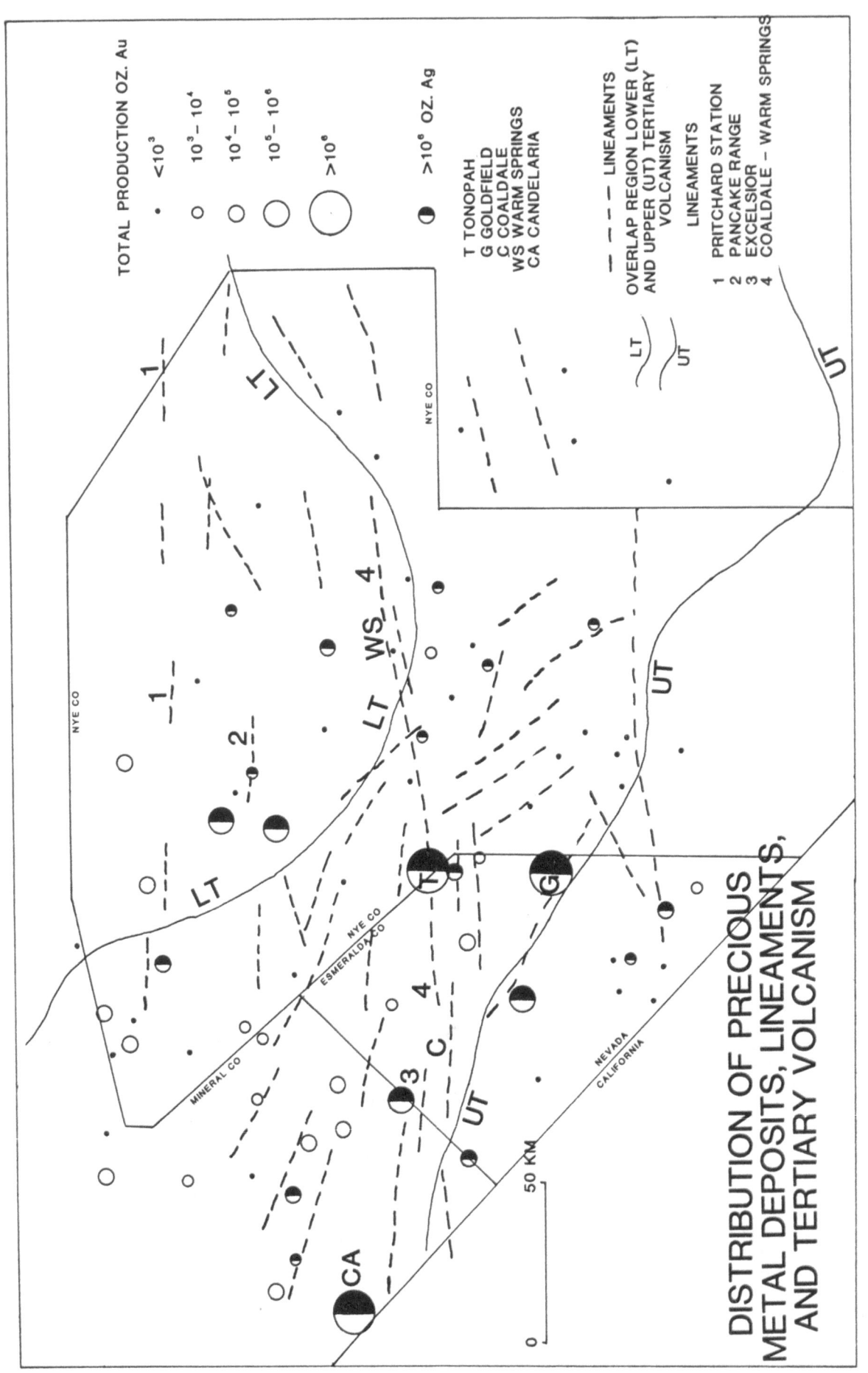

Fig 4 Map of the distribution of precious metal deposits, lineaments (from Landsat and geophysical data), and Tertiary volcanism.

examination of these data sets show the mapped lineaments and other structures form a zone (Fig. 5) that runs slightly south of west from Ely through the Mt. Hamilton area and the Mt. Hamilton area is a intersection of the Battle Mountain-Eureka Mineral Belt Trend and the newly described Ely-Mt. Hamilton Mineral Belt Trend.[11] More importantly the Landsat imagery clearly shows the extension of the structures and alteration of the Ely-Mt. Hamilton Mineral Belt to the areas west of Mt. Hamilton, which have yet to be explored in detail but are known to contain some precious metal anomalies. The aeromagnetic data clearly shows anomolous magnetic highs associated with these deposits and prospects.

YUCCA MOUNTAIN AREA, SOUTHERN NEVADA

The Yucca Mountain area, Nye county, of southern Nevada (Fig. 1) contains some precious metal producing districts. These are the Bullfrog and Fluorine mining districts to the west of Yucca Mountain which are known respectfully for their gold, silver, lead and copper deposits and gold, mercury and silver deposits. To the south of Yucca Mountain is the Fairplay mining district known for its gold, silver and lead deposits and to the east of Yucca Mountain is the Wahmonie mining district known for its silver and gold deposits.

Yucca Mountain, itself, and the area surrounding it were not primarily examined for their potential mineral wealth, although that is under consideration, but are being studied as a potential, future nuclear waste repository. As a result of this proposed repository a vast amount of remotely sensed data has been collected over the area. In addition the geographic information of lineaments and existing faults, roads, and seismic foci have been plotted and co-registered with various types of imagery.

While the acquisition, development and display of these digital data sets were primarily to help determine the feasibility of the repository, these data sets show how an exploration geologist could use these sets interactively in their exploration program. The aeromagnetic and gravity geophysical data along with the TM image lineaments and existing faults show structures never before mapped or considered in the feasibility studies of the repository. These data sets also show the relationship of regional mining districts, their ore deposits and regional structures.

CONCLUSIONS

While all the case studies mentioned above show a variety of uses for remotely sensed data and their usefullness when interactively used with geophysical and geographic information, for the most part this can only be considered a set of reconnaissance studies. Further, more focused studies need the availabilty of high spatial resolution and high multispectral resolution imagery over these areas. This data is now only commercially available from a number of sources in the form of aircraft scanner data but hopefully sometime in the not too distant future will be available from satellite or space station sensors.

References

1.   Sabins, F. F., Jr. Remote Sensing; Principles and Interpretation (2nd Edition). W. H. Freeman, New York, 1987, 449 p.

2.   Rowan, L. C. and P. Wetlaufer. Relation between regional lineament systems and structural zones in Nevada. American Association of Petroleum Geologist Bulletin 65, 1981, p. 1414-1432.

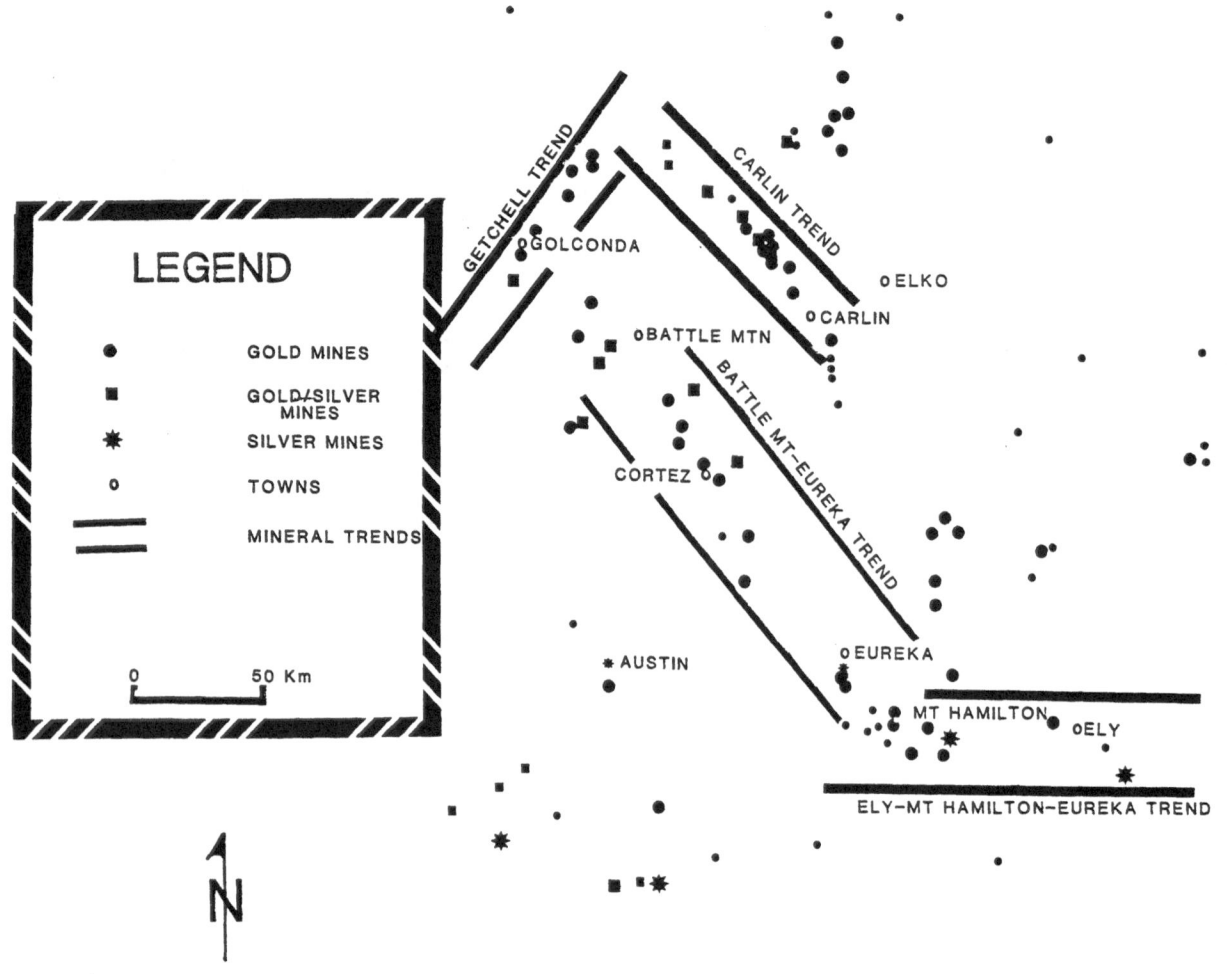

Fig 5 Regional map showing the location of mineral trends and gold/silver deposits in northern and eastern Nevada.

3.   Goetz, A. F. H., B. N. Rock, and L. C. Rowan. Remote Sensing for Exploration: An Overview. _Economic Geology_, Vol. 78, no. 4, 1983, 573-590.

4.   Drury, S. A. Image Interpretation in Geology. Allen & Unwin, 1987, London, 243 p.

5.   Kepper, J. C., T. P. Lugaski, and J. S. MacDonald. Discrimination of lithologic units, alteration patterns and major structural blocks in the Tonopah, Nevada area using Thematic Mapper data. _Proceedings of the Fifth Thematic Conference on Remote Sensing for Exploration Geology, Mineral and Energy Exploration: Technology for a Competitive World_. Reno, Nevada September 29-October 2, 1986, Vol. I, 1987, p. 97-115.

6.   Kepper, J. C., T. P. Lugaski and W. Aymard. Regional structural blocks, volcanism, and the precious metal deposits in the south-central Nevada structural zone. _IGARSS'89, Quantitative Remote Sensing: An Economic Tool for the Nineties_, Vancouver, Canada. July 10-14, 1989, Vol. 3, p. 1421-1423.

7.   Kepper, J. C., T. P. Lugaski and W. Aymard. 1990. Precious metal deposits, structural blocks, and volcanism in the south-central Nevada structural zone. (abst.) _Great Basin Symposium, Geology and Ore Deposits of the Great Basin_, Reno/Sparks, Nevada. April 1-5, 1990.

8.   Bonham, H. F. and T. P. Lugaski. Exploration for epithermal, volcanic-hosted, gold-silver deposits. (abst.),

GOLD 89 in Europe.

9. Bonham, H. F., T.P. Lugaski
and W. Aymard. Epithermal,
volcanic-hosted, gold-silver
deposit exploration techniques
useful in Nevada and other
areas. (abst), Great Basin
Symposium: Geology and Ore
Deposits of the Great Basin,
Reno/Sparks, Nevada April 1-5,
1990.

10. Lugaski, T. P., L. Garside
and H. F. Bonham. The use of
remote sensing to delineate gold
and other precious metal
deposits in Nevada.
International Symposium on Gold
Geology and Exploration, Oct.
1989. Shenyang, China, 6 p.

11. Larson, L. T., T. P.
Lugaski and W. Aymard.
Integration of geological,
geophysical, and Thematic Mapper
remote sensing data in relation
to the geologic occurrence of
precious and base metal deposits
in the Ely-Mt. Hamilton-Eureka,
Nevada area. Seventh Thematic
Conference on Remote Sensing for
Exploration Geology; Methods,
Integration, Solutions. ERIM.
Oct. 2-6, 1989. Calgary. 8 p.

12. Carpenter, J. A., R.R.
Elliot, and B. F. W. Sawyer The
History of Fifty Years of Mining
at Tonopah, 1900-1950.
University of Nevada Bulletin,
Vol. XLVII, no. 1. 153 p.

13. Elliot, R. R. Nevada's
Twentieth-Century Mining Boom,
Tonopah, Goldfield, Ely.
University of Nevada Press,
Reno, Nevada, 1966, 344 p.

14. Bonham, H. F. and L. J.
Garside. Geology of the Tonopah,
Lone Mountain, Klondike, and
Northern Mud Lake Quadrangles,
Nevada. Nevada Bureau of Mines
and Geology Bulletin 92, 1979,
142 p.

15. Bonham, H. F. and L. J.
Garside. Geochemical
reconnaissance of the Tonopah,
Lone Mountain, Klondike, and
Northern Mud Lake Quadrangles.
Nevada. Bureau of Mines and
Geology Bulletin 95, 1982, 68 p.

16. Stewart, J. H. Tectonics
of the Walker Lane Belt, Western
Great basin: Mesozoic and
Cenozoic Deformation in a Shear
Zone. In W. G. Ernst, ed.,
Metamorphism and Crustal

Evolution of the Western U.S.
Ruby Vol. VII, Prentice Hall, N.
J., 1988, p. 683-713.

Geological exploration in the western United States by use of airborne scanner imagery

W. P. Loughlin B.Sc.
National Remote Sensing Centre, Farnborough, Hampshire, England

SYNOPSIS

Airborne multispectral scanners have been operational in the western United States since 1985. Imagery has been acquired over known gold producing belts and from potentially prospective areas in Nevada, Oregon and adjacent states. Three scanners have been flown, manufactured by Daedalus, Geoscan and Collins; images from the first two of these have been used for the case histories in this review.

The western U.S. is a mature exploration province with a long history of mineral exploration and exploitation; many 'discoveries' in the last decade have been in the vicinity of old workings or prospects. In view of this a successful airborne scanner survey must be able to direct geologists to subtle alteration zones overlooked or disregarded by early prospectors, to mineralised zones within known alteration zones, or to previously undetected and potentially prospective geological structure. Case histories discussed and illustrated here demonstrate that new prospects are being discovered from interpretation of airborne scanner imagery, and new exploration plays can often be generated around existing prospects. Recommendations for the exploitation of archived image data and the planning of future surveys are presented.

This paper reviews methodologies for structural, geomorphic and lithological interpretation of pre-dawn acquired thermal infrared imagery and alteration mapping from mid-day acquired multispectral imagery.

An important new technique which depends on selective input of channels for Principal Component Analysis (PCA), and subsequent input of previously-generated Principal Component images for further PCA, has been developed during this study. The methodology is named the "Cròsta technique". The images generated by this method are reliable and definitive for alteration mapping from both airborne and satellite multispectral images over the Great Basin area.

INTRODUCTION

The purpose of this study is to review the geological applications of airborne scanner imagery for mineral exploration in and around the arid Great Basin area. The review summarises five years of operational surveying, and the collective experiences of many field geologists who have used the imagery as a prospecting aid.

Comparisons are made between different types of imagery (day-time acquired multispectral and pre-dawn acquired thermal), between imagery from different scanners (Daedalus AADS 1268, (ATM) and Geoscan AMSS mkII), and between airborne and satellite (TM) image data.

Direct comparison between airborne and Landsat TM imagery has been of considerable benefit in that image analysis techniques developed during this study are applicable to both.

Exploration in north America using remote sensing

In complete contrast to the situation in Australia[1], remote sensing is not yet fully accepted by north American exploration geologists. Field geologists have been understandably disillusioned by 'false' image anomalies and, in consequence, many exploration managers believe that traditional prospecting, especially prospect evaluation, is

FIGURE 1:
LOCATION MAP

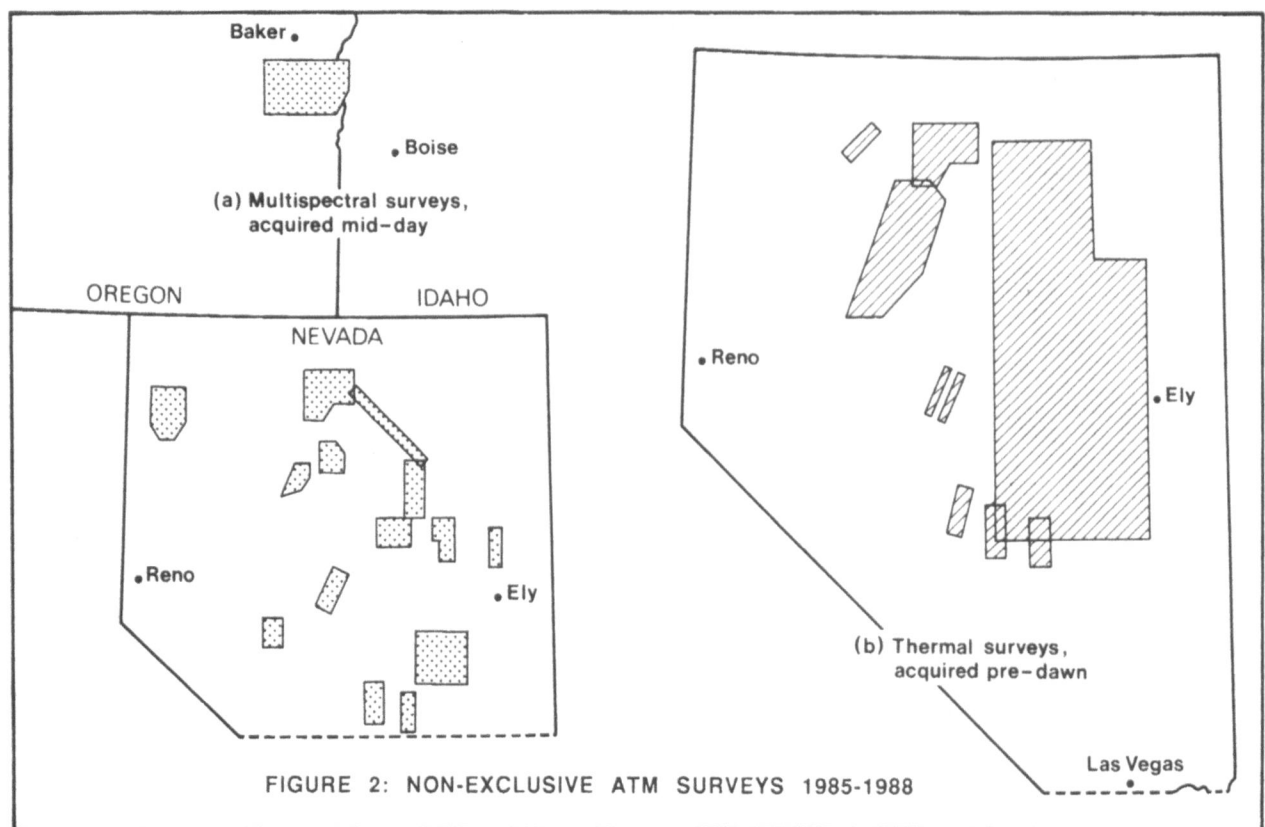

FIGURE 2: NON-EXCLUSIVE ATM SURVEYS 1985-1988

(a) Multispectral surveys, acquired mid-day

(b) Thermal surveys, acquired pre-dawn

more cost effective (and successful) than using remote sensing as a target finder.

Image processing techniques tested here have been selected to serve the needs of the prospectors and exploration geologists. Those techniques which require some detailed or expensively-acquired knowledge of the spectral properties of surfaces at each site (or which have been found unreliable in the Great Basin environment) have been avoided. The techniques can all be implemented on PC-based image processors.

'Test' sites chosen for the study are, or resemble, *bona fide* exploration prospects typical of those in the geological, geomorphic and climatic environments covered by available scanner surveys.

Large, well known and often highly disturbed mining districts (such as the Comstock and Cuprite districts) are deliberately omitted from this review; they are not representative of most of the recent precious metal discoveries in the region. Fortunately a number of poorly exposed deposits were imaged by either ATM or TM before or soon after discovery (or disturbance) and the image processing techniques developed during this study have been applied to these older images.

The scanners

The first fully operational survey was flown in 1985, by the Daedalus AADS 1268 (ATM). The Geoscan (AMSS mkII) scanner began USA operations in 1989 and covered some 100,00 sq. km. in the first season[2]. A prospect discovery from the first Geoscan campaign is reported elsewhere[3]. For this study Geoscan image data is available for one critical site.

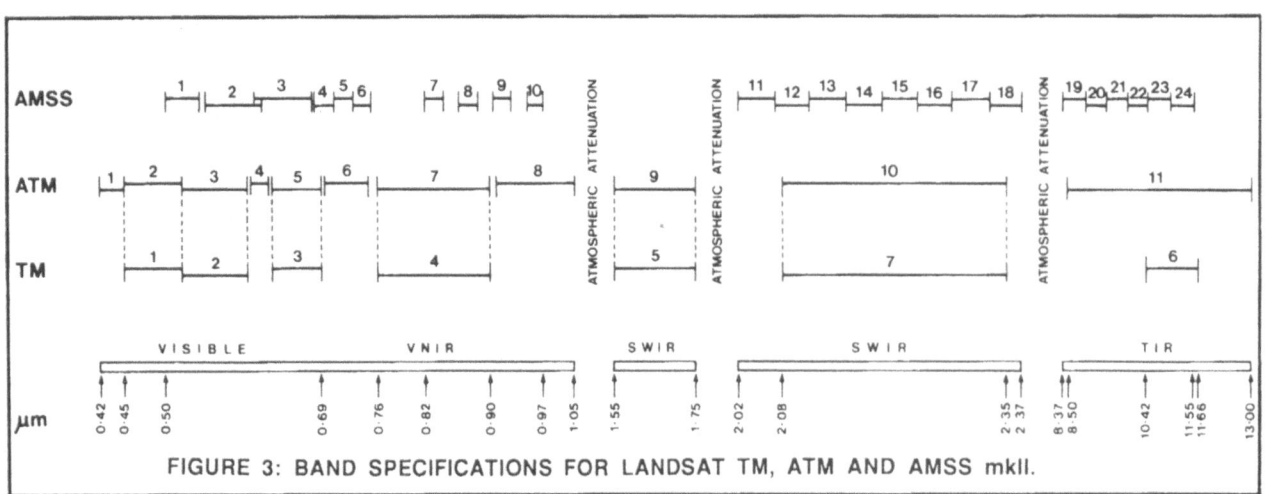

FIGURE 3: BAND SPECIFICATIONS FOR LANDSAT TM, ATM AND AMSS mkII.

The two basic classes of imagery described here (thermal images acquired before dawn and multispectral images acquired at mid-day) are discussed and reviewed according to image type. Where appropriate cross-references are made to the application of other types of imagery over each test site.

The Geophysical and Environmental Research (Collins) imaging spectrometer (GERIS) shows promise for detailed mineral mapping, but has yet to be used for large scale regional reconnaissance. A discussion of GERIS applications is therefore beyond the scope of this review. Recent GERIS case histories are reported by Taranik & Kruse[4], Kierein-Young & Kruse[5], and Akiyama et. al.[6].

THE GREAT BASIN & DISSEMINATED PRECIOUS METAL DEPOSITS

The Great Basin.

The Great Basin is an area of internal drainage within the Basin and Range physiographic province, characterised by typically north-south (or NNE-SSW) trending mountain and hill ranges and intervening alluvial valleys. The province has been formed by relatively recent crustal extension between the western edge of the Rockies (e.g. the Wasatch Mountains escarpment in Utah) and the Sierra Nevada - Cascade Range (in California/Nevada, Oregon and Washington). Valleys lie at an average elevation of around 1,350m; the ranges are much more varied in elevation, from low, undulating hills to lofty and craggy mountain chains with crests 1,500m above the valley floors.

Vegetation is largely dependent on moisture from winter snows. Lower and middle elevations are dominated by sagebrush and sparse grasses, with salt-tolerant scrub around playas. Woody species are more prolific on higher ground where conifers grow on moist ground protected from evaporation loss on north-facing slopes and junipers on the drier south-facing slopes. Highest elevations trap sufficient moisture to support good stands of timber and grassy meadows. Woodland has often been considerably thinned out around old mining camps as a source of mine timber and smelter charcoal.

Great Basin Geology

A long and complex geological history is due to repeated episodes of marine and continental sedimentation, uplift and erosion, orogenic compression and cratonization, igneous intrusion and crustal extension. The pre-Quaternary strata are well exposed in the recently uplifted mountain ranges; folded granite lenses within 1,450 m.y. old metamorphic rocks of southern Nevada have been dated at 1,740 m.y. old[7].

A number of major compressional events have been recognised. A Lower Palaeozoic (Antler) orogeny in late Devonian, early Mississippian, with greatest effect in central Nevada; an Upper Palaeozoic/Mesozoic (Sonoma) orogeny during late Permian and early Triassic, centred on west central Nevada; a late Mesozoic (Sevier) orogeny,

which is recognised in western Utah and easternmost Nevada. The compressions have resulted in the superimposition of allochthonous and autochthonous rock assemblages of contrasting facies, for instance along the Roberts Mountain thrust in central Nevada. Thrust packages are of major importance in providing a selection of suitable host rocks for later mineralisation.

Bulk-minable precious metal deposits.

Precious metals exploration has been going through a long and unusually sustained period of activity since the price increases of the late 'seventies and early 'eighties. This latter day gold rush has also coincided with (or has helped stimulate) a better understanding of Pacific Rim epithermal deposits. Heap leach mining methods have also been refined such that extremely low grade (e.g. 1ppm Au), but near surface, deposits can often be mined as ore.

The recently exploited Great Basin deposits are often loosely termed 'bulk-minable deposits of the Carlin type'. There are, in fact, many variants and the Carlin deposits are a particular type. The McLaughlin (California), Round Mountain (Nevada), Paradise Peak (Nevada), and Mercur (Utah) deposits are just a few examples of world class mineral producers within the region which differ in geology and mineralogy, yet share one common feature - they have been formed by epithermal processes within previously active, large hydrothermal systems.

The precious metal deposits are usually the result of a late Cretaceous to Tertiary age mineralising event. Host rock lithologies are sedimentary, volcanic or even plutonic in origin. Structural control is very important in forming channelways for the mineralising fluids. A "typical" Nevada deposit is illustrated on Figure 5.

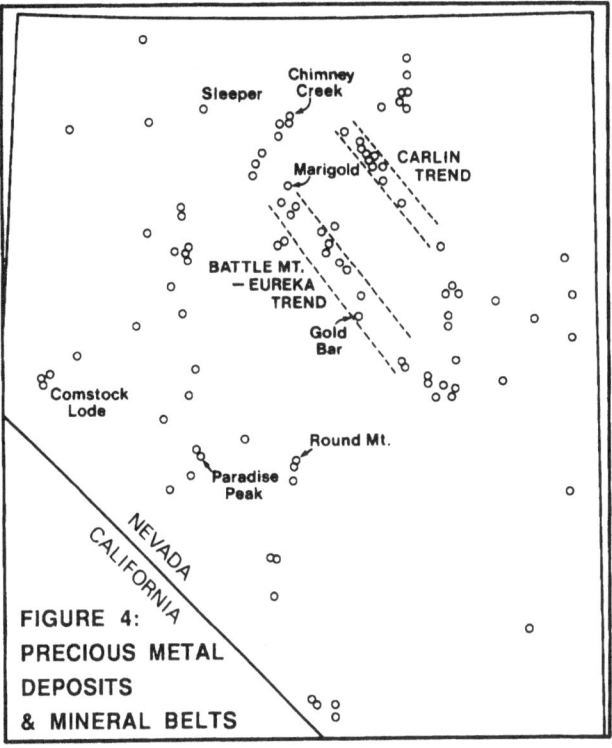

FIGURE 4: PRECIOUS METAL DEPOSITS & MINERAL BELTS

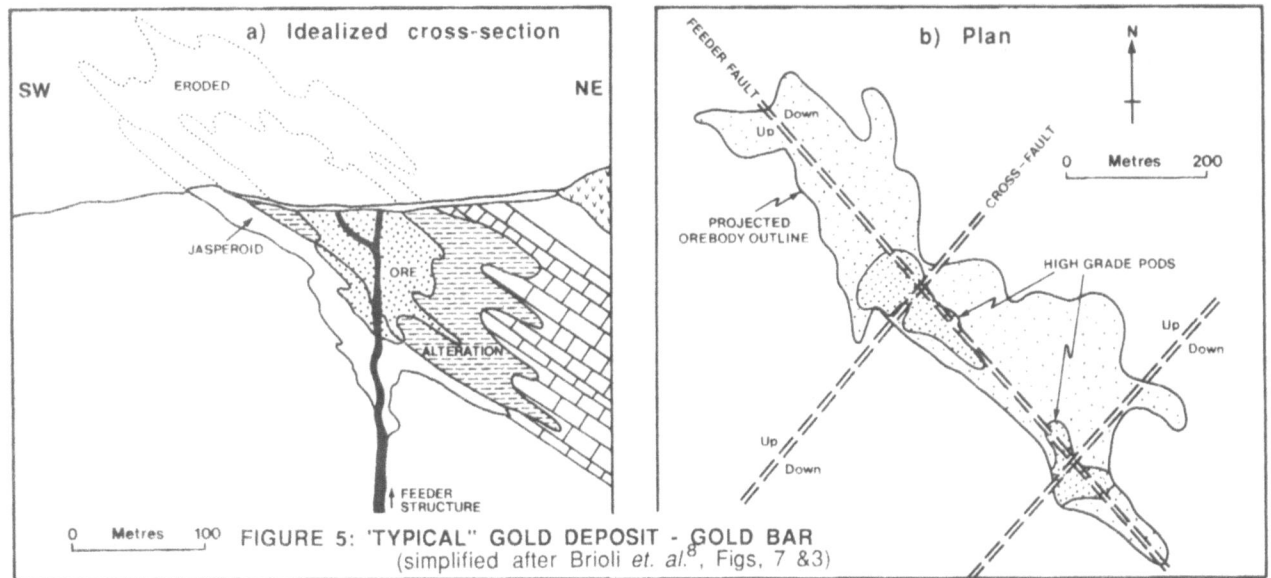

FIGURE 5: 'TYPICAL" GOLD DEPOSIT - GOLD BAR
(simplified after Brioli *et. al.*[8], Figs, 7 &3)

Pediment Plays

Nevada discoveries such the Sleeper, Gold Bar, and Marigold deposits are located between the main range front faults and basin boundary faults. They are largely (or completely, in the case of Marigold) obscured by piedmont fan deposits, and have led to an intensive search in similar geomorphic locations. This exploration target is known as a 'Pediment Play', and is explored for by sampling of every outcrop exposed through piedmont fan gravels, by soil and plant assays, by shallow seismic methods, by airborne and ground geophysics, and by interpretation of pre-dawn acquired thermal imagery.

Alteration and remote sensing characteristics

Alteration

The deposits are often associated with anomalous concentrations of mercury, arsenic, antimony and thallium, often with tungsten, flourite and molybdenum. Large alteration haloes are developed with zonation of alunitic to propylitic 'argillic' alteration, extensive silicification and pyritisation. Oxidation of iron-bearing sulphide minerals leaves iron oxide stains ('colors') on outcrops in the neighbourhood of the deposits. The recognition of these colours, in the vicinity of 'bleached', argillically altered, outcrop, led early prospectors to the discovery outcrops of most of the historic, and extant, productive orebodies.

The siliceous caprocks are known as "jasperoid"; in addition, the highest level silicification is occasionally preserved as opaline sinter which can also be seen surrounding active hydrothermal systems (hot springs) today.

Remote sensing characteristics

Images should be processed to facilitate the detection of 'argillic' alteration (hydroxyl-bearing minerals), silicification, and iron oxide staining ('FeOx') in association with favourable geology and structure.

The spectral characteristics of alteration minerals associated with mineral deposits have been discussed in detail in the remote sensing literature; for example, Hunt and Salisbury,[9] Hunt *et. al.*[10]. The 2.08-2.35 microns spectral window is of crucial importance for detection of absorption due to hydroxyl-bearing minerals, such as sheet silicates. The visible to near infrared covers diagnostic spectra for iron-oxide minerals.

Krohn[11] made spectral measurements of weathered rock samples from mineralised outcrop in Nevada and Idaho and found that many mineralised jasperoids also have a distinct hydroxyl-like absorption feature in the 2.08-2.35 microns spectral window. Krohn's work is of crucial importance for the understanding of mineralised jasperoid detection by remote sensing.

The thermal infrared has been promoted as the silica detection spectral region; however, there are problems in the Nevada environment, in that thermal images record variations in surface temperature (influenced by soil moisture, surface roughness and vegetation) and silica discrimination by inter-band comparison in the TIR has yet to be convincingly demonstrated in field conditions.

PRE-DAWN ACQUIRED THERMAL INFRARED IMAGERY

The use of thermal imagery for geologic investigations has been discussed in the remote sensing literature (e.g. Sabins,[12] Kahle[13] & Drury[14]). A number of images have also been published which indicate the potential of the technique (e.g. Drury[14], Fig. 6.7).

This is the first review of the systematic application of TIR imagery for regional scale mineral exploration. The test sites illustrated and described below are covered by the most extensive TIR image database in the public domain (Figure 2b) and a comprehensive description is provided which is drawn largely (and with permission) from the original study of the Osgood Mountains imagery[15].

226

Explanation of thermal infrared images

The thermal infra-red channel (of the ATM scanner) measures the radiant flux in the 8.5-13.5 micron spectral window. Within a good degree of approximation, this measurement is directly proportional to the radiant temperature of the ground surface and the brightness of the images is a good measure of the actual temperature of the ground at the time the imagery was acquired.

Variations of surface temperature in natural conditions are driven almost entirely by diurnal heating and cooling. The speed and extent to which natural materials warm up in the morning and cool down during the afternoon and at night are a function of the balance of heat fluxes at the ground surface and of the thermal properties of the ground itself.

The degree of ground warming is determined by the physical properties of the surface (albedo, moisture content, roughness) which controls the proportion of incident solar energy absorbed. The incidence angle of the solar illumination, which is a function of the time of day and local topography, also causes variations in surface temperature. These factors are the cause of many well known features of thermal IR imagery; for example, dark rocks with low albedo are invariably warmer than light rocks in daytime imagery; north-facing slopes are cooler than south-facing slopes (in the northern hemisphere); east-facing slopes warm up faster in the morning and reach their peak temperatures earlier in the day than west-facing slopes.

The warming effect of solar irradiation is offset by a number of processes which dissipate heat from the ground surface. These are the sensible and latent heat fluxes caused by conduction and evaporation to the atmosphere, radiation of thermal energy and conduction of heat into the subsurface. Evaporation from damp soils has a strong cooling effect. Evapotranspiration tends to keep vegetation canopies markedly cooler than surrounding ground surfaces on daytime thermal imagery. Trees are invariably warmer than surrounding surfaces on pre-dawn imagery.

During the afternoon and at night when the ground surfaces are cooling, temperature variations caused by the direct incidence of sunlight tend to become more subdued and patterns caused by the conduction of heat from the subsurface can be more clearly seen. The property of the ground which has the most direct bearing on the conduction of heat from the subsurface during the heating part of the cycle and its return to the surface in the afternoon and at night is the thermal inertia. This is a measure of the capacity of the ground to resist changes of temperature and is a function of its density, thermal capacity and thermal diffusivity. These are in turn functions of the composition of the rock or soil profile and its texture. Materials with high thermal inertia tend to stay cool during the day but remain relatively warm at night. For example, open bodies of water appear warm on night-time imagery because their thermal inertia is greatly enhanced by the convection to the surface of heat stored at depth during the daytime. Variation in the water content of soils affects their density, thermal capacity and thermal diffusivity and therefore has a strong impact on their thermal inertia. Many of the patterns visible on the Nevada thermal images in the pediment and valley areas are thought to be caused by variation in the water content of the soils acting through its effect on thermal inertia and surface evaporation. More comprehensive reviews of the thermal properties of rocks and soils are given by, for instance, Sabins[12,16] (pp119-175 & pp125-175), and Drury[14] (pp149-164).

Geological interpretation of pre-dawn TIR images.

Structural interpretation

Geological structures are detected on thermal images in a number of ways. Fractures are marked by cool zones due to damp soil and subterranean drainage, and warm zones due to woody vegetation (night-time imagery). Insulation by dry gravel or soil partially or completely covering buried fault scarps or rock pediment also produces temperature variations. Strong linear contrasts between dry and damp soils may reflect normal faults where subsurface drainage is freer and less retentive of groundwater on the upthrown side than on the downthrown side.

Well expressed lineaments (e.g., Fig. 14 below, structure NE of Mt. Hope) do not necessarily reflect more prospective structures than those with more subtle image appearance, especially in the tectonically active Great Basin environment. It is likely that important fault zones will have subtle expression in areas with thicker overburden, or where recent re-activation has been slight or non-existent.

Silicification detection by thermal contrast

Pre-dawn TIR images can often detect silicification and siliceous sinter, where a marked physical contrast exists between the siliceous materials and their host. The silicified rocks seem to retain heat much longer than surrounding rocks and soils, such that they appear as bright targets on the imagery.

Photogeological interpretation

The geological interpretation of single band, monochromatic, TIR images is entirely photogeological in approach, provided that the thermal properties, discussed above, are taken into account along with the general principles for interpretation (Fig. 6). Care should also be taken to recognise non-geologic phenomena such as wind streaks and smear (see Sabins[12] Fig. 5.13), warm tones from dense vegetation canopies, especially on north-facing slopes, and the 'chaotic' effects of rugged topography. The case histories provide examples of most of the features interpretable from pre-dawn TIR images.

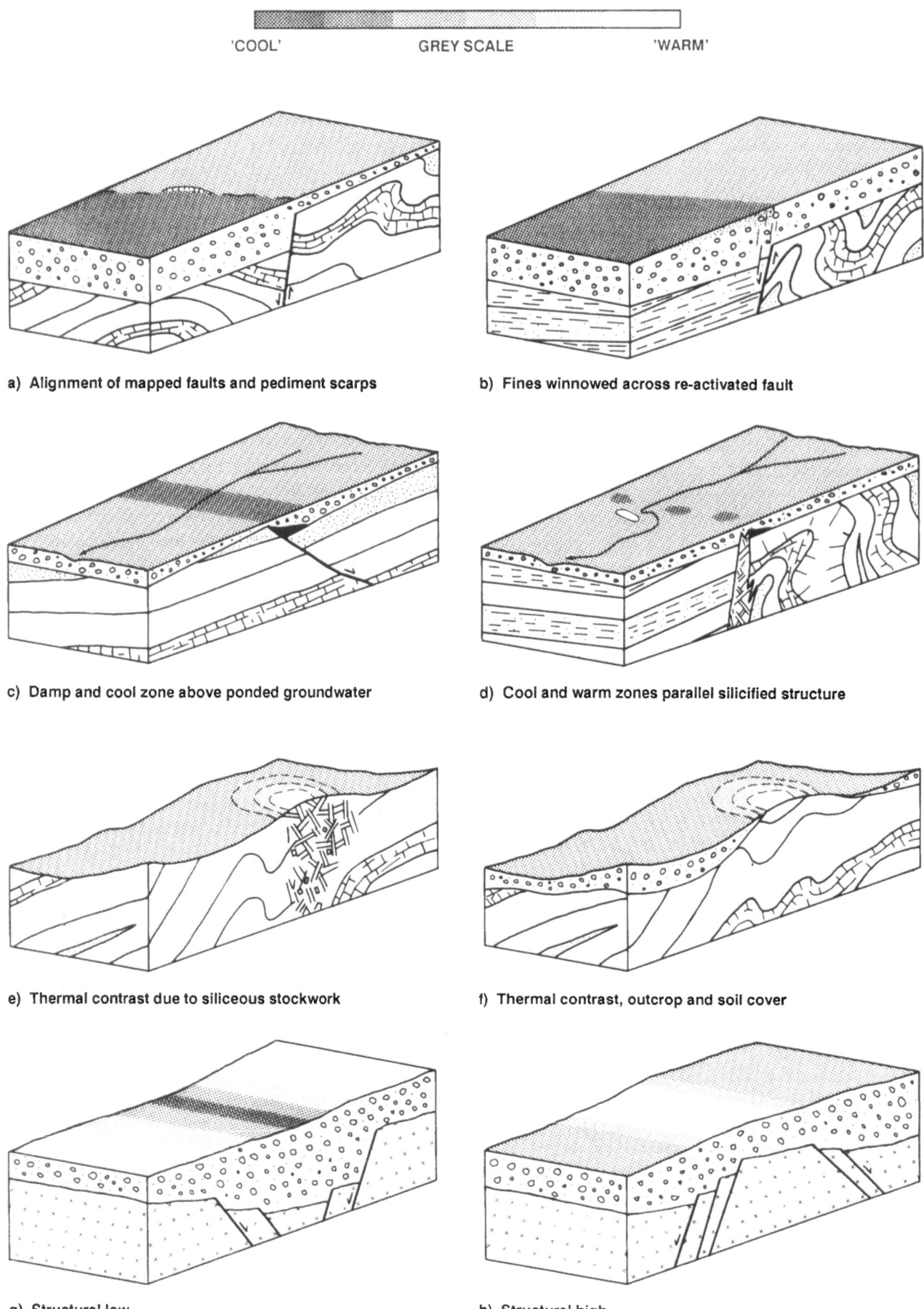

'COOL'   GREY SCALE   'WARM'

a) Alignment of mapped faults and pediment scarps

b) Fines winnowed across re-activated fault

c) Damp and cool zone above ponded groundwater

d) Cool and warm zones parallel silicified structure

e) Thermal contrast due to siliceous stockwork

f) Thermal contrast, outcrop and soil cover

g) Structural low

h) Structural high

FIGURE 6: PRINCIPLES OF (PRE-DAWN) THERMAL IMAGE INTERPRETATION

228

The ATM, Landsat TM and Geoscan AMSS mkII band passes are compared on Figure 3, above.

ATM imagery is offered by Hunting in two forms, hardcopy photographic products and digital imagery recorded on computer compatible tape (CCT). Hardcopy composite images of ATM channels 10,7,3 (equivalent to the TM composite 7,5,2; in RGB) provide an interpretation base for plotting the results of interactive processing of the digital image data. The hardcopy imagery gives good discrimination of lithology, with vegetation in bright green colours, and it often highlights hydroxyl-bearing minerals (cyan hues) and iron-oxide staining (reddish hues). It was never intended that the composites were to be used on their own for both geologic discrimination and mineral exploration purposes, and, as will be demonstrated, interactive processing is essential for extraction of reliable information from the ATM imagery.

The Geoscan imagery is routinely produced in hardcopy form with four different composites from each flight line laid up on one hardcopy sheet. This method removes the synoptic interpretation possibilities of hardcopy with adjacent image swath layout, but this is partially compensated by the opportunity to directly compare images from different spectral regions over the same site. The AMSS digital image data is recorded on optical disk and can thus be quickly accessed for interactive interrogation on the Geoscan-designed 'Gypsy' image processor.

Three band composites - false FeOx anomalies.

Landsat (or airborne) TM composites which include the 'Clay Band' as the red component (such as TM 7,5,2, TM 7,4,2, TM 7,5,1 or TM 7,4,1) very often highlight iron-oxide staining ('FeOx'), as reported by Legg[17]. The same composites are also prone to false FeOx anomalies, especially on satellite images where geomorphic interpretation is difficult due to lower resolution.

Many false FeOx anomalies are reported as areas of disturbed ground or recent soil erosion, boulder trains on steep mountain slopes, zones where vegetal cover has been removed (e.g. by brush fires), and surfaces strewn with pebbles and cobbles. In all of these vegetation cover is minimal and the surfaces are extremely dry in comparison with surrounding soils. It is a reasonable assumption that anomalous high reflectance in the mid infra-red from these areas is due to greater than background exposure of lithologic materials (rock fragments) which increase reflectance and less than average exposure of humic (and moist) soils and vegetal materials which, if present, would suppress reflectance in TM7.

Dry, loose, pebbly and bouldery materials become extremely hot in the desert sunshine, irrespective of surface albedo. This is especially true at mid-day in spring, summer and autumn when multispectral imagery is routinely acquired. Because of this there is a positive correlation between the TM7 region and the TIR for many 'false' FeOx anomalies.

Increased reflectance in TM7 should not be due to a thermal effect as thermal emmittance in TM7 only occurs above *about* 160 degrees centigrade[18] and radiance of EM energy reflected by the earth's surface falls rapidly towards zero beyond *about* 2.5 microns[14]. However the positive TM7/TIR correlation can also be seen on areas of recent brush fires. These surfaces are covered with blackened vegetal litter and are not dominated by strongly reflective (in TM7) lithologic materials. Black surfaces are effective heat absorbers. It is a strange coincidence that the reflective spectral behaviour of charred vegetation should include anomalously high reflectance in the TM7 region.

Examples of false anomalies

Boulder trains on a number of small mountains in the vicinity of the Bailey prospect (site 11 below) are reddish on the AMSS 18,8,1(RGB) composite as they are on coincident ATM 10,7,3 composites (TM 7,4,2). The boulder trains are anomalously 'hot' on both AMSS and ATM thermal imagery. The same positive TM7/TIR correlation can be demonstrated at other test sites covered in this review. At Malheur City (site 8) an area of disturbed bare soil which was first interpreted as a particularly strong FeOx target on ATM 10,7,3 imagery is in fact no more, and no less, iron-stained than surrounding residual soils over pyritised bedrock. The anomaly shows positive TM7/TIR correlation. Genuinely 'false' TM7 anomalies elsewhere also show the same positive correlation with the thermal region. At Copper Creek (site 9) a false anomaly is due to subsoil debris in a landslip from the Eldorado water ditch (constructed in the last century to supply placer mines around Malheur City) and at Ironside (site 10) TM7 false anomalies are areas of bare soil near vehicle tracks.

Recognition of false FeOx anomalies

In every case the false anomalies could have been eliminated by examination of the thermal imagery (ATM 11) on which they are the warmest zones present. Field verification of these 'anomalies' has been costly in time and assay expenditures. The recognition of false anomalies is extremely important and interpreters of simple three band composites should be fully aware of the problem. To effectively eliminate them the imagery should be interactively processed for detection of iron oxide using interband comparison, for instance in the visible part of the spectrum (TM3>TM1).

Multispectral imagery - alteration mapping

The band ratio approach (for hydroxyl or iron-oxide detection) has been found to be totally unsuitable for field prospecting at most of the test sites studied in this review. Whilst the method works reasonably well over sites where rocks are well exposed and the 'argillic' materials have high albedo in the visible to mid infra-red, such as Copper Creek (site 9) and Portuguese Mountain (site 7), the information can just as reliably be extracted from simple three band composites (airborne TM) of these sites.

Attempts to 'correct' imagery (to 'improve' band ratios) for atmospheric, topographic and vegetation effects are inappropriate for two reasons. The methods do not equally compensate for atmospheric path radiance from sunlit and shadowed terrain; Great Basin climatic environments produce contrasting vegetation cover on north- and south-facing slopes (due to contrast in soil moisture content).

A number of 'predictive' techniques for alteration mapping have been described and summarized.[19] Early attempts to use logarithmic and least squares residual techniques[20] on the Oregon ATM data had disappointing results, occasionally with less success than band ratios. The residual images often highlight vegetation-dominated pixels as well as hydroxyls. Removal of vegetation effects requires baseline-type techniques. The application of residual, baseline techniques or, indeed, any of the predictive methods is expensive in computer time and inappropriate for geologists with PC-based image processors.

Pairwise Principal Components (pairs of only two images input for Principal Component Analysis[21]) have also been tried over multispectral sites in this study. The images give some indication of the location of known 'argillic' alteration or iron-staining, but are not (spatially) definitive enough to be useful in the field.

Principal Component Analysis and the "Crôsta technique"

Principal Component Analysis (PCA) is applied to raw data and, most important, does not require radiometric correction of aircraft data for off-nadir effects.

During processing of eleven band ATM imagery of the Malheur City prospect (site 8) it became clear that Principal Component Analysis (PCA) was best at uniquely discriminating known pyritised areas and mapping the full extent of pyritisation. PC images were compared with iron-oxide ratio images (e.g. ATM 5/2, for areas of known pyritisation) and found to be more specific than the latter. Subsequent PCA on progressively fewer, but selected, ATM bands produced even better results at Malheur City, due to reduction of 'redundant' bands, especially TM7

Principal Component Analysis over all multispectral (TM or ATM) sites studied in this review is equally successful. The reduction of input bands to suit FeOx or hydroxyl discrimination (e.g. TM 1,3,4 & 5 for FeOx, and TM 1 or 2 or 3,4,5 & 7 for hydroxyl) generates PC images which show hydroxyl and FeOx outcrops known at each site, and previously unsuspected mineralised outcrop.

The methodology works well on Landsat TM imagery as described below and illustrated on Figure 7.

'Feature Oriented Principal Components Selection' (FPCS) has been described by Crôsta and McM. Moore.[22] Their method of examining the eigenvector loadings (from each input band in relation to the output PC images) allows a prediction of which PC image contains the required information and it indicates whether the information is in negative form (darkest pixels) or positive form (brightest pixels). The PC image selection process can be automated, as will be clear from the examples below.

The methodology described here is a modified and developed version of the Crôsta and McM. Moore technique. The modification is that only pertinent bands are input for PCA ("selective" rather than "standard" PCA[21]). The development is that PCA is performed again on previously generated PC images. The methodology is here called the "Crôsta technique".

The Crôsta technique

The Landsat TM image of the Mount Tobin mercury district reproduced on Figure 7a has been processed to detect hydroxyl-bearing minerals by applying PCA to uncorrected data from TM bands 1,4,5 & 7. TM3 has been deliberately omitted to avoid mapping iron minerals. The following eigenvectors loadings are produced:-

	TM1	TM4	TM5	TM7
PC1	32%	30%	79%	43%
PC2	74%	47%	-48%	-01%
PC3	-51%	82%	03%	-26%
PC4	-31%	10%	-39%	86%

Table I. Eigenvector loadings for PCA on TM bands 1,4,5 & 7 of Mount Tobin subscene.

The PC image which discriminates those pixels dominated by hydroxyl-bearing minerals is that from Table I which has the greatest magnitude contribution, irrespective of its' sign (positive or negative) from the 'clay band', TM7. The contribution from TM5 should be opposite in sign. PC4 meets these conditions. As the loading due to hydroxyl absorption (which should reduce image brightness in the 'clay' band) from TM7 is positive (+86%) in PC4, that image shows hydroxyl-dominated pixels as the darkest in the scene; the PC4 image was consequently negated to produce the "Crôsta hydroxyl" (C.H.) image reproduced in Figure 7a. If the 'clay band' eigenvector loading had been negative the image in Figure 7a would not have required negation. The C.H. image contains less than 1% of the image information from TM bands 1,4,5 & 7 and is therefore extremely specific to the spectral differences between bands 7 and 5.

The same reasoning for the selection of the Crôsta hydroxyl PC image also allows a prediction that PC3 from Table I will show vegetation (highly reflective in TM4) as the brightest pixels in that image; if there is a desire

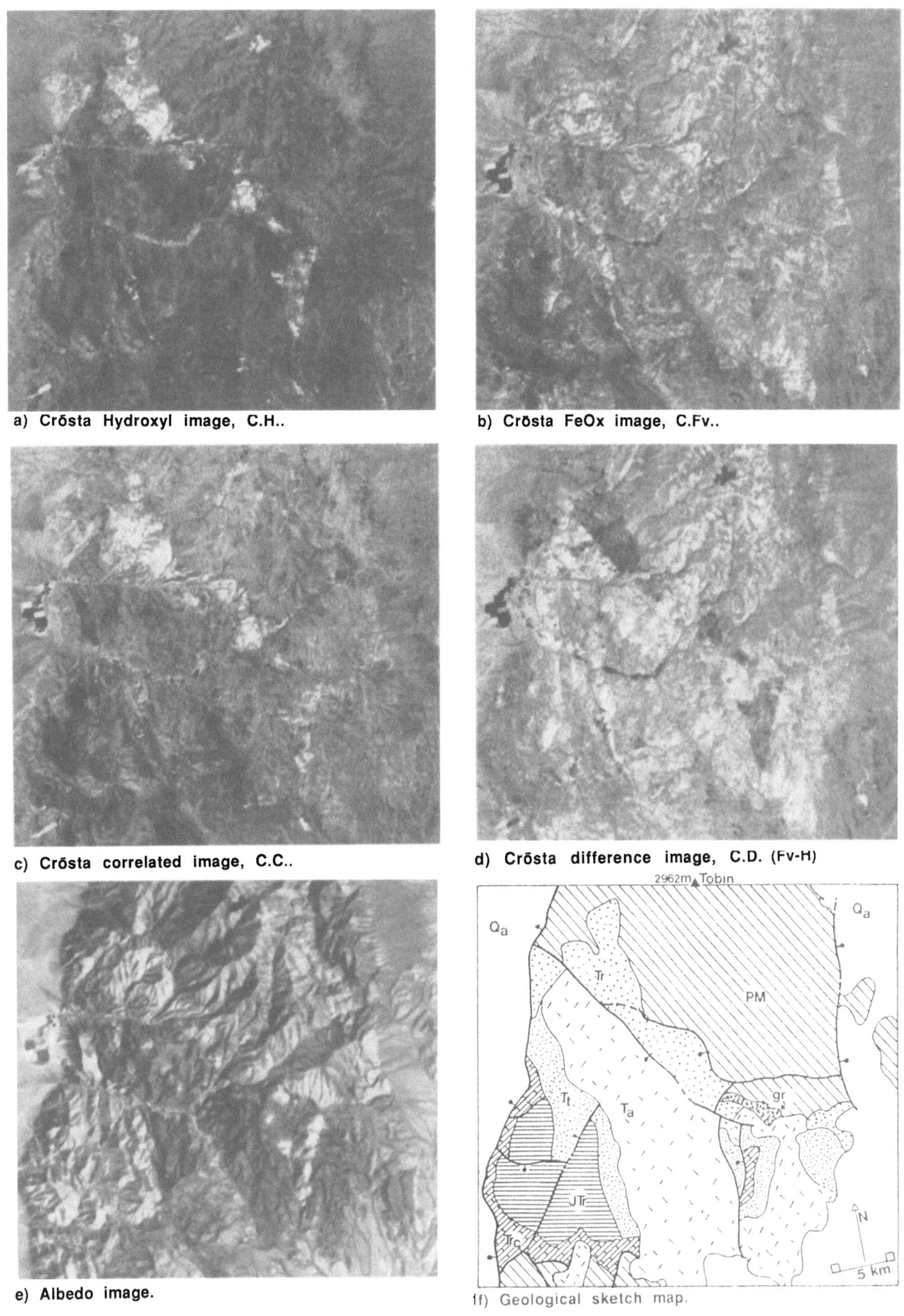

a) Crõsta Hydroxyl image, C.H..

b) Crõsta FeOx image, C.Fv..

c) Crõsta correlated image, C.C..

d) Crõsta difference image, C.D. (Fv-H)

e) Albedo image.

1f) Geological sketch map.

FIGURE 7: MOUNT TOBIN Hg DISTRICT. Landsat TM (p42/r31, 29 July,1984)

'remove' vegetated pixels from the Crósta hydroxyl image (to compensate for vegetation reflectance in TM5) then it is a straightforward procedure to subtract the Crósta vegetation image from the Crósta hydroxyl image. In practice none of the multispectral test sites required removal of vegetated pixels.

The Tobin mercury district is also covered by ATM imagery. An ATM Crósta anomaly (of around five or more 7.5m resolution pixels) can be generated for every anomalous TM pixel (the converse is obviously not true). Because of this, and judging from known field information it is clear that isolated anomalous TM pixels detected by the Crósta hydroxyl image are genuinely altered ; they are not due to 'noise', 'speckle', vegetation or any other spurious phenomena.

The iron-oxide detection image ("Crósta FeOx", C.Fv.) reproduced on Figure 7b was produced by applying PCA to raw data from TM 1,3,4 & 5. Iron oxide spectra have diagnostic 'shoulders' (strongly increased reflectance from shorter to longer wavelength bands) across TM 1&3 and across TM 4&5. In this case TM7 has been omitted to avoid mapping hydroxyls. Eigenvector loadings:-

	TM1	TM3	TM4	TM5
PC1	34%	39%	32%	79%
PC2	61%	47%	23%	-59%
PC3	-34%	-18%	91%	-13%
PC4	63%	-77%	10%	07%

Table II. Eigenvector loadings for PCA on TM bands 1,3,4 & 5 of Mount Tobin subscene.

By looking for the largest magnitude eigenvector loading from either of TM3 or TM5 (in both of which iron-oxide reflectance is at a maximum for each respective spectral region), and by checking that the contribution from TM1 or TM4 is opposite in sign (i.e. looking for positive versus negative eigenvector loadings from either of TM3 versus TM1 or TM5 versus TM4) PC4 was selected as being the PC image to distinguish FeOx dominated pixels; its loading of 77% from TM3 is greater than the 59% loading from TM5 in PC2. As the loading to PC4 from TM3 is negative (-77%) the PC4 image shows FeOx-dominated pixels as darkest pixels and has been negated for reproduction in Figure 7b. This C.Fv. image contains half of 1% of the image information contained in TM bands 1,3,4 & 5. It is therefore extremely specific to spectral differences between bands 1 and 3.

On other test sites the Crosta FeOx images are sometimes selected by the increase in iron-oxide reflectance across TM4 and TM5 ('C.Fm', i.e. FeOx from MIR). This is especially true where "mixed" pixels are present, for instance at Portuguese Mountain (site 7, Fig. 15). At this site hydroxyl-bearing jasperoid outcrops are also heavily iron-stained (and therefore more likely to be mineralised, which they are). The presence of both FeOx and hydroxyl has the effect of increasing TM5 reflectance much more than if only one of the two categories (FeOx or hydroxyl) was present. At Portuguese Mountain two FeOx images are produced by the Crósta technique. One relies on TM3/TM1 (C.Fv., FeOx from visible)) and indicates 'background' iron-staining (iron-rich Tertiary volcanic rocks), whilst the TM5/TM4 image (C.Fm.) detects the pyritised jasperoid outcrops. In actual fact the Crósta FeOx image for Portuguese Mountain is specific to iron-stained hydroxyl-bearing outcrop (mineralised jasperoid), and not merely to "iron-stained" surfaces in general. At a few test sites FeOx images are selected where both TM3>TM1 and TM5>TM4 are similarly opposite in sign, that is TM5 and TM3 are both positive (or negative) and TM4 and TM1 are both negative (or positive). These FeOx images are labelled 'C.Fv&m', since FeOx is selected due to high reflectance in both of the visible and mid-IR spectral regions.

It is clear from the images reproduced in Figures 7a and 7b that colour compositing is not necessary to discriminate mineralised zones within the Mount Tobin TM subscene. It is a simple procedure to map or classify increasing intensities of either hydroxyl or FeOx by, for instance, density slicing of the monochrome Crósta images. It should also be a relatively straightforward process to produce thematic maps (e.g. on a 'Versatec' plotter) with contoured intensities of hydroxyl-type alteration and iron-staining.

Colour composites from Crósta images

To produce colour composites which show both hydroxyl and FeOx it is best to perform a pairwise PCA with the Crósta hydroxyl image (e.g. C.H. image of Fig. 7a) and the Crósta FeOx image (C.Fv., Fig. 7b) as input. The following eigenvector loadings are produced:-

	C.Hydroxyl	C.FeOx
PC1	-70%	72%
PC2	72%	70%

Table III. Eigenvector loadings for pairwise PCA on Crósta Hydroxyl image and Crósta FeOx image. Mount Tobin TM subscene.

The images derived from the above pairwise PCA are reproduced in Figures 7c and 7d. As indicated by the eigenvector loadings PC2 (Fig. 7c, C.C. image, i.e. correlation between C.H. and C.Fv.) shows areas where both hydroxyl-absorption and FeOx are anomalous as bright pixels, and PC1 (Fig. 7d, C.D.(Fv.-H), FeOx bright & hydroxyls dark) depicts the two categories as darkest and brightest pixels respectively.

The Crósta hydroxyl image (C.H., Fig. 7a), the Crósta FeOx image (C.Fv., Fig. 7b) and the Pairwise correlation hydroxyl plus FeOx image (C.C. image of Fig. 7c) can now be input for colour compositing. All three of the components contain specific information directly related to 'argillic' alteration, iron-oxide staining or both. A composite of (for instance) common hydroxyl/FeOx in red, Crósta hydroxyl in green and Crósta FeOx in blue will distinguish "alteration" (FeOx and hydroxyl as brightest areas) and show hydroxyl absorption in 'reddish-yellowish-greenish' hues, and FeOx staining in 'bluish' hues. The photogeologist can then select areas of iron-staining within larger 'argillic' alteration zones, or suspected areas of genuine 'argillic' alteration near or within iron-stained areas. The actual ranking of associations (for field priority) will obviously depend on the perceived exploration model.

An albedo image (PC1 of PCA on TM bands 1,3,4, &5) is reproduced in Figure 7e as a key to the Crósta images. Formations on the geological sketch map (Fig. 7f) are as follows:- PM, Pennsylvanian-Mississippian sediments; Tr, Triassic andesitic and rhyolitic rocks; Trc, Triassic carbonate rocks; JTr, Triassic-Jurassic sediments; gr, granitic intrusive; Tt Tertiary tuffaceous rocks; Ta, Tertiary andesite flows; Qa, valley alluvium. Most of the mercury production was from the mines along the NW-SE trending structures between the Palaeozoic rocks and the younger sequences to the south (i.e, within the main hydroxyl zone of Figure 7a) with a few mines and prospects to the SW (i.e., near the small anomalies in the SW part of Fig. 7a).

Other applications of the Crósta technique

The Crósta technique is successful on raw data from airborne scanner data (as well as Landsat TM), and no correction is necessary for the extreme off-nadir effects which are characteristic of airborne scanners. The same method should therefore be useful for information extraction from future generations of off-nadir SPOT imagery. These applications need not be restricted to geologic studies.

TEST SITE CASE HISTORIES

Test sites chosen for this review

Many sites are covered by a variety of imagery, indicated thus:- Pre-dawn thermal (**TIR**), mid-day acquired ATM multispectral (**ATM mss**), Geoscan multispectral (**AMSS**) and Landsat TM (**TM**). Certain targets were either located by the application of particular image types, or the geological understanding of the sites was improved by the imagery. This is shown, where appropriate, by an asterisk(*) following the image type:- e.g., the Mulligan Gap outcrops (site 6) were first found on TIR imagery and Landsat TM images were consulted after the discovery.

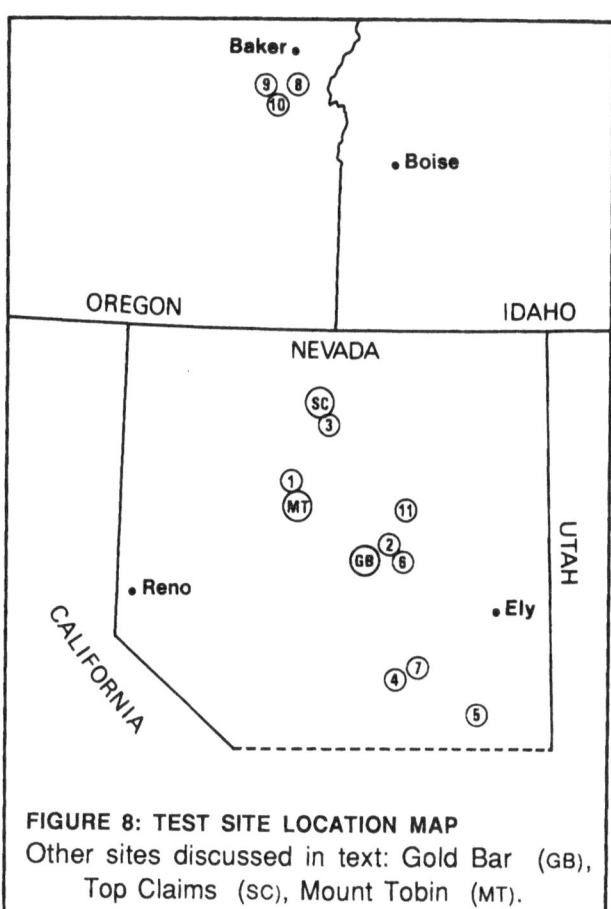

FIGURE 8: TEST SITE LOCATION MAP
Other sites discussed in text: Gold Bar (GB), Top Claims (SC), Mount Tobin (MT).

Colour air photography is also available for some sites. The resolution of the aircraft multispectral imagery, in Nevada, is approximately 7-8m ; the nominal resolution of pre-dawn TIR imagery is between 12.5-15m. Oregon sites are covered by ATM 11-channel data, 10m resolution, acquired mid-summer 1987.

Test site case histories - TIR imagery

Site 1 Goldbanks-Pollard Canyon. T30N, R37E (Township & Range grid reference). Pershing County, Nevada. An area to the west and northwest of the Mount Tobin range, TIR*, TM*. Figure 9.

A sinuous structure detected on pre-dawn TIR imagery. The structure crosses Grass Valley from the vicinity of Goldbanks Hills, northern Stillwater Range, to Pollard Canyon in the northern Tobin Range, and continues further north. The 'dogleg' shape of the structure may be important in that the southern inflection in the lineament trace occurs in the vicinity of the old Goldbanks vein gold district, and the reverse inflection, in the Tobin Range, is in an area where silicified and pyritised rhyolites are located. The rhyolites are apparently barren of precious metals, apart from one small outcrop of silicified fault breccia (Au to 140ppb) which is located directly on the TIR structure.

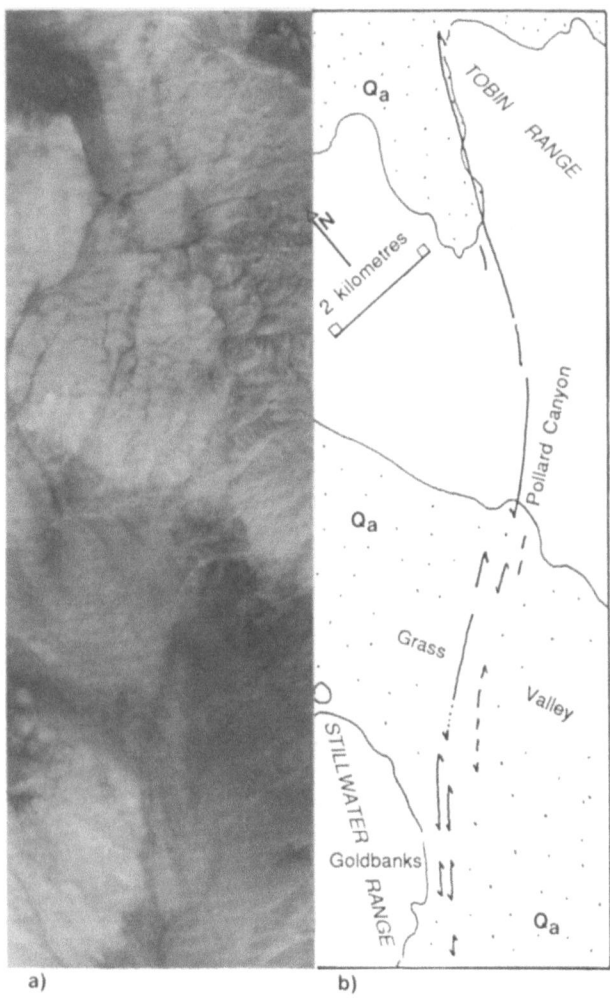

a)   b)

Figure 9: Goldbanks-Pollard Canyon

On enhanced Landsat TM imagery the southern part of the TIR structure crossing Grass Valley is not detected. The northern end of the structure, in the Tobin Range, is easily seen as the local range front fault.

Site 2. Garden Valley. T23N, R52E. Eureka County, Nevada. A pediment play originally located by soil and plant assays. TIR*, ATM mss*, TM. Figure 10.

The Garden Valley area was a property submittal to Billiton Minerals in 1989. It is located midway between the 'Battle Mountain - Eureka' and 'Carlin' Trends, Figure 4. The Gold Bar deposit (Fig. 5) is located some 27 kilometres to the southwest. The prospect ('Bigwun Claims') had previously produced anomalous pathfinder elements in assays from soil samples and apparently anomalous gold values in (ashed sagebrush) plant material. These geochemical and geobotanical anomalies are located on two adjacent topographical highs, in a pediment area of Garden Valley.

Examination of the pre-dawn acquired TIR images indicates that the anomalous areas coincide with areas of probable thin soil cover over bedrock, similar to the hypothetical occurrence illustrated in Figure 6f. The TIR

also provided specific structural information (Fig. 10b) with which to target a drilling programme. A classification of the TIR structures can be inferred, according to probable age and affinity; oldest structures trending NW-SE, parallel to both the 'Carlin Trend' and the 'Battle Mountain - Eureka Trend' ; next youngest, the NE-SW trending structures, with possible affinities to cross faults which have had some control over localisation of high grade ore at Gold Bar (see Fig. 5b above); the youngest structures, trending N-S, are related to the local range-forming faults.

The topographic highs have high albedo on both TM and ATM mss three band composites, though with no cyan hues indicative of 'argillic' alteration. Interactive processing seems to confirm that hydroxyl-bearing minerals are not present at the site. Billiton have now completed exploratory drilling (early 1990) and have decided to undertake no further work.

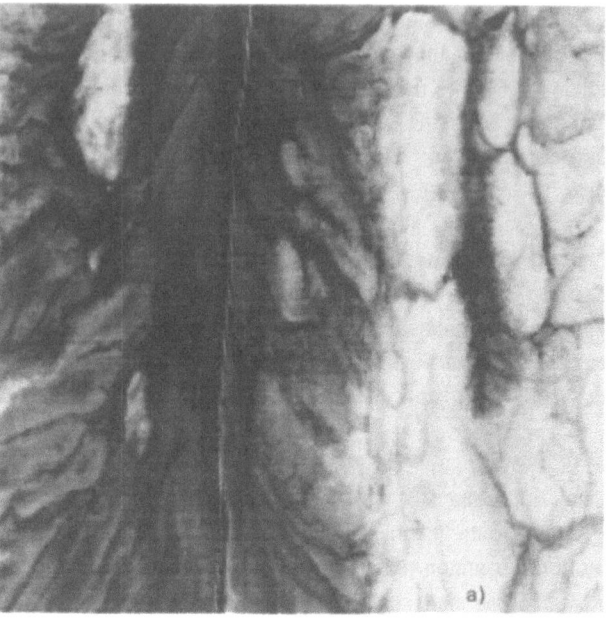

a)

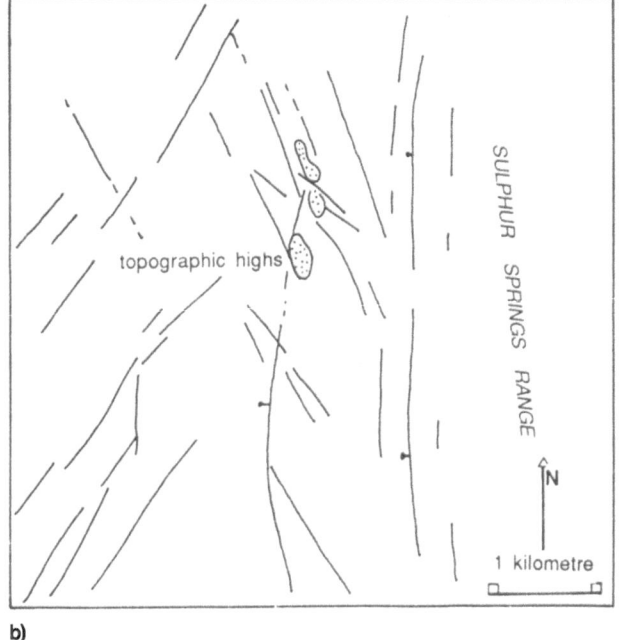

b)

FIGURE 10: GARDEN VALLEY

234

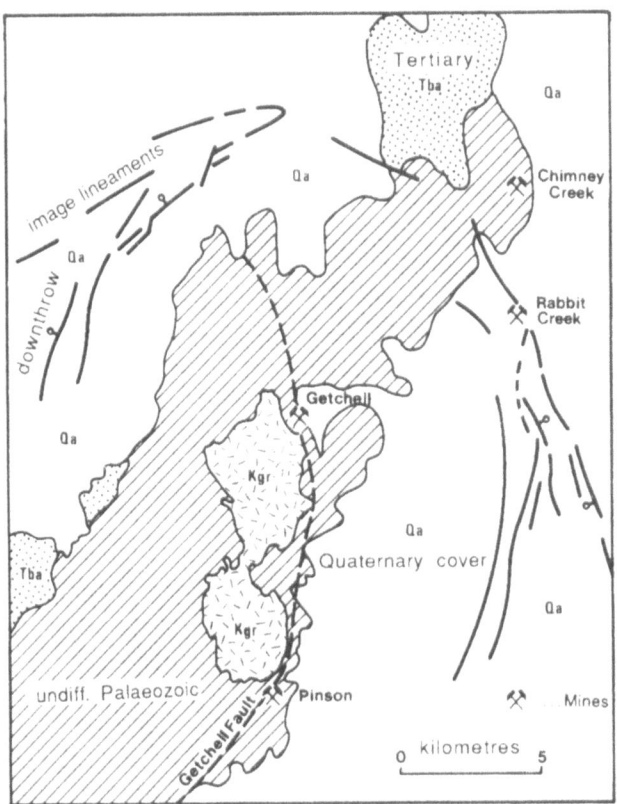

FIGURE 11: OSGOOD MOUNTAINS

Site 3. Osgood Mountains. Humboldt County, Nevada.
The area around the Getchell, Chimney Creek and Rabbit
Creek deposits. Pediment structure from TIR*, ATM mss,
TM. Figure 11.

Pre-dawn TIR images are produced (by Hunting
Technical Services Ltd. and The Robertson Group) as
monochromatic hardcopy image sheets with adjacent swaths
laid up in appropriate geographic position. These images can
be examined synoptically; the salient structural features
(such as those on Figs. 9 and 10 above) identified and
followed across adjacent image swaths for compilation onto
topographic maps.

The map reproduced on figure 11 is a compilation from
the Osgood Mountains survey.[15] A possible explanation for the
curvilinear structures which pass near the Chimney Creek
orebody and through the area of the Rabbit Creek orebody is
that the structures represent the surface trace of fractures
developed around the subsurface margins of the Osgood
pluton, 'Kgr.'. It is known that the granite has an extensive
thermal aureole in a northerly direction away from its'
exposed contact with country rocks,[23] and it is therefore
possible that the presence of a much larger pluton at depth,
and its relative crustal buoyancy, has resulted in fracturing
around the pluton margin during recent tectonism.

Site 4. Heart Hills. T9N, R52E, Nye County, Nevada.
Silicified fault in Tertiary volcanic rocks. TIR*. Figure 12.
This site was located by photogeological interpretation.

As can be seen from Figure 12 the interpretation depends on
three factors:- 1 - recognition of anomalously warm zones
in a pediment setting (cf. Fig. 6e), 2 - alignment along the
extension of a mapped fault in Tertiary tuffaceous rocks,
and 3 - the recognition of 'ponding' of subsurface moisture
on the upthrown side of the fault (cf. Fig. 6d). The zone is
indeed a silicified fault which has been overlooked during
the latest gold rush (no sample flags present). The
jasperoids are barren of both gold and toxic pathfinders and
are of no further interest.

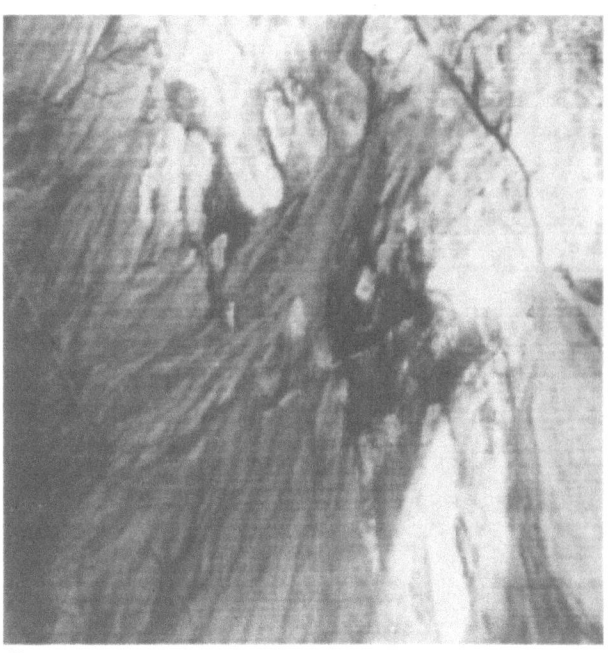

a)

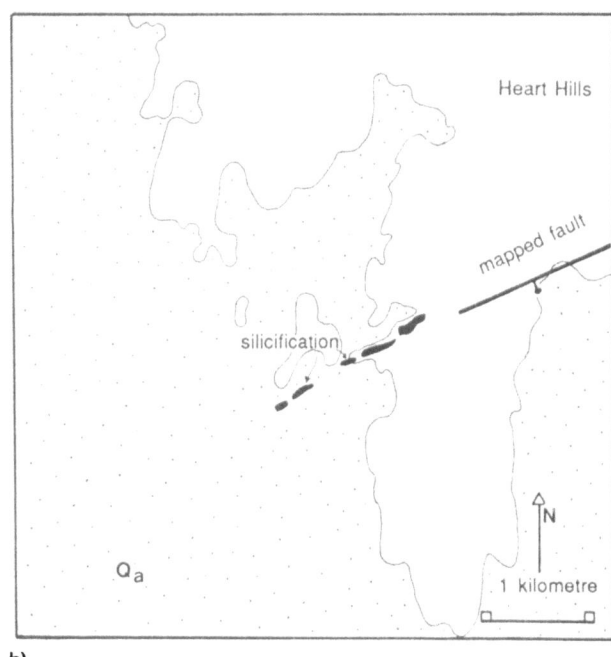

b)

FIGURE 12: HEART HILLS

a

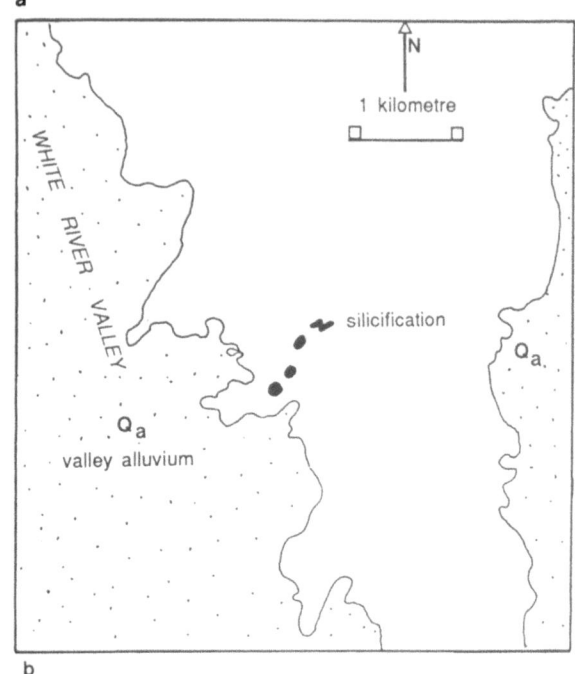

b

FIGURE 13: ADAMS-McGILL

Site 5. Adams-McGill. T5N, R61E, Nye County, Nevada. Silicified volcanic rocks, pediment setting. TIR*. Figure 13.

This site is an example of silicification detection on pre-dawn TIR imagery through the retention of heat by the silicified rocks in contrast to the lower temperatures of enclosing rocks, in this case ash-fall tuffs. The locality is on undulating pediment, underlain by tuffs, with very thin, or absent, residual soil cover. The anomalous heat from the target outcrops is therefore entirely due to the thermal contrast between silicified and unsilicified rock, and not due to dry rock contrasting with surrounding 'moist' soils, which could be the case for Mulligan Gap (site 6, Fig. 14) and is definitely the cause of the 'warm' tones over the Garden Valley topographic highs (site 2, Fig. 10).

Site 6. Mulligan Gap   T22N, R52E, Eureka County, Nevada. Silicified Palaeozoic carbonates, in a pediment setting. TIR*, TM, (ATM mss not consulted). Figure 14.

The Mulligan Gap outcrops were located by field checking pre-dawn TIR anomalies. They lie on a previously unrecognised SW-NE trending cross-structure, and in a pediment area where outcrops are not recorded on the published geological map of Eureka County[24]. The outcrops are very subtle features, limited in aerial extent, and rise only one or two metres above the local terrain. They consist of partly to extensively silicified Palaeozoic carbonate rocks.

Landsat TM imagery has been consulted for this site. On simple three-band colour composites (e.g. TM 7,5,1 & TM

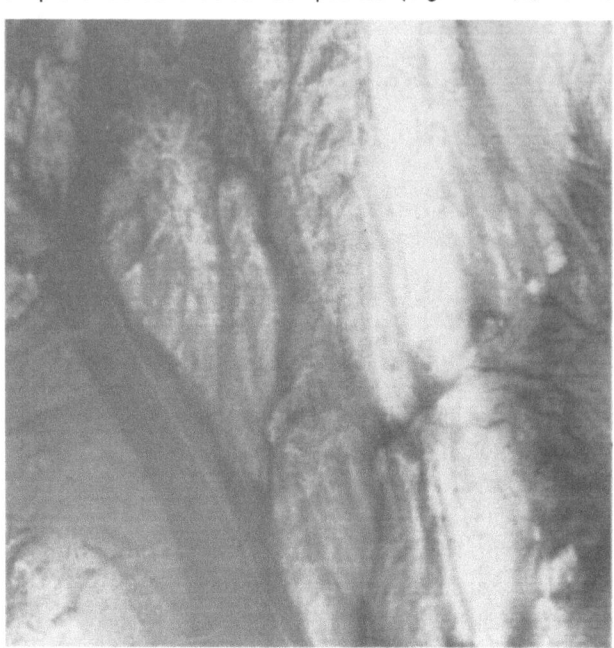

a

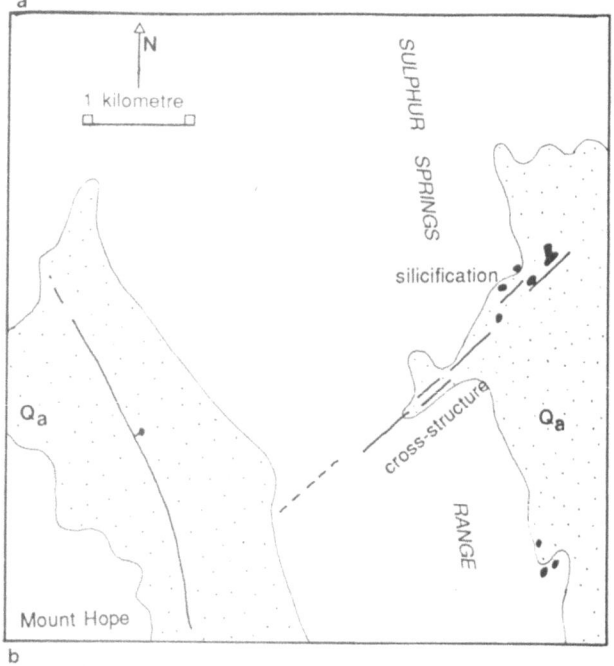

b

FIGURE 14: MULLIGAN GAP

236

7,4,1) the outcrops appear as bright zones indistinguishable from areas of similar appearance which are due to fresh exposures of bare soil on nearby piedmont gravel fans; they would not be recognised as anomalies from the colour composites alone. By applying the Cròsta technique a strong (hydroxyl) anomaly appears on the TM imagery on outcrop some distance from the TIR anomalies but not on the silicified outcrops themselves. It must be presumed that hydroxyl-bearing minerals are not present in the silicified outcrops at all, or, if they are present, they are in insufficient amounts within each 30metre TM pixel to cause a detectable absorption effect in TM7.

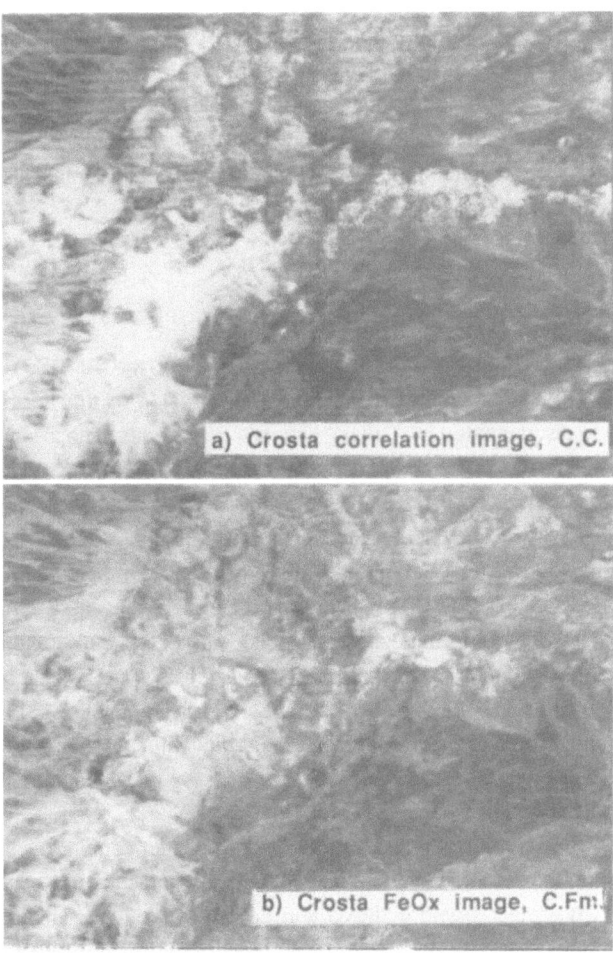

a) Crosta correlation image, C.C.

b) Crosta FeOx image, C.Fm.

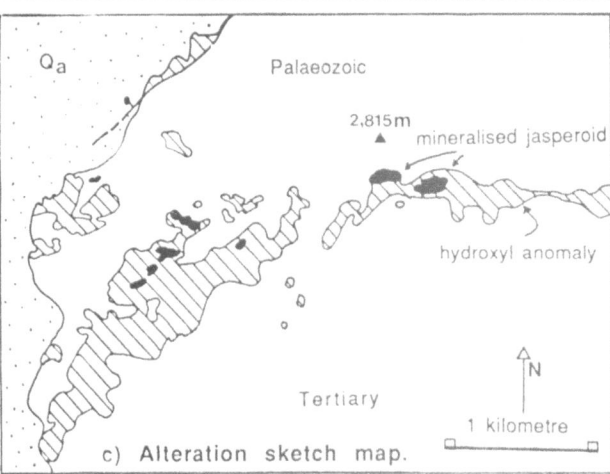

c) Alteration sketch map.

FIGURE 15: PORTUGUESE MOUNTAIN.

Test site case histories - multispectral imagery

Site 7. Portuguese Mountain T10N, R54E, Nye County, Nevada.

Classic iron-stained jasperoid, located on a fault, surrounded by a halo of bleached and altered Palaeozoic carbonates. ATM* mss, TIR. Figure 15.

The Portuguese mountain anomalies were first detected during image processing of the 1986 Pancake Range ATM survey. Hardcopy imagery (ATM 10,7,3 in RGB) offered by Hunting detects iron-oxide stained outcrops on an east-west trending fault structure within a larger halo of 'argillic' alteration. Field-checking of the anomaly (in 1989) revealed that the iron-staining is on pyritised jasperoid outcrop, and the 'argillically' altered zone is heavily bleached palaeozoic limestone (shown on the Nye County geological map[25] as Tertiary ash-flow tuffs, which are often very pale in colour). The jasperoid is mineralised, with pathfinder elements well above background levels. Outcrops of jasperoid in the vicinity carry gold to 140ppb.

The locality is an interesting one in that the pyritised jasperoids give mixed pixels in the TM7 region, with hydroxyls tending to reduce DN values and iron oxides tending to increase them. The jasperoids consequently produce much stronger FeOx anomalies on ATM 9,7,3 composites. The methodology used for the production of the Cròsta imagery (Fig. 15) is discussed in the description of the Cròsta technique above.

Site 8. Malheur City Prospect T13S, R41E, Malheur County, Oregon. ATM mss*. Potential disseminated gold deposit. A pyritised porphyritic intrusive and its envelope of (?)Jurassic or Triassic age volcano-sedimentary country rocks are host to mineralisation. Not illustrated.

This prospect is presently farmed out to Billiton Minerals for detailed exploration. The prospect is unusual in that:- there is no significant development of 'argillic' alteration, the mineralisation is unaccompanied by silicification, the host is an albitised (G.J. Breen, *pers. comm.*) granitoid intrusive with widely disseminated gold (to around 0.3 oz/t), and, in complete contrast, the adjoining property (to the east) has probable syngenetic, volcanic-exhalitive gold mineralisation intimately and exclusively associated with narrow sulphide lenses and disseminations within a thin unit of felsic volcaniclastic rocks. Mafic and ultramafic rocks north and east of the mineralised area mark a probable crustal suture. The gold mineralisation produced the placer deposits mined from local creeks by the original settlers of Malheur County.

Bedrock is masked by residual soils. These give good 'FeOx' signatures on ratio images and, especially, on PCA images selected by the Cròsta technique. The Cròsta FeOx images permit confident mapping of the full extent of surficial expression of pyritised bedrock to guide exploration. PCA of all eleven ATM bands (including the TIR) provides useful structural information.

Site 9. Copper Creek T13S, R38E, Baker County, Oregon. Large SE Oregon fossil hot spring system. Altered Mesozoic sedimentary and ultramafic rocks, and Tertiary volcanic rocks. ATM mss[*]. Not illustrated.

This fossil hot spring system is typical of many in the SE Oregon-NW Nevada region in that it has many features of gold-bearing systems but is apparently barren of that metal. The anomaly was found during field-checking of the Oregon ATM mss survey. It has been used to test image processing techniques. A 'false' TM7 anomaly has been discussed above.

Site 10. Ironside Anomaly T14S, R39E, Malheur County, Oregon. ATM mss[*].

SE Oregon fossil hot spring. Altered Mesozoic and Tertiary host rocks. Not illustrated.

Unlike the Copper Creek site, which Ironside resembles in extent and style of alteration, this anomaly is not easily detectable other than on the multispectral imagery. The site has been used for trials of image processing techniques; false (TM7) anomalies have already been discussed. Very little prospecting has been carried out on the anomaly. Five grab samples of jasperoid and siliceous sinter are barren of precious metals.

Site 11. Bailey Prospect T27N, R53E, Elko County, Nevada. In southern part of Sulphur Springs Range (Pinon Range). Potential limestone-hosted gold deposit. Located, sampled, and drilled prior to image acquisition or processing. ATM mss[*], AMSS[*], TIR (not studied in detail), TM[*]. Figure 16.

The prospect was originally located by Atlas Precious Metals Inc.. Samples from jasperoid outcrops gave assays to 385ppb gold. A drilling program (eleven reverse circulation holes, deepest hole 200feet) failed to upgrade Atlas' target concepts (Fig. 16c).

The property was subsequently acquired by Breen Minerals Inc.. Breen Minerals' model postulates a narrower target zone located along a N-S or NNW-SSE trending structure. This is based on:- the linear nature of the Palaeozoic-Tertiary contact (i.e. possibly fault controlled) and the obvious increase in stockwork intensity. Although the Breen model would increase Tertiary overburden in the prospective eastern portion of the area, the location of higher grade ore on a feeder structure may compensate for this, especially in light of recent deep discoveries beneath some of the Carlin Trend orebodies. The Breen model is shown on Figure 16d.

Geoscan AMSS mkII imagery, Bailey Prospect

Geoscan AMSS mkII imagery was flown in an attempt to locate the postulated fault. The Geoscan hardcopy imagery (provided by Breen Minerals for this study) is presented as four different composites on one image sheet. These are:- 'true colour' (bands 3,2,1 in RGB); 'geological

discrimination' (18,8,1); 'Al clays' (11-14, 11-15, 11-16 band differences in the hydroxyl region) and 'silicates' (23-20, 22-20, 21-20 band differences in the TIR region. Band passes are given on figure 3, above.

Of these composites the 'Al clays' image distinguishes known jasperoid uniquely and unequivocally, and the image gives a precise indication of a possible extension of the postulated feeder fault, almost exactly as shown on Figure 16b (zone 3) as well as a possible parallel structure to the east (zone 2, Fig. 16b). It does not differentiate different classes of "hydroxyl"; the anomalies are white targets indicating equal intensities in all three band subtractions. The other images are of little use for mineral exploration. The 18,8,1 image suffers from topographic shadowing, probably due to atmospheric correction considering the time of image acquisition (June 6, mid-day). The TIR band difference image is totally dominated by surface heat effects with no discrimination of silicate and carbonate minerals, as mapped in the field (Fig. 16a).

ATM imagery, Bailey Prospect.

The mineralised jasperoids detected by the AMSS 'Al clays' image, and the possible fault extension to the north, are discriminated by Crŏsta hydroxyl PC images. These can be generated from as few as four (e.g. ATM 2or3or5,7,9,10) or from all eleven ATM bands, provided that ATM 9 and 10 are used. The Crŏsta hydroxyl zones are shown on the sketch map, Figure 16b. The correlation between the ATM Crŏsta hydroxyl image(s) and the AMSS 'Al clays' image is almost exactly pixel for pixel, due to the close match in resolution between the two images (7-7.5m). ATM three band composites do not show the mineralised zone extension, although the ATM 10,9,3 (RGB) image has a weak cyan hue over the known jasperoid outcrop, Figure 16b, zone 1.

Crŏsta FeOx images for the Bailey prospect compare well with the location of mineralised grab samples; a pairwise PCA on the Crosta hydroxyl and FeOx images (C.C. image, Fig. 17) reinforces the selection of the mineralised jasperoid and possible extensions (zones 1,2 &3 of Fig. 16b). The C.C. image highlights the mineralised outcrop as the brightest areas on the scene.

Landsat TM imagery, Bailey Prospect

On the TM image of the prospect the main jasperoid zone is detected by Crŏsta PCA analysis. The northern extension of the structure, as shown on Figure 16b, zone 3, is not detected. This is a direct function of an approximately sixteen-fold difference in resolution between the airborne and satellite images.

The jasperoid zone had been located by traditional ground prospecting prior to this remote sensing investigation. It could have been detected on landsat TM. The spatial

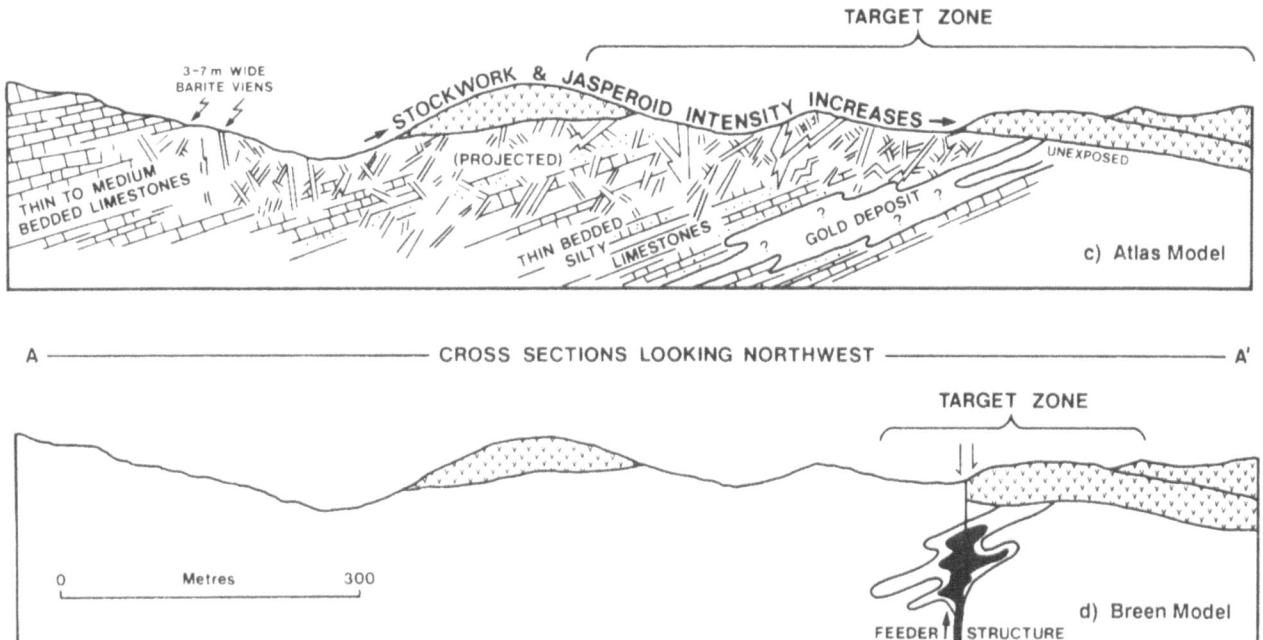

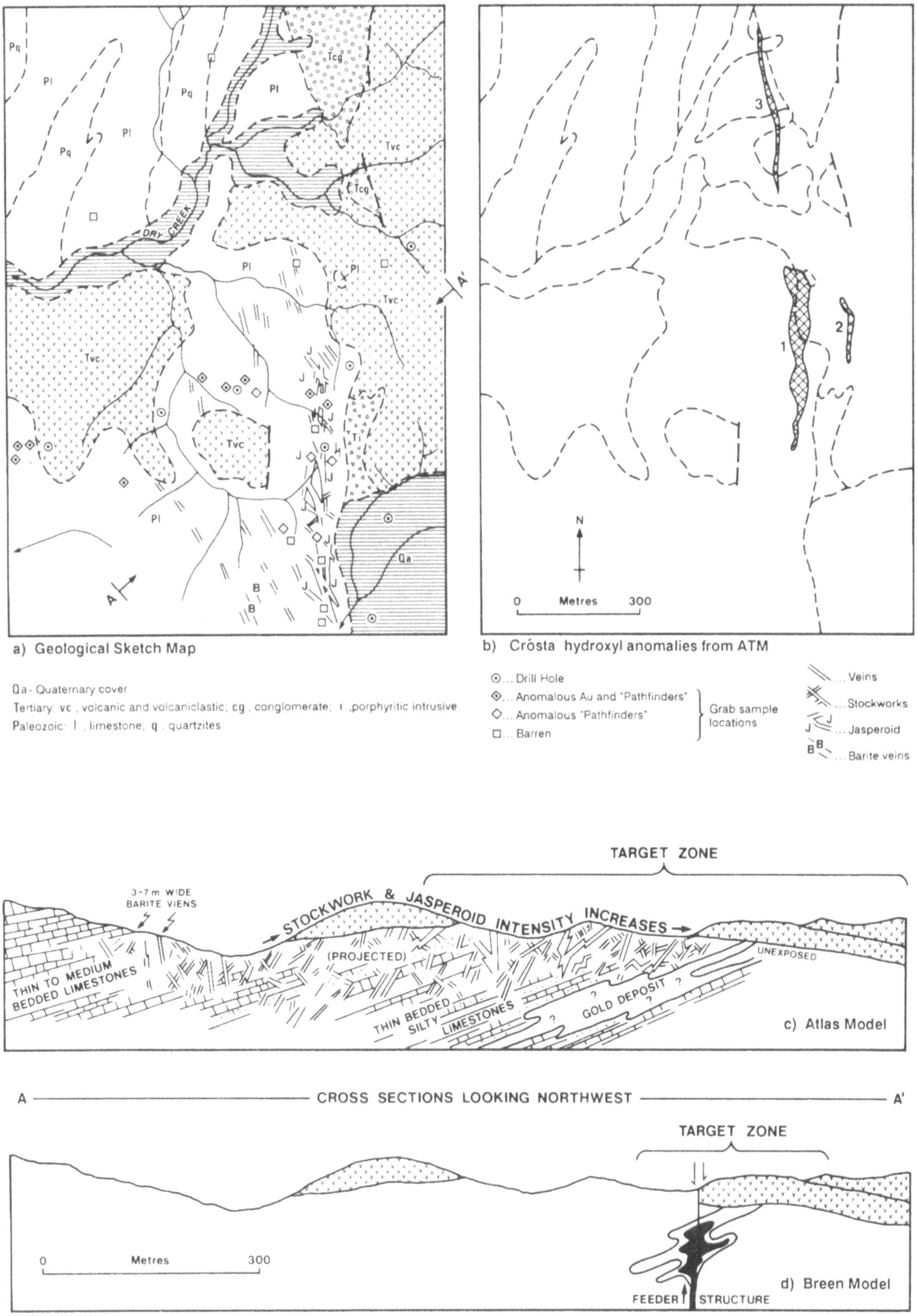

a) Geological Sketch Map

Qa - Quaternary cover
Tertiary: vc , volcanic and volcaniclastic; cg , conglomerate; ı ,porphyritic intrusive
Paleozoic: l , limestone; q , quartzites

b) Cròsta hydroxyl anomalies from ATM

⊙... Drill Hole
◈... Anomalous Au and "Pathfinders"   } Grab sample
◇... Anomalous "Pathfinders"   } locations
☐... Barren

╲╲╲ ... Veins
⤬ ...Stockworks
⧸⧹ ...Jasperoid
B╲ᴮ ... Barite veins

TARGET ZONE

3-7 m WIDE BARITE VIENS

STOCKWORK & JASPEROID INTENSITY INCREASES →

THIN TO MEDIUM BEDDED LIMESTONES

(PROJECTED)

THIN BEDDED SILTY LIMESTONES

? GOLD DEPOSIT ?

UNEXPOSED

c) Atlas Model

A ———— CROSS SECTIONS LOOKING NORTHWEST ———— A'

TARGET ZONE

0   Metres   300

FEEDER STRUCTURE

d) Breen Model

FIGURE 16: THE BAILEY PROSPECT, SULPHUR SPRINGS (PIÑON) RANGE, ELKO COUNTY, NEVADA.

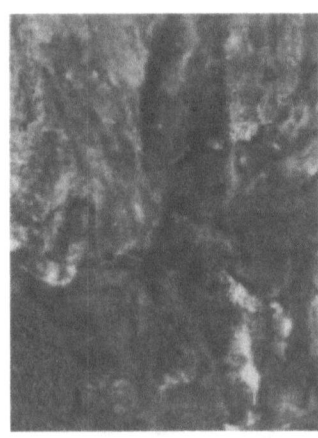

Crosta Correlation image (C.C.), H+Fm. Both Hydroxyl & FeOx are bright. Compare image with Fig. 16b.

FIGURE 17: BAILEY PROSPECT (ATM)

resolution of the airborne scanners has been crucial for testing and apparent confirmation of the Breen model at this prospect; Landsat TM would not have helped justify the Breen model.

Landsat TM and ATM comparison at other sites in Nevada

The image interrogation methodologies developed for some of the above sites were then tried on imagery (ATM mss and oldest available Landsat TM) over a few selected, and known, sites elsewhere in Nevada (Fig. 8); the Mount Tobin Mercury district (illustrated in Fig. 7 above), the original 'Top Claims' 600 metres north of the Chimney Creek orebody, and the Gold Bar and Gold Bar extensions near and within the Roberts Mountains. All of these sites are 'anomalous' on both the airborne and satellite images (processed using the Crōsta technique), though subtle features and extensions are only detectable on the airborne imagery, in a similar manner to those at the Bailey prospect. For instance, the jasperoids at the 'Top Claims' are detectable on both (1986) airborne and (pre-discovery, July 1984 acquisition) Landsat TM; whilst the small outcrops above the Chimney Creek orebody (on the same structure further south) are only anomalous on the airborne imagery, and would not have been detected directly from the satellite TM data.

It seems to be a general rule that airborne anomalies encompassing some 5-12 pixels can often be detected by one or two satellite TM pixels, depending on the intensity of the alteration. Deep shadow on satellite images, especially on spring or autumn scenes, is an obvious problem.

CONCLUSIONS

Thermal imagery acquired pre-dawn.

Pre-dawn acquired thermal infrared images are valuable for structural mapping, silicification detection and geomorphic interpretation. They are particularly applicable over low hill ranges, and the pediment and valley margin areas of the Great Basin region.

Alteration mapping.

The Crōsta technique for alteration mapping using

Principal Component Analysis, as developed and modified during this study, is equally applicable to both airborne and satellite TM imagery. The technique doesn't require radiometric correction, is unaffected by albedo and most topographic effects, and can be carried out on low-cost PC-based image processing systems.The method is not prone to 'false' anomalies.

Future scanner surveys for mineral exploration

The comparison between eleven-channel ATM imagery and the (24-channel) Geoscan AMSS mkII hardcopy imagery is interesting in that only five bands (2or3,5,7,9,10 of the ATM) are required to map both hydroxyl-bearing minerals (in identical manner to the AMSS mkII 'Al. clays' band subtraction image) as well as iron-stained outcrop. It should be possible to re-process the Geoscan imagery, probably by the Crōsta technique, to subdivide the hydroxyl region. The lack of a "TM5" band pass on the AMSS mk.II (as flown in the USA in 1989) may be a disadvantage for mineralised jasperoid detection, especially in view of the results from the Portuguese Mountain test site. The Geoscan multi-spectral TIR channels will need further evaluation (different processing, night surveys) before their utility can be fully evaluated for mineral exploration.

The spatial resolution of both airborne scanners is a distinct and important advantage over Landsat TM, especially in a heavily prospected environment, and this is very well demonstrated at the Bailey prospect and in the vicinity of the 'Top Claims' and Chimney Creek. Future scanners should include dedicated 'geological' channels (say around 14-20, depending on whether multispectral TIR is considered necessary) but with <u>increased spatial resolution</u>.

Satellite TM images can be corrected for geometric fidelity to extract a specific grid reference for individual anomalous pixels. This should be useful for rapid reconnaissance field checking in areas where airborne imagery doesn't exist or is unavailable.

An 'ideal' remote sensing campaign should start with Landsat TM image interpretation to locate alteration trends in favourable geological environments. The TM interpretation would then be used for the cost-effective selection of areas and specifications (especially of spatial resolution) for flying airborne surveys.

Image processing systems are essential for access to the imagery at all stages, and should be under the control of the exploration geologists. Hardcopy imagery should be treated as a mere compilation base for plotting results of interactive processing and interpretation.

ACKNOWLEDGEMENTS

This study would not have been possible without the co-operation and assistance of the NRSC and, in alphabetical order, Billiton Minerals, Breen Minerals, Global Earth Sciences, Hunting Technical Services, Sergio Pastor, Rio Tinto Zinc and The Robertson Group. The help of all the geologists who collected field data and those in the western USA whose comments contributed to the work (knowingly or otherwise) is also gratefully acknowledged. G.H. Griffiths edited the text. A. Thwaites & P. Giles drew the Figures.

References

1.   Bailey, J.D. The Australian Landsat TM Story: a Commercial Success. Proceedings of the 7th (ERIM) Thematic Conference: Remote Sensing for Exploration Geology, Calgary, Oct. 2-6, 1989. p.1359-1366.

2.   Lyon, R.J.P., and Honey, F.R. Relating Ground Mineralogy via Spectral Signatures to 18-Channel Airborne Imagery Obtained with the Geoscan mkII Advanced Scanner: a 1989 Case History from the Leonora, Western Australia, Gold District. Proceedings of the 7th (ERIM) Thematic Conference: Remote Sensing for Exploration Geology, Alberta, Canada, Oct. 2-6,vol.I 1989. p.331-348

3.   Coopersmith, H.G. A Geoscan AMSS Discovery at Sentinel, Humboldt County, Nevada. To be presented at: Society for Mining, Metallurgy and Exploration GOLDTech 4. Reno, Nevada. September 10-12, 1990.

4.   Taranik, D.L., and Kruse, F.A. Iron Mineral Reflectance in Geophysical and Environmental Research Imaging Spectrometer (GERIS) Data. Proceedings of the 7th (ERIM) Thematic Conference: Remote Sensing for Exploration Geology, Alberta, Canada, Oct. 2-6, 1989. p.445-458.

5.   Kierein-Young, K.S., and Kruse, F.A. Comparison of Landsat Thematic Mapper Images and Geophysical and Environmental Research Imaging Spectrometer Data for Alteration Mapping. Proceedings of the 7th (ERIM) Thematic Conference: Remote Sensing for Exploration Geology, Alberta, Canada, Oct. 2-6, 1989. p.349-359

6.   Akiyama, A., Komai, J., Yokoyama, T., and Okada, K. Digital Processing and Analysis of Airborne Multispectral Data for Mapping Hydrothermal Alteration at Yerington, Nevada. Proceedings of the 7th (ERIM) Thematic Conference: Remote Sensing for Exploration Geology, Alberta, Canada, Oct. 2-6, 1989. p.969-980.

7.   Stewart, J.H. Geology of Nevada: a discussion to accompany the Geologic Map of Nevada. Nevada Bureau of Mines & Geology. Special Publication 4, 1980. 136pp.

8.   Brioli, C., McN. French, G., Shaddrick, D.R. and Weaver, R.R. Geology and Gold Mineralization of the Gold Bar Deposit, Eureka County, Nevada. In: Bulk Minable Precious Metal Deposits of the Western United States. (ed. Schafer, R.W. et. al.) 1988. p.51-72.

9.   Hunt, G.R. and Salisbury, J.W. Visible and Near-Infrared Spectra of Minerals and Rocks: I.Silicate Minerals. Modern Geology, vol.1,1970. p.283-300.

10.   Hunt, G.R., Salisbury, J.W., and Lenhoff, C.J. Visible and Near-Infrared Spectra of Minerals and Rocks: III. Oxides and Hydroxides. Modern Geology, vol.2, 1971. p.195-205.

11.   Krohn M.D. Spectral Properties (0.4 to 25 microns) of selected rocks associated with Disseminated Gold and Silver Deposits in Nevada and Idaho. U.S. Geol. Surv. Open-File Report,   85-576,1985. 37pp.

12.   Sabins, F.F. Remote Sensing: Principles and Interpretation. W.H. Freeman, New York. 1978. 426pp.

13.   Kahle, A.B. Surface thermal properties. In: Remote Sensing in Geology. Siegal, B.S., and Gillespie, A.R. (editors). Wiley, New York. 1980. p.257-273.

14.   Drury, S.A. Image Interpretation in Geology. Allen and Unwin, London. 1987. 243pp.

15.   Hunting Geology and Geophysics Ltd. A Non-Exclusive Study of Night-Time ATM Imagery of the Osgood Mountains Area, Nevada. Unpublished Report, 1986. 16pp.

16.   Sabins, F.F. Remote Sensing: Principles and Interpretation. 2nd Edition.W.H. Freeman, New York. 1986.

17.   Legg, C.A. Applications of the Airborne Thematic Mapper. 1. Alteration Zones at Jabal Sa'id. Directorate General Mineral Resources (Saudi Arabia) Open-File Report DGMR-OR-05-47,. 1984. 5pp., 4 Figs.

18.   Rothery, D.A. A Re-Interpretation of Landsat TM Data on Chernobyl. International. Journal of Remote Sensing, vol.10, number 8, 1989. p.1423-1427.

19.   Lamb, A.D. and Pendock, N.E. Band Prediction Techniques for the Mapping of Hydrothermal Alteration. Proceedings of the 7th (ERIM) Thematic Conference: Remote Sensing for Exploration Geology, Calgary, Oct. 2-6, 1989. p.1317-1329.

20.   Fraser, S.J., Gabell, A.R., Green, A.A. and Huntington J.F. Targeting Epithermal Alteration and Gossans in Weathered and Vegetated Terrains using Aircraft Scanners: Successful Australian Case Histories. Proceedings of the 5th (ERIM) Thematic Conference: Remote Sensing for Exploration Geology, Reno, Nevada, September 29th-October 2, 1986. p.63-84.

21.   Chavez, P.S. and Kwarteng, A.Y. Extracting Spectral Contrast in Landsat Thematic Mapper Image Data using Selective PC Analysis. Photogrammetric Engineering and Remote Sensing, Vol. 55,No.3, 1989. p.339-348.

22.   Crôsta, A.P. and McM.Moore, J. Enhancement of Landsat Thematic Mapper Imagery for Residual Soil Mapping in SW Minais Gerais State, Brazil: A Prospecting Case History in Greenstone Belt Terrain. Proceedings of the 7th (ERIM) Thematic Conference: Remote Sensing for Exploration Geology, Calgary, Oct. 2-6, 1989. p.1173-1187.

23.   Hotz, P.E. and Willden, R. Geology and Mineral Resources of the Osgood Mountains Quadrangle, Humboldt County, Nevada. USGS Professional Paper 431, 1964. 132pp.

24.   Roberts, R.J., Montgomery, K.M. and Lehner, R.E. Geology and Mineral Resources of Eureka County, Nevada. Nevada Bureau of Mines Bulletin 64. 1967. 152pp.

25.   Kleinhampl, F.J., and Ziony, J.I. Geology of northern Nye County, Nevada. Nevada Bureau of Mines and Geology. Map to accompany Bulletin 99A. 1985.

Gold exploration in greenstone belts by use of Landsat TM

N. Pendock B.Sc., B.Sc.(Hons.), Ph.D.
Department of Computational and Applied Mathematics, University of Witwatersrand, Johannesburg, South Africa

INTRODUCTION

So geographers, in afric maps,
With savage pictures fill their gaps,
And o'er uninhabitable downs
Place elephants for want of towns.
J. Swift

SYNOPSIS

The ability of the Landsat Thematic Mapper (TM) to measure mineral spectra in the 2.2 μm region of the electromagnetic spectrum has enabled the remote detection and mapping of rock and soil 'clay' signatures. This has resulted in improved lithological discrimination for geological mapping and has also had direct economic implications for the detection of hydrothermal alteration haloes associated with exposed or near surface porphyry or epithermal ore deposits.

One application for the satellite detection of hydrothermally altered rocks is gold exploration in Archean greenstone belts. Current geological thinking is that gold was deposited in the greenstone belts from hydrothermal solutions. The wall rocks of the mineralized veins are typically altered by the hydrothermal fluids and the mineral products of this alteration and subsequent weathering products occur in the soil profiles above the gold deposits.

The TM satellite collects data in a number of spectral wavebands ranging from the visible to near infrared. The various reflectance properties of the minerals present in hydrothermally altered rocks and soils developed above them enable detection of hydrothermal alteration zones in TM imagery. These zones are thus potential areas of gold mineralization in granite and greenstone terranes.

We consider several techniques for the detection and enhancement of 'clay' anomalies in TM imagery and use them to the detect hydrothermal alteration zones in a TM image of the Murchison greenstone belt, Northern Transvaal, South Africa.

Landsat Thematic Mapper imagery is a useful tool for regional geological exploration. Various types of information may be extracted from the data :

- The synoptic view provided by the satellite provides information on basic regional geology and the nature of the terrain as well as drainage and lineament patterns.

- For the large parts of the world where maps are lacking or inadequate, TM plays a useful cartographic role for ground exploration activities and may be used as a base map for airborne geophysical investigations.

- Detailed geological information of surface types and stratigraphy are available due to the relatively good image spatial resolution.

- In arid areas, mineralogical information of concentrations of clay and iron-type minerals may be extracted by the processing of the image spectral bands.

The TM satellite collects data in six spectral wavebands ranging from the visible to near infrared. A thermal channel is also collected but is of little use for mineral exploration due to the coarse spatial resolution as well as the usual problems with daytime thermal data. The various reflectance characteristics of the minerals present in the surface and subcropping rocks, as well as the soils developed above them, enable the detection of spectrally anomalous zones in TM imagery. These anomalous regions may be of importance for geological exploration.

243

For example, rocks which have been hydrothermally altered may exhibit alteration haloes detectable in satellite imagery. Such zones are potential areas of gold mineralization in granite and greenstone terranes since gold may have been deposited in the rocks by the hydrothermal solutions.

The extraction of mineralogical information in vegetated areas is more problematic, although some information may be gleaned by using geobotanical methods, for example, stressed vegetation may give a clue to the nature of the soil in which the vegetation is growing.

THE GEOLOGICAL ENVIRONMENT

Over six thousand gold workings are scattered throughout the greenstone belts of Africa. Nearly 350 gold mines or prospects are known to occur in the Barberton area of South Africa alone. Current geological thinking [Anhaeusser and Viljoen, 1986] is that the primary source of gold was the volcanic rocks of the greenstone belts. Various physical processes, including sedimentation and biological mechanisms, have been postulated to explain the concentration of gold into economically viable deposits. The major concentrating agent was probably fluid penetration associated with igneous activity : in most cases, an intrusive granitic body generated heat in turn mobilizing the gold which was transported as hydrothermal fluids. These mineralizing fluids precipitated the gold into structurally controlled traps. The wall rocks of the mineralized veins were typically altered by the hydrothermal fluids and the mineral products of this alteration and subsequent weathering products occur in the soil profiles above the gold deposits. These alteration haloes are usually more extensive than the ore zone itself and are thus useful indicators for exploration activities.

The mineralogical nature of an alteration halo depends on the composition of the hydrothermal fluids and wall rocks, as well as the hydrothermal conditions (temperature and pressure) prevalent during the deposition of the gold. Landsat TM imagery may be used to detect 'clay' alteration minerals, including *kaolinite, montmorillonite, mica, talc* and *chlorite*. TM is thus a cost-effective means of identifying potential alteration zones in greenstone terrains. In addition, detection of 'clay' anomalies in regions with soil cover may indicate the presence of subcropping greenstone remnants.

IDENTIFYING SPECTRAL ANOMALIES

Spectral anomalies in multispectral imagery are groups of pixels with a spectral response different from what we would expect. One method of identifying these pixels is to use a set of image bands to predict the spectral response of another band or set of bands. Spectral anomalies will be badly predicted. Hook and Munday [1988] review other approaches including *band ratios* and *directed principal components*. In this particular application, we are interested in identifying clay anomalies in TM imagery and so wish to predict the $2.2\mu m$ band. Lamb and Pendock [1989] mention three ways of doing this : *multiple linear regression, polynomial regression* and a *data adaptive linear filter*. They apply the techniques to a TM image of a Pre-Cambrian porphyry system in Namibia and delineate the phyllic and propylitic alteration zones with some accuracy.

Multiple regression estimates the expected value of an image band as a linear combination of the other bands using a single set of partial regression coefficients, while polynomial regression fits a polynomial of a specified degree to each spectrum (pixel). Midway between these two approaches in terms of the number of different transformations used is the *data adaptive linear filter*. Another important difference between this technique and the regression methods, is that image spatial information plays a rôle in the calculation of the adaptive filter weights.

Multiple Linear Regression

Multiple linear regression may be applied to determine the linear combination of the first five TM spectral bands $S = (S_1, \ldots, S_5)$ which best describes the clay band S_6. We assume a multiple linear regression model of the form

$$E(S_6 | S_1, \ldots, S_6) = w_0 + \sum_{k=1}^{5} w_k S_k$$

$E(\cdot)$ is the expected value operator. Least squares estimates of $w_0, \ldots, w_5$ may be obtained by minimizing

$$E = \sum_{all\ pixels\ j} \left(s_{6j} - w_0 - \sum_{k=1}^{5} w_k s_{kj} \right)^2$$

where s_{ij} is the jth spectral value from spectral band S_i. Setting the partial derivatives of E with respect to $w_0, \ldots, w_5$ to 0 results in a system of *normal equations* $CW = X$:

$$\begin{pmatrix} N & \sum s_{1j} & \cdots & \sum s_{5j} \\ \sum s_{1j} & \sum s_{1j}s_{2j} & \cdots & \sum s_{5j}s_{2j} \\ \vdots & \vdots & \ddots & \vdots \\ \sum s_{5j} & \sum s_{1j}s_{5j} & \cdots & \sum s_{5j}^2 \end{pmatrix} \begin{pmatrix} w_0 \\ w_1 \\ \vdots \\ w_5 \end{pmatrix} = \begin{pmatrix} \sum s_{6j} \\ \sum s_{1j}s_{6j} \\ \vdots \\ \sum s_{5j}s_{6j} \end{pmatrix}$$

where N is the number of pixels in the image. The above system has solution $W = C^{-1}X$.

In order to identify regions with anomalous clay spectra, we should ensure that, as far as possible, we exclude clay regions when we collect statistics for the covariance matrix. Fraser *et al* [1986] use this technique to target epithermal alteration and gossans in weathered vegetated terranes from airborne scanner imagery. The main drawback with this technique is that a single transformation is used to predict the clay band. The method estimates the optimal linear prediction coefficients for the whole scene, but may be suboptimal for a specific part of the image. In addition, no image spatial information is used. In fact, we would obtain the same prediction coefficients if we jumbled up the image pixels! We would hope to be able to improve our prediction by using the prior information that the alteration areas are spatialy homogenous and have a minimum size. This leads us to consider data adaptive filters.

Data Adaptive Linear Filter

If we are given the reflectance values for the first five spectral bands of a TM pixel $s = (s_1, \ldots, s_5)$, we are interested in finding a set of weights $w = (w_1, \ldots, w_5)$ such that the predicted response of the clay band $\hat{s}_6 = ws'$ is as close as possible to the observed value s_6. The error between the output of the filter w and our desired response is

$$e = \hat{s}_6 - s_6$$

In order to minimize the *mean square* error we must consider

$$e^2 = ws'sw' - 2ws_6s' + s_6{}^2$$

which has expected value

$$E(e^2) = wE(s's)w' - 2wE(s_6s') + E(s_6{}^2)$$

Now $r = E(s's)$ is the 5×5 autocorrelation matrix of the vector s and $p = E(s_6s)$ is the cross-correlation vector of s with s_6 so we have

$$E(e^2) = wrw' - 2wp' + E(s_6{}^2)$$

with derivative with respect to w

$$\nabla = 2rw' - 2p'$$

The mean square error is minimum when $\nabla = 0$, i.e. when

$$w' = r^{-1}p'$$

The above equation is a system of five *normal equations* and may be solved in O(25) multiplications and divisions by the *Levinson-Durbin* algorithm [Silvia and Robinson, 1979].

This is an unsatisfactoy solution for two reasons :

- The algorithm is computationally expensive for a whole TM scene. In addition, the computational complexity grows with the square of the number of predictor

bands. With the trend in new and proposed remote sensing systems to more spectral bands, this approach will be impractical for these high spectral dimension images.

- We wish to *couple* a pixel to neighbouring pixels since if we know the predictor weights for a pixel, the next pixel on the scan line will probably have similar predictor weights, given the continuous nature of remotely sensed scenes. Thus we would expect to be able to do less work to determine the new weights. In addition, we should like the predictions made from neighbouring pixels to influence our predictions for a pixel. This will be useful in noisy imagery to avoid spurious anomalies and will provide us with some control over the size of predicted anomalies.

We may achieve this by considering an adaptive algorithm : the *next* weight vector w^{i+1} is equal to the *present* weight vector w^i plus a change proportional to the gradient of the present prediction error

$$w^{i+1} = w^i + \alpha \nabla^i$$

where α is some *convergence* factor which regulates by how much prediction weights may change between adjacent pixels and ∇^i is the gradient of the error at the ith pixel. This gradient may be estimated by $\hat{\nabla}^i = 2e^i(s_1^i, \ldots, s_5^i)'$ With this estimate of the gradient, we have a *data adaptive clay prediction algorithm :* $w^{i+1} = w^i + 2\alpha e^i(s^i)'$. Initial prediction weights are simply chosen as $w^0 = (1/5, \ldots, 1/5)'$ and a value is chosen for the convergence parameter α. This choice will depend on the amount of noise in the data and the spatial size of the anomalies we wish to identify.

The filter advances pixel-by-pixel and when it reaches the end of the first scan line, we change direction and proceed along the next scan line in the reverse direction to avoid an unnessary *jump* in the prediction process. The amount of computation is linear in the number of bands used for predicting the output.

The other simple prediction filter we shall consider consists of fitting polynomials to each spectrum and then evaluating the polynomial to estimate the value in the clay band.

Polynomial Fitting

Given a spectrum of six values $s_1, \ldots, s_6$ we may fit a unique polynomial of degree five to these points so that when this polynomial is evaluated, $s_1, \ldots, s_6$ may be retrieved exactly. This may be achieved using the method of *Newton-Gregory* or *Lagrange* [Gerald, 1970]. If we choose a lower degree polynomial, we would expect, in general, to find some error between our observed reading and the values calculated from the polynomial. We would also expect this error to increase as the degree of the interpolating polynomial decreases.

We may consider our six spectral values as being six (x, y) points, the y coordinate the spectral value and the x value the spectral wavelength. The provision of the spectral wavelength gives a weighting to the spectral values according to their relative positions in the spectrum. If this weighting is disregarded, we may simply treat the points as $(1, s_1), \ldots, (6, s_6)$ or, since the clay band is the last band, number the points in reverse order $(5, s_1), (4, s_2), \ldots, (0, s_6)$. This latter representation has the advantage of avoiding the evaluation of the interpolating polynomial, since the interpolated clay value is now simply the constant term of the interpolating polynomial.

If we choose n as the degree of the interpolating polynomial ($n \leq 5$), then our model is $\hat{y} = \sum_{i=0}^{n} w_i x^i$ with errors $e_i = \hat{y}_i - y_i$. Minimizing $\sum_{i=1}^{6} e_i^2$ results in $n + 1$ *normal equations* which we may write in matrix form as $(X'X)W = X'Y$ where

$$X = \begin{pmatrix} 1 & x_1 & x_1^2 & \cdots & x_1^n \\ 1 & x_2 & x_2^2 & \cdots & x_2^n \\ \vdots & \vdots & \vdots & \ddots & \vdots \\ 1 & x_{n+1} & x_{n+1}^2 & \cdots & x_{n+1}^n \end{pmatrix} \text{ and } Y = \begin{pmatrix} y_1 \\ y_2 \\ \vdots \\ y_{n+1} \end{pmatrix}$$

The solution W is given by $W = (X'X)^{-1} X'Y$. Since the expression $(X'X)^{-1} X'$ is independent of the spectral values it may be precomputed.

CASE STUDY : THE MURCHISON GREENSTONE BELT

The Murchison greenstone belt is situated in the North Eastern Transvaal, South Africa, and hosts a variety of mineral deposits including gold, antimony, mercury, copper, zinc and emeralds. Several producing gold mines are located in the Murchison Range, including Consolidated Murchison, the fifth largest gold producer in Archaean terranes in South Africa [Pearton and Viljoen, 1986].

According to Söhnge [1986], hydrothermal remobilization at the time of granite emplacement produced antimony ore bodies in the Murchison greenstone belt. The wall rocks are carbonated talc shists (altered komatiite) and contain mercury, lead and gold, which were probably precipitated by volcanic exhalations. There is extensive outcrop, so we have a possibility of detecting alteration zones using remotely sensed data.

Figure 1 is a TM band 5 image of a $30 \times 30km$ region around the Consolidated Murchison Mine. The southern most river in the image is the Selati, in which alluvial

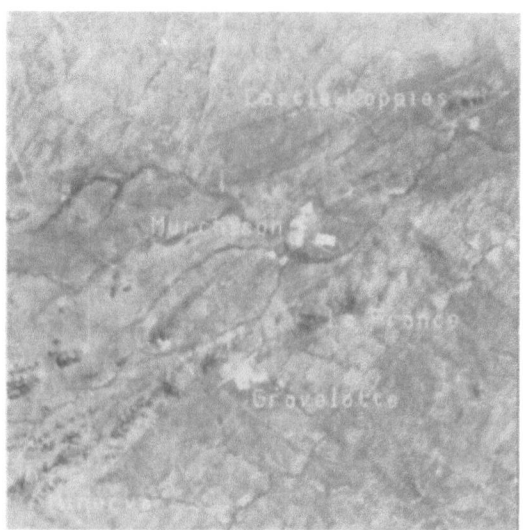

Fig. 1 TM band 5.

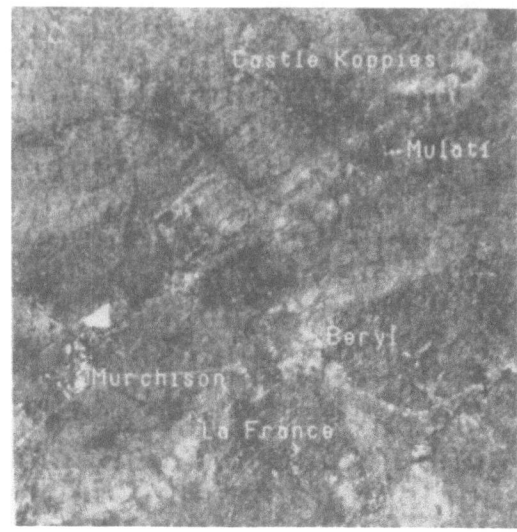

Fig. 2 clay anomalies.

gold was found in 1869. This was the earliest recorded discovery of gold in the area.

'Clay' anomalies were computed using the algorithms described above. Best agreement with the mapped geology was obtained using the *data adaptive clay prediction algorithm*. Figures 2 and 3 are clay anomaly images with white features most anomalous through to black (no anomaly).

The Spitskop area (Figure 3) provided a large amount of the early gold mined in the area. Gold occurs at the Minerva mine (Figure 3) in disseminated quartz-pyrite zones in deformed albite porphyry bodies. High temperature pyrite-arsenopyrite veins host the gold at Castle Koppies (Figure 2). The La France formation (Figure 2) is characterized by metamorphic minerals with a high alumina content. A new gold mine is currently being developed

Fig. 3 clay anomalies.

in this region by South Witwatersrand Exploration, Ltd. The imagery presented here was acquired and processed prior to the announcement of this new mining venture.

CONCLUSIONS

'Clay' anomalies may be identified in TM imagery of the Murchison greenstone belt. These satellite spectral anomalies show good correspondence with mapped gold deposits. This approach shows promise for the exploration of

• gold in greenstone belts

• gold hosted in shear zones in granite bodies

• subcropping greenstone remnants indicated by spectral anomalies in the soils developed above them.

References

1 Anhaeusser C.R. and Viljoen M.J. Archaean metallogeny of Southern Africa in Mineral Deposits of Southern Africa. The Geological Society of South Africa, 1986.

2 Fraser S.J. et al. Targeting epithermal alteration and gossans in weathered vegetated terrains using aircraft scanners : successful Australian case histories in Proceedings of the V th Thematic Conference : 'Remote sensing for exploration geology'. Reno, Nevada, 1986.

3 Applied numerical analysis. Gerald C. Addison-Wesley 1970.

4 Hook S.J. and Munday T.J. Preprocessing and analysis of airborne visible near and shortwave infrared data for the detection of alteration in weathered vegetated terrain in Proceedings of the IGARSS '88 Symposium. Edinburgh, 1988.

5 Statistics and experimental design in engineering and the physical sciences. Johnson N.L. and Leone F.C. Wiley, 1977.

6 Lamb A. and Pendock N. Band prediction techniques for the mapping of hydrothermal alteration in Proceedings of the VII th Thematic Conference : 'Remote sensing for exploration geology'. Calgary, Canada, 1989.

7 Pearton T.N. and Viljoen M.J. Antimony mineralization in the Murchison greenstone belt - an overview in Mineral Deposits of Southern Africa. The Geological Society of South Africa, 1986.

8 Söhnge A.P.G. Mineral provinces of Southern Africa in Mineral Deposits of Southern Africa. The Geological Society of South Africa, 1986.

9 Deconvolution of geophysical time series in the exploration for oil and natural gas. Silvia M.T. and Robinson E.A. Elsevier, 1979.

Mineral exploration—case histories

Remote sensing of a biogeochemical anomaly associated with a base-metal deposit in the Spanish Pyrite Belt

C. Banninger
Institute for Image Processing and Computer Graphics, Joanneum Research, Graz, Austria

SYNOPSIS

An analysis of multitemporal Landsat Thematic Mapper (TM) data from a heavy metal stressed pine forest in south-western Spain defined TM bands and transformations best suited for detecting the presence of anomalous concentrations of copper, lead, and zinc in the underlying soil. Based upon Thematic Mapper data from January and August acquisition dates, spectral bands and transformations utilising TM bands in the visible wavelength region for the winter scene and those in the shortwave infrared wavelength region for the summer scene proved most successful in defining metal–related anomalies in the pine forest. The results of the study support not only the employment of a biogeochemically based Thematic Mapper survey as a surrogate approach to a soil geochemical survey for the search for orebodies in the Pyrite Belt, but also show the importance of selecting the proper spectral bands and transformations with respect to the aquisition dates of the Landsat data.

INTRODUCTION

As one of the principal metallogenic provinces in Europe, the 7800 km² Spanish–Portuguese Pyrite Belt has been the focus of continuous mineral exploration activity since pre–Roman times, which has resulted in the discovery and development of numerous base metal deposits in the region. Most deposits with obvious surface expressions (e.g., gossans, stockworks) have, however, been found, but the likelihood remains that a considerable number of buried deposits are still to be discovered, even at shallow depths. The massive, polymetallic nature of the Pyrite Belt orebodies, with their high copper–lead–zinc content, gives rise to extensive, although often subtle, geochemical and geophysical expressions of their presence, but only a small fraction of the region has undergone detailed geochemical or geophysical survey work.

The Mediterranean climate of the Pyrite Belt, with its hot, dry summers and mild, moist winters, favours the establishment of deep–rooting, water–seeking plant species that are able to sample large volumes of the subsurface over a wide site area for the presence of geochemical anomalies. For plants rooted in soils containing toxic concentrations of heavy metals, the uptake of the metals into the plant system results in physiological and morphological changes to the plant that can be detected by remote sensing means. The large area encompassed by the Pyrite Belt makes it ideal for the application of remote sensing technology to mineral exploration, especially that related to spaceborne sensor systems such as the Landsat Thematic Mapper (TM).

The Spanish government, in cooperation with NASA, has established a programme to evaluated the usefulness of Landsat Thematic Mapper data for mineral exploration in Spain. One aspect of this programme involves the application of TM data for the detection and identification of biogeochemical anomalies associated with base metal deposits in the Spanish Pyrite Belt. The results of one segment of this study are reported herein.

REGIONAL SETTING

Stretching across southwestern Spain and southern Portugal for nearly 230 km (Figure 1), the Pyrite Belt comprises an assemblage of volcanic and sedimentary rocks that underlie a gentle upland terrain of low to moderately high hills that rarely exceed 250 m in height above the valley floors.[1] Sulphide mineralisation is confined to a 20 – 35 km wide exposure of predominately felsic pyroclastic rocks that have been strongly folded during the Hercynian Orogeny and which are overlain to the south by a thick sequence of slates and greywackes. Mineralisation consists of stratiform, lenticular–shaped, massive polymetallic pyrite bodies up to 4 – 5 km in length and several kilometres in width, but generally less than 10 m in thickness, containing chalcopyrite, sphalerite, and galena.[2,3,5,6] Numerous open–cast mines occur throughout the Pyrite Belt, including the important Rio Tinto Mine.

The predominately dry Mediterranean climate favours a vegetation cover characterised by communities of evergreen oak, pine, and eucalyptus trees and low–growing shrubs consisting of maquis and garrigue.

TEST SITE DESCRIPTION

The Pyrite Belt contains several mine sites that have stands of undisturbed vegetation occurring in close proximity to existing or former orebodies that make them suitable for the study of heavy metal induced biogeochemical anomalies by Landsat Thematic Mapper data. One such site (no. 20 in Figure 1) situated in the south-central part of the Spanish Pyrite Belt adjacent to the village of Sotiel Coronada comprises a densely wooded stand of naturally occurring pine trees (*Pinus pinea* L.) lying adjacent to an abandoned open-cast mine and growing in soils derived from the fine–grained tuffaceous rocks that are host to the area's cupriferous pyrite bodies. No obvious signs of stress, such as advanced chlorosis or needle

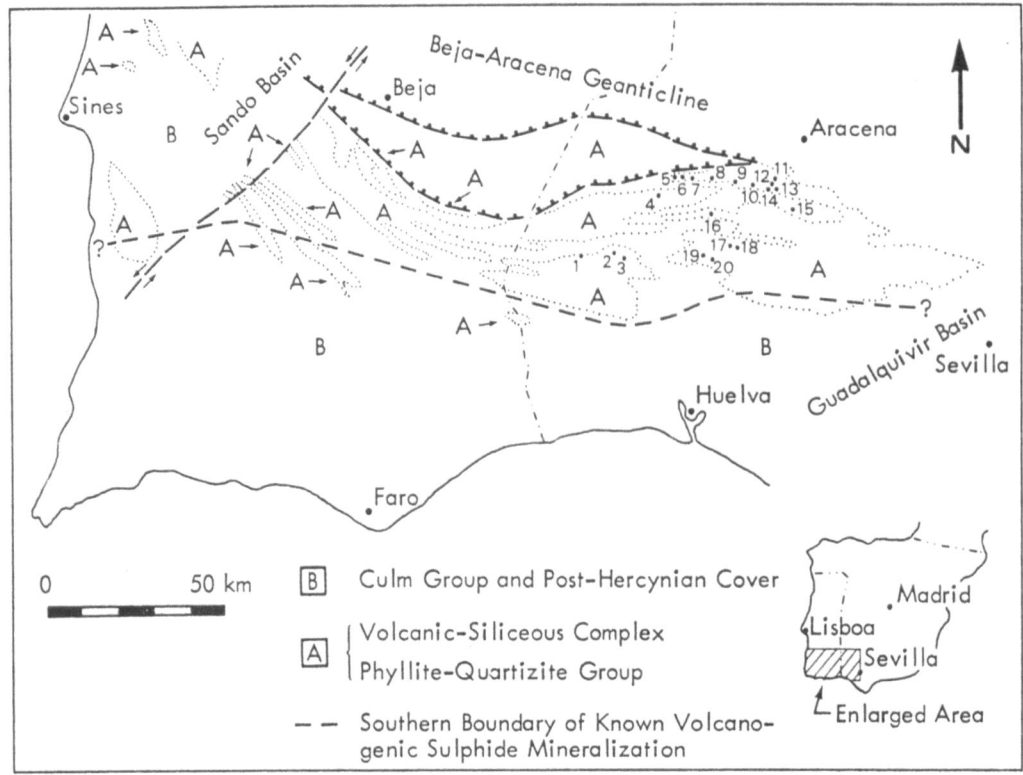

Figure 1: Geological Map of the Spanish–Portuguese Pyrite Belt Showing the Location of Present and Former Mine Sites.[4,6]

loss, are evident in the pine tree canopy, although those trees situated closest to the open pit have a more open crown structure compared with those farther removed from the orebody.

SPECTRAL RESPONSE OF PLANTS TO HEAVY METALS

The presence of high concentrations of heavy metals in the soil substrate can adversely affect plant growth and result in physiological and morphological changes that are detectable by remote sensing instruments. These changes range from previsual stress symptoms related to internal plant functions to marked decreases in plant biomass, which separately or together can alter the spectral response characteristics of a plant canopy. Stress–induced physiological changes are normally characterised by a reduction in leaf pigment and water content and morphological changes by a decrease in leaf size or density. This generally results in an increase in a plant canopy's reflectance in the visible (400 – 700 nm) and short-wave (1300 – 2500 nm) wavelength regions and a decrease in the near–infrared (700 – 1300 nm) region.

LANDSAT TM BANDS FOR VEGETATION ASSESSMENT

In contrast to the Landsat MSS spectral bands, those of the Thematic Mapper are configured to measure specific plant properties that can be used to assess the state of health of a vegetation canopy (Table I). The reflective TM bands cover the visible to shortwave infrared wavelength region from 450 – 2350 nm in six discrete channels, with bandwidths varying

between 60 and 80 namometres in the visible and between 140 and 270 nanometres in the infrared spectral regions. Table I lists the six TM reflective bands (Band 6 is in the thermal region) and their wavelength intervals and gives a brief description of the plant properties measured by each band.[7] With a swath width of 185 km, a spatial resolution of 30 m, and 256 quantised radiometric intervals,[8] Landsat Thematic Mapper data have the potential to detect geobotanical anomalies associated with ore deposits of small size and subtle biogeochemial expression.

GROUND AND LANDSAT TM DATA COLLECTION

Ground data comprise soil samples collected from the base of the B-horizon (15 – 20 cm depth) at 50-m grid spacings and analysed for copper, lead, and zinc by atomic absorption spectrophotometry (80-mesh soil fraction used). Soil metal values range from 26 – 153 ppm for copper, 30 – 325 ppm for lead, and 39 – 205 ppm for zinc. These concentration levels are sufficiently high to be toxic to vegetation growing on the test site, particularly with respect to lead. The conversion of the point source soil metal values to their respective soil metal isopleth maps provided a suitable base for the integration of these data with the Landsat Thematic Mapper data.

Landsat TM scenes used in the study are from January 1983 and August 1984. Data processing consisted of the removal of the haze component in all six reflective bands for each of the scene dates by the darkest object subtraction method[9] before their employment in any computational analysis. The number of pixels corresponding to the test site for the January and August TM scene dates are 85 to 89, respectively, and

252

Table I: Landsat Thematic Mapper Bands for Vegetation Monitoring.

TM Band	Bandwidth (nm)	Plant Canopy Relationship
TM 1	450 – 520	Region of maximum chlorophyll and carotenoid absorption. Strong relationship between spectral reflectance and plant pigment concentration.
TM 2	520 – 600	Green vegetation reflection peak. Slight sensitivity to plant chlorophyll concentration.
TM 3	630 – 690	Strong chlorophyll absorption region. Strong relationship between spectral reflectance and and plant chlorophyll concentration.
TM 4	760 – 900	High vegetation reflectance. Strong relationship between spectral reflectance and vegetation density and vigor.
TM 5 and TM 7	1550 – 1750 2080 – 2350	Reflection peaks within the water absorption region of the shortwave infrared. Both bands are highly sensitive to leaf water content.

Table II: Landsat TM Bands and Transformations Employed in the Analysis.

Band 1   (TM1)   Landsat TM Band
Band 2   (TM2)
Band 3   (TM3)
Band 4   (TM4)
Band 5   (TM5)
Band 7   (TM7)

Band 4 – Band 1   (BD1)   Band Difference
Band 4 – Band 2   (BD2)
Band 4 – Band 3   (BD3)
Band 4 – Band 5   (BD5)
Band 4 – Band 7   (BD7)

Band 4/Band 1   (R41)   Simple Band Ratio
Band 4/Band 2   (R42)
Band 4/Band 3   (R43)
Band 4/Band 5   (R45)
Band 4/Band 7   (R47)
Band 2/Band 3   (B23)
Band 3/Band 1   (R31)
Band 5/Band 7   (R57)

(Band 4 – Band 1)/(Band 4 + Band 1)   (ND1)   Normalised Difference
(Band 4 – Band 2)/(Band 4 + Band 2)   (ND2)
(Band 4 – Band 3)/(Band 4 + Band 3)   (ND3)
(Band 4 – Band 5)/(Band 4 + Band 5)   (ND5)
(Band 4 – Band 7)/(Band 4 + Band 7)   (ND7)

0.3037(TM1) + 0.2793(TM2) + 0.4743(TM3)   (TMB)   Brightness Index[10]
+ 0.5585(TM4) + 0.5082(TM5) + 0.1863(TM7)

− 0.2848(TM1) + 0.2435(TM2) − 0.5436(TM3)   (TMG)   Greenness Index[10]
+ 0.7243(TM4) + 0.0840(TM5) − 0.1800(TM7)

0.1509(TM1) + 0.1973(TM2) + 0.3279(TM3)   (TMW)   Wetness Index[10]
+ 0.3406(TM4) − 0.7112(TM5) − 0.4572(TM7)

Table III: Upper Ranked Landsat TM Bands and Transformations for Geobotanical Remote Sensing in the Pyrite Belt.

January	August
TM4	TMW
TMB	TM7
TM1	BD7
BD2	ND7
TM3	BD5
PC1	TM5
TM2	R31
TM5	PC1
TMG	ND5
BD1	R45
TM7	R47
BD3	R23

were co-registered with the soil metal isopleth maps to facilitate the merging of the two data sets. Table II lists the thirty single and transformed Thematic Mapper bands employed in the study, based on their ability to discriminate stress in plant canopies.

DATA INTEGRATION AND ANALYSIS

A comparison between the ground and Landsat data sets required their integration to a common data format. This was achieved by calculating the weighted average of the copper, lead, and zinc soil values for each TM pixel (the Landsat pixel metal value) corresponding to the test site, using a 36-point dot grid scaled to the 30 × 30 m ground dimensions of the TM resolution cell in conjunction with the respective isopleth map for each metal. A linear regression of Landsat band and transformed radiance values against logarithmically transformed Landsat pixel metal values defined the best TM bands and transformations for detecting geochemically induced anomalies in the pine tree canopy and, by implication, the presence of heavy metals in the underlying soil.

Based on the correlation coefficients derived for each pairwise combination of the regressed data sets, a ranking of the thirty TM bands and transformations for both Landsat scene dates was obtained by calculating the weighted frequency sums for each of the twenty-one data sets used in the analysis. A list of the top ranked bands and transformations with correlation coefficients greater than 0.60 at the 99 per cent probability level is given in Table III.

DISCUSSION OF RESULTS AND CONCLUSIONS

The results of the statistical analysis presented in Table III appear, at first glance, to be a potpourri of spectral bands and transformations, with no apparent commonalty amongst them other than their incorporation of Thematic Mapper bands. A closer inspection, however, reveals a consistency within each scene date with respect to the major stress conditions prevailing at the time of year the data were acquired. For the January scene date, nine of the twelve top-ranked TM bands and transformations include bands 1, 2, and 3 and their transformations. These bands measure radiation in the visible part of the spectrum, where a strong relationship exists between plant physiological processes and plant spectral characteristics. This implies that differences in the spectral response between metal-stressed and non-stressed pine trees are predominately related to differences in needle pigment content (particularly the chlorophylls) at this time of the year. A similar relationship exists between the upper-ranked TM bands and transformations for the August scene date exits, but with respect to bands 5 and 7. These bands are positioned in the water absorption region of the shortwave infrared, and hence give a measure of water stress in a plant canopy. As August is one of the hottest and driest months of the year in the Pyrite Belt, water stress would be severe in the pine tree stand at this time and even more acute in the metal-stressed trees, due to metal-induced impairment of their physiological system.

As to the choice of TM bands and transformations to be employed in a geobotanical survey, any of those in the top three positions would be suitable, as their discriminating power in detecting metal stress in the pine tree canopy is comparable. These same bands and transformations, however, may not be equally effective for other vegetation types or at other times of the year than those given.

Too often, the focus of remote sensing studies is on the determination of the best bands or transformed bands for employment in a vegetation survey, with little regard given to their usefulness during different times of the year. The results of this study, although limited to two scene dates, show the importance of taking into consideration seasonally related changes in canopy reflectance in a geobotanically based remote sensing survey, especially in the Pyrite Belt.

This study has demonstrated that Landsat Thematic Mapper data have the capacity to detect and delineate geobotanical anomalies in vegetation stands related to the presence of heavy metals in the soil substrate, and, therefore, can be used as an alternative approach to a soil geochemical survey. The selection of the appropriate TM bands and transformations for discriminating metal stress in vegetation canopies is crucial to the success of a remote sensing based survey and is dependent upon the time of the year and season in which the spectral data are acquired. An understanding of the environmentally related physiological and morphological changes expected to be encountered in a plant canopy at a particular time of the year is also necesarry for the proper identification and interpretation of geobotanical anomalies defined by a remote sensing survey.

ACKNOWLEDGEMENTS

I thank NASA Ames Research Center in Mountain View, California, and the Instituto Nacional de Tecnica Aerospacial in Madrid, Spain, for providing the Landsat Thematic Mapper CCT's used in the study.

REFERENCES

1. Schermerhorn, L.J.G. An Outline Stratigraphy of the Iberian Pyrite Belt. *Boletin Geologico y Minero*, vol. 82–83–84, 1971, pp. 239 – 268.

2. Schermerhorn, L.J.G. The Deposition of Volcanics and Pyrite in the Iberian Pyrite Belt. *Mineral. Deposita (Berl.)*, vol. 5, 1970, pp. 273 – 279.

3. Schermerhorn, L.J.G. Pyrite Emplacement by Gravity Flow. *Boletin Geologico y Minero*, vol. 82–83–84, 1971, pp. 304 – 308.

4. Schermerhorn, L.J.G. Spilites, Regional Metamorphism and Subduction in the Iberian Pyrite Belt: Some Comments. *Geologie en Mijnbouw*, vol. 54, no. 1, 1975, pp. 23 – 35.

5. Strauss, G.K., and Madel, J. Geology of Massive Sullphide Deposits in the Spanish–Portuguese Pyrite Belt. *Geologische Rundschau*, vol. 63, 1974, pp. 191 – 211.

6. Strauss, G.K., Madel, J., and Alonso, F.F. Exploration Practice for Strata-Bound Volcanogenic Sulphide Deposits in the Spanish–Portuguese Pyrite Belt: Geology, Geophysics, and Geochemistry.in: *Time- and Strata-Bound*, D.D. Klemm and H. – J. Schneider (eds.), Springer-Verlag, Berlin, 1977, pp. 55 – 93.

7. Tucker, C.J. A Comparison of Satellite Sensor Bands of Vegetation Monitoring.*Photogrametric Engineering and Remote Sensing*, vol. 44, no. 11, 1978, pp. 1369 – 1380.

8. Slater, P.N. *Remote Sensing: Optics and Optical Systems*, Addison–Wesley, Reading, Mass., 1989, 575 p.

9. Crane, R.B. Preprocessing Techniques to Reduce Atmospheric and Sensor Variability in Multispectral Scanner Data.*Proc. 7th International Symposium on Remote Sensing of Environment*, University of Michigan, Ann Arbor, 1971, pp. 1345 – 1355.

10. Crist, E.P. The Thematic Mapper Tasseled Cap-A Preliminary Formulation. *Proc. 9th International Symposium Machine Processing of Remotely Sensed Data*, Purdue University, West Lafayette, Indiana, 1983, pp. 357 – 363.

Effective methods of remote sensing applied to prediction of and prospecting for solid mineral resources

L. M. Natapov

'Aerogeologia', V/O Zarubezhgeologia, Ministry of Geology, Moscow, U.S.S.R.

Remote sensing is used successfully to predict and prospect for precious, rare and base metals and for kimberlite pipes that may be host to diamond deposits.

A few examples are given in this brief presentation.

1. Tin mineralisation in Yakutia

The geology of the region of interest comprises terrigenous sediments of Upper Palaeozoic age that are thrust and folded and cut by granites. The terrain is mountainous with sparse taiga and bare summits. The exploration task was to reveal poorly exposed and buried tin deposits in the region where cassiterite-quartz and cassiterite-other metal sulphide ores are well known. The known occurrences concentrate in narrow bands transverse to the strike of the folding.

Prediction was carried out using an automatic computer system "Region". An analysis of aerial photographs and high-resolution (5m) space images provided input into the prediction.

The results of the prediction were prepared as 1:200 000 scale maps using a basic cell of 2kmx2km. Target areas of 4sq km minimum, about the size of an average orefield, can be identified using this cellsize.

The first phase of the analysis considered 40 criteria of which 12 are derived from photo-interpretation: lineaments, ring structures, buried granitoid intrusions regions of hydrothermally altered rocks et al. An heuristic modelling algorithm resulted in the recognition 19 factors influencing ore formation.

The validity of the prediction derived from these factors was checked. A quantitative evaluation of the results determined the critical factors and ranked their importance. The zone of influence of each factor was also computed. Factors derived from photo-interpretation, in particular major faults and breaks identified on satellite images, were important.

Nine competitive versions of prediction maps were prepared using various combinations of experts participating in the analysis. Ore targets were divided into several classes, 40% being reserved as control samples.

A taxonomic "special point" algorithm is applied simultaneously. The region is divided into zones which are analogous to a reference ("special point") orefield of defined mineral or ore type. The result is a qualitative composition for each ore target and a prediction of potential resource. The version in which the reference type was, only deposits predicted potential larger targets.

Promisingly prospective areas were predicted for 7% of the total area of 30 000 sq km and areas of probable mineralisation were predicted locally within areas of more general potential.

Prospecting and validating the prediction was carried out by the authors of the prediction who were using the system. Field traverses, sampling, trenching, magnetic surveying,and tin analysis by X-ray radiography were carried out in 14 areas. Of these, 3 have poor tin mineralisation, 4 contain the root section of eroded tin deposits and 7 contained newly discovered deposits, 2 being of high potential.

It is emphasised that lineaments observed on satellite images were found to be more significant than fractures depicted on geological maps.

2. Gold mineralisation in west Tien Shan

The region comprises intensively folded shales of early Palaeozoic age underlying an arid mountainous terrain.

Composite multispectral air photographs at a scale of 1:10 000 were studied to reveal undiscovered zones of metasomatic alteration around gold deposits. Alteration zones are observed on false-colour image composites on the western flanks of one deposit, displaced along a fault by 500m from the ore deposit.

3. Kimberlites of the Siberian Platform

The geology of the region of interest is gently dipping carbonate rocks overlain locally by thin terrigenous sediments of Jurassic age. There are many Triassic age sills, dykes and diabases. The terrain is flat with sparse taiga.

The study included manual and automatic photo-anomaly interpretation of panchromatic and multispectral air photographs. The method is effective for delineating kimberlite. Many dykes and diatremes, some with diameters several hundred metres across, were revealed. The photo-interpretation is more effective than aeromagnetic prospecting in this region of crystalline shield in which the magnetic field is complex.

There seems to be a clear relationship between lineaments seen on the photographs and kimberlite pipes. The lineaments seems to represent narrow strips of fractures that widen over the kimberlites.

Tests were carried out to determine how well kimberlite pipes could be identified directly on the images. A set of images of an area including 96% of known kimberlite bodies of the Siberian Shield. Dark anomalies coincided with 28.7% of the kimberlites. The anomalies represented thicker vegetation, in particular Alder trees and bushes. Elsewhere drainage patterns or local interruption of ridges of hard rock or lineament intersection coinciding with photo-anomalies revealed a great number of kimberlites at very little cost.

Manual interpretation, however, can result in error and false anomalies. A method has been developed to identify these errors and minimise them. Photo-anomalies can be qantified on digital images and assigned a symbol. There are 3 groups: photometric; topological; and textural. Furthermore automatic recognition of "pipe-like" spots can be achieved. Photo-anomalies are categorised for their form and edge contrast. A priori dimensions and shape were used to describe the form of the pipes; dimensions of 50m to 500m and a ratio of short axis to long axis from 1:1 to 1:5 were selected. The edge contrast was determined by comparison to known kimberlite bodies.

Two versions of the anlysis were carried out depending on the strength of the edge of the pipe-like features that were correlated with the results of independent manual interpretation carried out by 5 interpreters. The automatic and manually interpreted results showed good general agreement. Some anomalies were better or only seen on the automatic version whilst others, those with poorly defined edges, were only recognised during the manual interpretation.

258

Remote sensing in mineral exploration—really a practical tool?: image processing and GIS applications in exploration projects

Enrique Ortega
Jesús Artieda
Rupert Haydn
Peter Volk
TdA—Teledetección Aplicada S.A., Madrid, Spain

ABSTRACT

During recent years Remote Sensing and Image Processing have become progressively more important in mineral exploration.

In some cases the results obtained are convincing. However, this does not reflect the general situation.

The reliability of the results is strongly controlled by several factors: climatic conditions, vegetation cover, geological background information, scale of the survey and genetic type of deposits. In this paper a critical review of the real capabilities of this technique is presented, commenting on two case histories for prospecting on Sn/W/Au and Cu/Ag/Au mineralizations in Spain and South America.

Economic factors concerning conventional methods and a remote sensing approach are discussed.

1. INTRODUCTION

In recent years, the use of remote sensing and image processing techniques in mineral exploration has steadily increased. This increase is results from the combination of two basic factors:

a) The actual problems to be solved by mineral exploration activities are different from those several decades ago. This is a consequence of higher exploration costs, the need to search for buried deposits and also the need to manage and analyze more and more data.

b) On the basis of dropping hardware prices and the development of image processing software on powerful personal computers, image processing has become an affordable tool.

As a consequence, the number of image processing systems operated all over the world rose dramatically and many of them are devoted to mineral exploration, being installed in official entities, consulting companies or directly in mining companies.

This must be basically considered a positive development, since it enables access to new data and new information for more users. Nevertheless, often Remote Sensing and Image Processing Techniques are not very accepted in the field of exploration, since the results obtained are not convincing or the methodology has not been able to detect mineralizations.

Often, these negative results are due to an overestimation of the method's capability or to an inadequate utilization of the data.

The number of possibilities to combine and to analyze the data is manifold (mainly in GIS data bases where multivariate data sets are stored). Therefore it is necessary to know exactly the problem to be solved and to choose very carefully among possible combinations and processing strategies.

The use of Remote Sensing, Image Processing and GIS-techniques supports the understanding and interpretation of geological and metallogenetic processes. They have several advantages compared to conventional exploration activities and can provide new information, but they cannot fill gaps in the geological knowledge of an area. The processing and interpretation of the data will be only as good as the level of geological background knowledge.

Nevertheless, the correct use of Remote Sensing and Image Processing and GIS- techniques can provide a useful tool to strongly increase the amount and quality of information obtainable from the original data.

The aim of this paper is to describe some aspects to be considered before starting a Remote Sensing project, the logical approach to solve existing problems and the kind of results to be expected. Two case histories, where these techniques have been successfully applied will be presented.

2. GENERAL CONSIDERATIONS FOR SUCCESSFUL APPLICATIONS

The analysis of remote sensing data for the mineral exploration is a highly interpretative process and it can only be successfully executed if several constraints and limiting factors are considered.

There are **restrictions in scale** even for high resolution satellite imagery. TM images can be used at scales down to 1:50.000 and SPOT-pan data are exploitable in scales down to 1.25.000. However, airborne scanner data do not have this limitation in principal, but they lack the synoptic aspect on a regional scale.

The **climatic conditions** of the area to be investigated are a decisive factor for the application of remote sensing. In the humid tropics and the moderate climate of the higher latitudes extensive cloud coverage makes it difficult to acquire cloudfree image data. Furthermore, the **dense vegetation cover** obscures any spectral response from rock and soils. In a hyperarid environment, sand covered bedrock cannot be detected using multispectral data; in such cases imaging radar sometimes has the potential to penetrate loose sediments and reveal tectonic details.

One of the most important aspects for successful remote sensing is the elaboration of a **geological model** and based on it, an ore-genetic concept. They must lead to the definition of spectral or structural indicators, which are related to mineralization targets searched. E.g. it is useless to process and analyze data for the delineation of hydrothermal alterations, if the target is a structurally controlled Pb-Zn vein type deposit, formed under low temperature hydrothermal conditions.

Based on the geological model and a preliminary idea of a working approach, **suitable sensor data** must be selected and subsequently a **processing concept** has to be developed. For mapping of hydrothermal alterations the short-wave infrared bands of TM are perfectly suited. Tectonic mapping should be performed by the aid of satellite data acquired with low sun elevation and high resolution. In some cases lithological features show spectrally a weak response. Therefore it is mandatory to select data with acquisition dates close to the sun solstice. This guarantees an optimum signal/noise ratio in the data and hence the best potential spectral separation. However, the processing strategy applied to the digital data finally determines the value of an image. It is vitally important to have a thorough experience in image processing strategy. We want to give a simple example:

Paleo-placers containing different valuable heavy minerals are developed within tertiary limonitic gravels. The high contents in Fe^{3+} minerals contrasts with the other rocks exposed in the area. A ratio of TM band 2 and 3 excellently marks this lithological contrast due to specific spectral absorptions in TM band 3. This ratio information is stretched and colourcoded to improve the visual separation and interpretation.

The **thematic interpretation** of this processed image data on one hand requires a knowledge of the geological conditions and on the other hand a thorough experience in remote sensing techniques. Visual interpretation is basically performed using criteria developed for aerial photographs. But these criteria have to be applied under consideration of the processing methodology, e.g. to exactly know the limits of every contrast enhancement performed.

Summarizing, it is obvious that only a detailed knowledge of all possibilities and constraints can lead to success. In many cases, where the preconditions quoted are not considered carefully, projects may fail completely.

3. VALUE-ADDING BY MEANS OF GIS-TECHNIQUES AND DATA INTEGRATION

It is a well-known fact in mineral exploration, that often orebodies are not found due to an individual significant response in one data set, but rather by the superposition of several, sometimes not significant anomalies in different data sets. The concept of superposing geological, spectral, geophysical maps has been used for a long time by exploration geologists. Geographic Information Systems (GIS) now provide an universal tool to store and analyze different data sets without the restrictions of printed maps (various scales, display and printing quality etc.).

There exist two different GIS concepts:

1) **Vector-based GIS**
   (All data are digitized and stored in a vectorized form).

2) **Raster-based GIS**
   (The data are stored in raster format, maps have to be digitized and converted to raster format).

Rasterbased GIS are faster and more versatile for interactive superposition and analysis. In general, they are easier to handle and thus better suited for interactive work. Additionally, processed satellite data can be integrated without major modifications and directly compared to all other information.

As a further development, the concept of an Exploration Data Base (EDB) has been established. This means a rasterized data base of all exploration-relevant data, preferably with a standardized raster cell size and in map sheet format. Minas de Almadén for example executed a large EDB covering all their concession areas based on the 1:50.000 map sheet division and including valuable auxiliary information (old mines, field observations, stream sediment samples etc.).

To display the correlation of different data sets and to use this information during field surveys, image products of EDB analyzed data have to be created. For this purpose RGB- and IHS-coding, contouring and Synthetic Stereo algorithms are excellent procedures to create highly interpretative products. For former projects combined IHS/Synthetic Stereo-coded images have been proven a kind of standard for the display of correlations between spectral, geophysical and geochemical data sets.

All steps of integrated data analysis described above can be seen as an extension of the analysis process for an individual data set (e.g. remote sensing) and they yield additional insights into the geological conditions of target area. Once the Exploration Data Base is created, it can be easily and continuously updated when new data are available or new exploration concepts have been elaborated. Even small-to mediumsized companies with limited exploration budgets can benefit from this concept, since the costs for hard- and software are the same as the price of 500 m of diamond core drilling.

In the following, the ideas expressed will be presented for two practical examples.

4. CU/AG/AU HYDROTHERMAL MINERALIZATIONS, CENTRAL ANDES, SOUTH AMERICA

The axial zone of the Andes orogenic belt is a very promising area for mineral exploration, and it could be considered very suitable for the application of Remote Sensing techniques as a consequence of several factors:

- no spectral interference with vegetation
- poorly developed soils
- reduced atmospheric absorption due to high altitude

Additionally, the nature of the mineralization is very favourable for spectral detection. Most of the deposits are located in hydrothermal alteration aureoles around volcanic or subvolcanic centers, expressed in important mineralogical changes (silification, argilization, ... etc.) which spread over wide areas. The near and middle infrared bands of the Landsat TM are very sensitive to these mineralogical changes.

Ratioing processes (involving the TM bands 3,4,5 and 7), followed by a carefully enhancement of the ratios allowed the discrimination of the iron oxide aureole and the clay alteration aureole, the first one being usually wider.

The study of the lineament pattern suggests a strong tectonic control of the volcanic complexes, being preferably located near or over the intersections of major lineaments.

The selected areas have been successfully checked in the field, and promising grades (several ppm gold) have been found in several places. Mostly the high grades are located in the core of the aureoles.

If we consider the difficulties related to field work in such areas (more than 3.500 meters altitude), the advantages of this technique are obvious. It optimizes a reliable, fast and economic selection of final target areas.

5. SN/W/AU MINERALIZATIONS RELATED TO GRANITIC CUPOLAS, CENTRO-IBERIAN ZONE, SPAIN

In the southern part of the Centro-Iberian zone there exist several Sn/W/Au mineralizations, always related to alkaline differentiations in calc-alkaline granitic bodies. These bodies belong to the hercynian orogen and the ore is hosted in precambrian or paleozoic sediments.

Most of the known deposits are not of economic interest, because their size is small and their grade is low. But it is well known that these types of deposits are usually hosted in the apical zone of the granitic cupola. Therefore, if a granitic body is outcropping, probably the area with the highest potential for mineralization has been already eroded.

But if a fertile granitic cupola is located at shallow depth below the surface, the richest mineralized zone may be preserved. Taking this basic idea as a guideline for prospection, Landsat TM data have been selected to outline areas where the contact metamorphic zone could be detected as a consequence of changes in the reflectivity due to thermal intrusion effects.

A methodology has been developed to enhance the images and to pick-out areas affected by metamorphism.

On the basis of RGB combinations of TM infrared bands (4,5,7), weak colour zonations in the host rock of outcropping granites can be observed. Consequently, interactive processing revealed an improved discrimination in histogram adapted Ratios. To display this zonation, an IHS -> RGB transformation was applied. The resulting colour image excellently outlines contact metamorphism by yellow to red colours, whereas the not-affected precambrian host rock is represented by a TM 4 image having simple grey tones. Despite the presence of vegetation interference effects, the obtained results have been very positive and some interesting areas have been selected, finding previously unmapped contact metamorphic zones. Normally, the mapping of these aureoles is based on the presence of some key minerals (e.g. andalusite), but the thermal and recrystallization effects are developed far away from the key minerals isogrades. By the aid of the near-infrared TM bands, it is possible to recognize these weak thermal effects and to delineate unknown aureoles.

The selected targets have been checked with other information available in the already mentioned database maintained by MAYASA (mainly with aeromagnetic, geochemical and gravimetric data), selecting the most promising zones to be studied in detail.

It is specially interesting to remark on the strong coincidence and fit between the geophysical information (gravimetry and aeromagnetic data) and the spectral response. The obtained information in both cases is coincident, but the costs and time necessary for processing of the remote sensing data are significantly lower. It is also remarkable that the use of the georeferenced database, integrating the information graphically by means of a GIS system, improved dramatically the thematic value of the original data.

6. CONCLUSIONS

It has been demonstrated that remote sensing and image processing techniques are powerful tools if they are used correctly.

If possibilities and constraints of remote sensing data are checked in the light of an exploration task, experts in image processing and interpretation may derive valuable geological information. Of course, sometimes the project conditions or limitations inherent in remote sensing do not allow a successful application. But in general this technology must be regarded today as an operational alternative or supplement to traditional methods.

The incorporation of GIS capabilities (EDB) and data integration methods improves the information compared to the original data. This enables the enhancement of interesting anomalies not detected by conventional methods.

The results obtained are strongly controlled by the reliability of the geological model, the exploration concept, the careful selection of data sets to be combined and the integration method.

Due to its economic feasibility and the actual costs, any mining company could afford the installation of such a spatial data base.

In addition to the technical improvements, the following economic implications can be pointed out:

- Saving of time for data processing and interpretation.
- Reduction of fieldwork and manpower.
- Possibility to perform exploration projects over large areas with a reduced team in a   short time.
- Easy reinterpretation of old campaigns and comparison with recent data.

Exploration logistics

Use of bathymetry derived from Landsat Thematic Mapper to plan seismic surveys in the Red Sea

R. W. Cleverly M.A., D.Phil.
BP Exploration, London, England

SYNOPSIS

Landsat Thematic Mapper (TM) have been used in the Red Sea, offshore Ethiopia, to help in the planning of a seismic survey. The area has unreliable bathymetric charts and highly variable water depth.

The Landsat TM data were geocoded to the local datum and map projection to an accuracy of about 100m. Depth sounding data were digitised from the bathymetric charts and used to calibrate the Landsat data.

TM band 1 was found to give the best results, giving good water penetration and being less noisy than band 2. The calibration with depth data suggested that maximum penetration was about 20m. This is about the limit for safe operation of the seismic vessel.

These results were used to lay out the planned seismic lines in around the shoals and islands, saving much time in the design stage of the survey and removing the need for an expensive bathymetric survey.

This technique has also been used successfully with SPOT data in the Gulf of Suez.

Introduction

Seismic acquisition in and around the Dahlak Islands, offshore Ethiopia, poses many problems due to the large number of islands and shoals and highly variable water depth. Many of the shallows are developed on the tops of salt diapirs, and this may mean that they vary from year to year depending on the relative rates of diapirism and dissolution of the salt. Available Admiralty charts were based on widely spaced 19th century surveys, and were found to be unreliable.

The BP concession offshore Ethiopia includes all the Dahlak Islands and extends approximately 120km offshore. The port of Massawa would be the main operational base.

Work done by BP in the Abu Zenima area of the Gulf of Suez demonstrated clearly the applicability of satellite data for mapping areas of shallow water. In this case, SPOT XS data were used and calibrated with a small (and expensive) ship-borne bathymetric survey. The SPOT data revealed details of the carbonate reefs that fringe the shore and enabled shallow seismic lines to be run through the reef.

This example suggested that satellite data could be used (in areas of clear water) for the definition of shallow water and used for seismic survey design and planning.

DATA ACQUISITION

Two Landsat Thematic Mapper scenes were acquired over the BP Dahlak Island concession. The data were rotated and geocoded to the local datum and projection to enable the satellite data to be directly compared with the various maps and charts for the area.

The westerly scene unfortunately was found to be badly affected by atmospheric dust and haze and was unsuitable for bathymetric work. The easterly scene was very good quality, with virtually no atmospheric effects and little digital noise.

DATA PROCESSING

Band 1 (visible blue) was found to give the best results as it gave the the maximum water penetration with minimum noise. The water appeared uniformly clear and there were no areas of turbid water that would affect the bathymetric estimation. A few dark oil slicks related to pollution from the shipping lanes were also noticed, but these were allowed for when digitising the water depth contours.

For the final display and hard copy a split processing technique was used. A land/water mask was calculated using band 7. This was used to split the land areas from water. The land areas were processed using standard satellite processing techniques. Bands 1, 3 and 7 were used as they have been found to give the optimum geological information. They were combined in blue, green and red respectively, contrast-stretched, colour saturation increased and edge enhanced. For the sea areas, band 1 on its own was used to display the water depth as a colour density slice. The land and sea masks were then recombined to give the final hard copy.

DATA CALIBRATION

Depth data were digitised from the Admiralty charts for the area. These were rather scattered point measurements, many dating back to 1880, and not precisely located. For each depth measurement, the DN values were extracted from the satellite image using the benefits of geocoding. The depths and DN vales were cross-plotted and correlations calculated.

The correlations were reasonable considering the inaccuracies inherent in the data. Absolute accuracies were in the range ±5m at 15m, but about ±1m at 5m water depth. Maximum penetration is probably about 20m, although the data would not be reliable at that depth due to possible variations in water turbidity.

The satellite processing contours generally compared well with those on the maps, although there were significant differences in some areas, suggesting that the bathymetric charts are inaccurate due to lack of measurements, or out-of-date due to changes in the sea floor due to recent salt movement.

Water depth data from the seismic survey were used to refine the Landsat calibration as a basis for planning future surveys in the area, although there were only a very few readings less than 20m.

CONCLUSION

These images were used for the refinement of the original seismic plan, as many of the original sketched line locations crossed areas of very shallow water. The revised plan could be used for more detailed costing and logistical planning and in fact there were few changes between the planned survey and the final acquisition. Plans for acquiring a bathymetric survey over the area were cancelled with a cost saving of many thousands of dollars.

Despite all this preliminary work however, the Landsat data were found to be insufficiently precise for confident shooting and the seismic chase boat was used to survey each line in advance of the main vessel. None of these precautions could prevent sharks severing the cable on four occasions!

266

Operational remote sensing in Saudi Arabia: Landsat data and logistical support for the earth science community

K. P. Ferguson B.A., M.A., F.R.G.S.
Earth Observation Satellite Company, Lanham, Maryland, U.S.A.

SYNOPSIS

For more than 15 years, Landsat data have been used operationally to support geological investigations, mineral exploration, and mapping in Saudi Arabia. In 1977, the Deputy Ministry for Mineral Resources (DMMR) established a remote sensing center in Jiddah, Saudi Arabia, to provide earth scientists with information derived from Landsat data. Earth scientists with DMMR, have used Landsat data for planning field work, for navigating helicopters in poorly mapped areas, as bases for compiling new maps, and as a source of information for revising old maps. Landsat data were used to produce three series of geometrically corrected image maps at scales ranging from 1:250,000 to 1:1,000,000. These maps, which provide national coverage, meet U.S. Geological Survey standards for map accuracy. At the height of this mapping program, the Environmental Research Institute of Michigan (ERIM) produced a new image map every three days.

BACKGROUND

In the Kingdom of Saudi Arabia, remote sensing has always been considered a logistical and technical support technology rather than just another esoteric research tool. Between 1974 and 1985, many significant technological achievements, particularly in cartography and image processing for geology, were pioneered in Saudi Arabia using Landsat data. In Saudi Arabia, remote sensing applications are considered to be a means of achieving expedient and cost-effective solutions to many technical and logistical problems encountered by operational programs of the Saudi Arabian government and cooperating foreign organizations.

Research with remotely sensed data was of little or no interest to users during the first years of application activities in the Kingdom. When research was conducted, it was to find ways to make remote sensing products more effective in supporting operational programs. Operational and logistical requirements have pulled the technology into the forefront of geo-scientific investigations in the Kingdom. The technology did not have to introduce or even sell itself. It was invited into the country to satisfy pressing needs.

REMOTE SENSING TECHNICAL SUPPORT FOR OPERATIONS

Since 1944, the U.S. Geological Survey (USGS) has had a mission in Saudi Arabia to investigate the mineral potential of the western part of the Kingdom. Since little was initially known about the region, preparation of basic geological and geographical maps was necessary. Between 1950 and 1963, a series of 1:500,000 scale geological and geographical maps covering the entire Kingdom was published in cooperation with the Arabian American Oil Company (ARAMCO). These maps were compiled on bases created from aerial photographs obtained during two separate photographic acquisition programs between 1950 and 1957.

In 1969, as investigations and exploration became more focused, the USGS began a 1:100,000 scale mapping program. These maps were based on the same aerial photography used to produce the earlier 1:500,000 scale map series. In 1973, a year after Landsat 1 was launched, nearly 50 of these quadrangles had been published. In that year, topographic engineers discovered that the photographic bases for two quadrangles in the southwestern part of the country could not be edge matched because they straddled areas on the edges of the two previously mentioned photographic missions. Photos from the two missions possessed different geometric distortions and were incompatible. Field work and publishing schedules did not permit time to acquire new aerial coverage. A common denominator was needed, and it was found in Landsat Multispectral Scanner System (MSS) data.

Full MSS coverage of the Kingdom was obtained during 1973. (See Fig. 1.) MSS Band 7 images were used to prepare the 1:100,000 scale bases for the two 30-minute quadrangles. These bases, produced at the DMMR photo lab in Jiddah, were the first of their kind in the world. Digital processing technology was in its infancy; MSS scanner lines, which become patently obvious at 1:100,000 scale, were obscured using a number 2 lead pencil. The desired results were achieved,

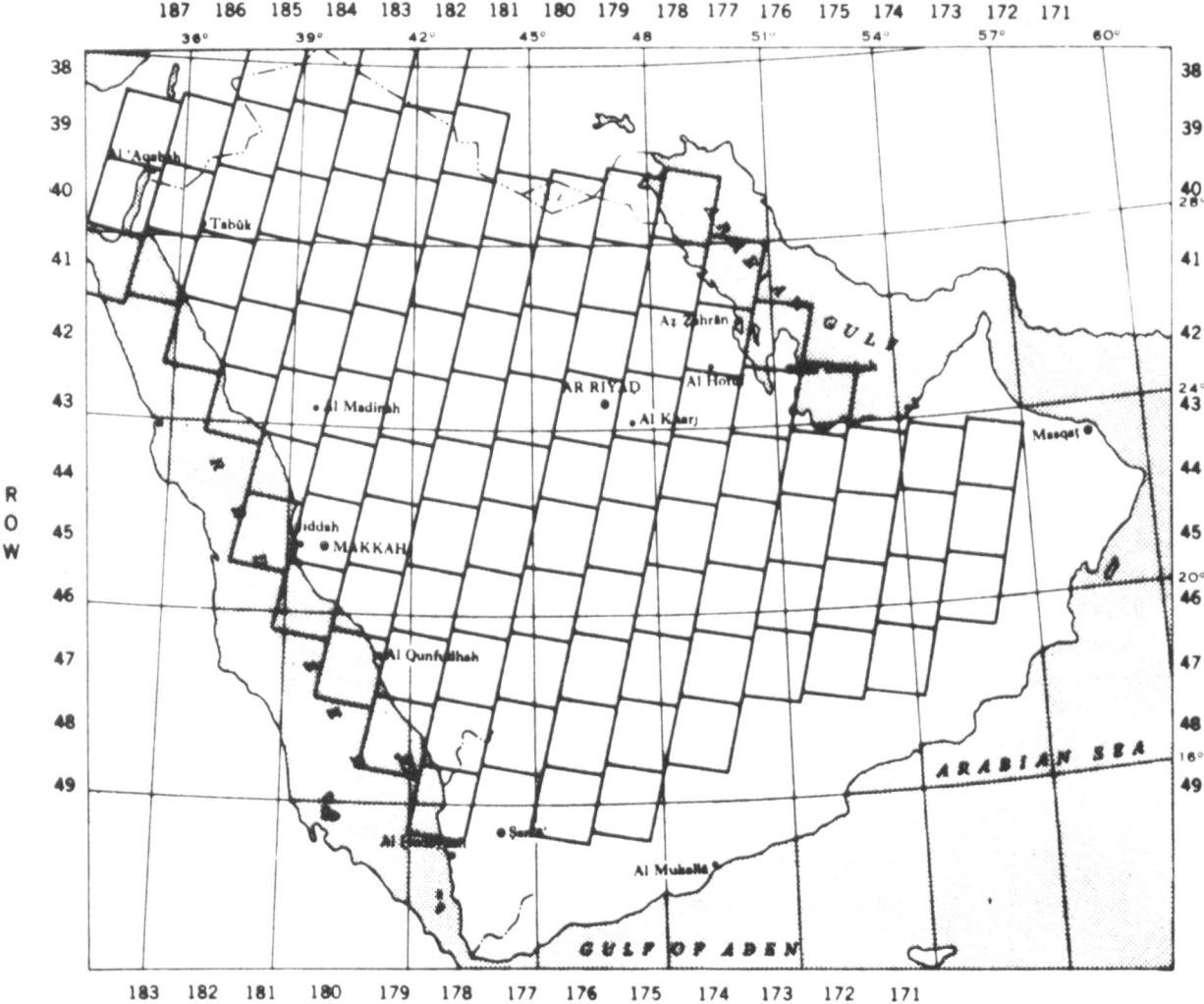

Fig. 1 - Index of Landsat Images, Kingdom of Saudi Arabia

field work proceeded without interruption, the maps were published on time, and the edges of neighboring quadrangles were properly matched. This modest success story set the stage in Saudi Arabia for an ambitious remote sensing program which has continued unabated to the present.

Base mapping program

Remote sensing activities in Saudi Arabia began in earnest in 1977 when the USGS assigned the author as a full-time remote sensing specialist to the USGS Saudi Arabian Mission in Jiddah. One of the initial tasks was to coordinate production of digitally-mosaicked, 1:250,000 scale, geometrically-corrected Landsat MSS base maps. These maps were to be used for compiling a series of geological and geographical maps which synthesized earlier work published at 1:100,000 scale.

The Arabian environment has two characteristics which are conducive to the production of images and mosaics with maximum geological and topographical detail. The first characteristic is predictable and prolonged periods of low cloud-cover; the second is minimal vegetative cover to obscure geological features. Unfortunately, extensive areas of highly reflective sand, silt and sedimentary rock saturate the MSS detectors and reduce tonal variations in the data.

Standard false color composite and black and white Landsat images initially produced for the Saudi program were heavily saturated and, to a large extent, useless except in the western third of the country which is dominated by darker toned Precambrian igneous and metamorphic rocks. The eastern two thirds of the country dominated by sedimentary Phanerozoic cover rocks were washed out. Quadrangles covering areas along the eastern edge of the Precambrian region and the western edge of the Phanerozoic cover were of no interpretative value to compilers and were cartographically unaesthetic.

The problem was only partially solved using images acquired during the annual period of low sun elevation. A satisfactory solution was reached at the USGS Flagstaff Facility by introducing a sun elevation correction procedure when the images were initially processed. This program created images with a uniform appearance over both light and dark areas. Band 7 data, chosen for its superior geological information content, were processed using a high pass filter which accentuated topographic shadows and linear features. Up to four MSS images were then digitally mosaicked using geodetic control data generated in the field by Doppler satellite positioning receivers to produce 1 by 1 1/2 degree quadrangles. (See Fig. 2.) A production pipeline was developed by trial and error which eventually led to one quadrangle being produced every three days at the Environmental Research Institute of Michigan (ERIM). (See Fig. 3.) All of these quadrangles exceeded the USGS accuracy standard of better than 300 metres root mean squared (RMS) error for maps at 1:250,000 scale. Between 1977 and 1984, a total of 111 quadrangles was published. (See Fig. 4.) The same production methods, MSS data, and geodetic control were used to produce 1:500,000 and 1:1,000,000 scale geometrically corrected color image maps for much of the country in 1984 and 1985. (See Fig. 5.)

Another Landsat mapping program was initiated in 1977 and completed in 1984 with the publication of a 1:2,000,000 scale geographic map of the Arabian Peninsula. This map replaced an out-of-date traditional geographic map published in 1963. Three years were required to assemble from 256 Landsat images what is still the largest film mosaic in the world. An additional three years were needed to compile the geographical overlay in cooperation with ARAMCO which eventually totalled 21 separate printing plates. Since many of the images used in the mosaic were acquired in 1973 and 1974, they did not show the extensive highway and other infrastructure developments begun by the government in 1976. Therefore, new Landsat Return Beam Vidicon (RBV) images were acquired over most of the country. The increased resolution of the RBV data facilitated determination of the location of roads and villages. This information, supplemented with information gathered in the field, was transferred manually to the master compilation mosaic.

Image processing for geological mapping

Since the 1:100,000 scale maps had been compiled by different geologists using black-and-white aerial photography over a period of years, rock unit boundaries and structural feature did not always join from sheet to sheet. Therefore geologists re-compiling these sheets on 1:250,000 scale quadrangles used false color MSS images (Bands 7 in red, 5 in green and 4 in blue) at the same scale to help identify where joins should occur. The color images were processed using a high pass filter, and a sun elevation adjustment algorithm. Scanner lines were digitally removed.

In addition to these images, nineteen other types of image products were processed to facilitate special mapping projects. The Bureau de Recherches Geologiques et Minieres (BRGM) used ratioed images to map the lithology of part of the Tuwaiq Escarpment in central Saudi Arabia. Riofinex Ltd., a subsidiary of Rio

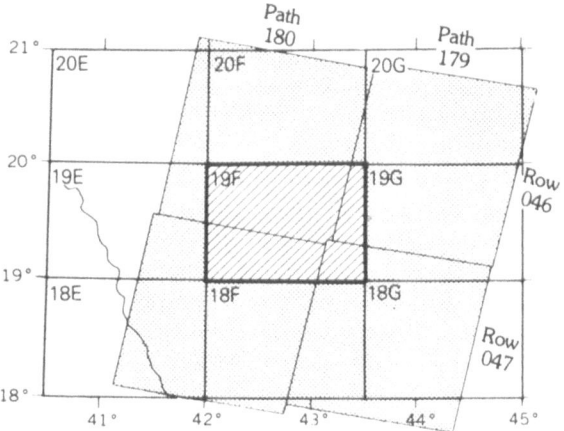

Image Index

Path/row	Image ID	Date acquired
179/046	1118–07003–7	Nov. 18, 1972
179/047	2330–06464–7	Dec. 18, 1975
180/046	1155–07060–7	Dec. 25, 1972
180/047	1155–07063–7	Dec. 25, 1972

Fig. 2 - Diagram showing typical orientation of Landsat images on a 1:250,000 scale geometrically corrected image map.

Tinto Zinc, used ratios to locate and map the extent of duricrust in the north central part of the country as part of a phosphate exploration program. The USGS used ratios to map the geology of Jabal Aja, the largest pluton in the peninsula.

Until 1985, all image processing was accomplished at facilities outside Saudi Arabia because it was more time- and cost-effective to rely on existing facilities with experienced staff. The USGS Flagstaff Facility, ERIM, and the Commonwealth Scientific and Industrial Research Organization (CSIRO) in Australia processed data according to specifications supplied by Saudi Arabia. In-country field geologists did all of the interpretation.

REMOTE SENSING SUPPORT FOR FIELD OPERATIONS

Landsat data products were heavily used by geologists and support staff to help conduct safe, cost-effective field work. Landsat data were used for planning field work, camp site selection and for land, sea and air navigation.

Planning field work

Geologists used color Landsat images to identify areas requiring closer examination. The images were annotated to show areas where work was completed and to show where more work was needed. Once a geologist had located features of interest, he could site the position of his camp. Camp siting was influenced by location of villages, where water and other consumable goods could be obtained, the location of roads to facilitate transportation of camp equipment, and

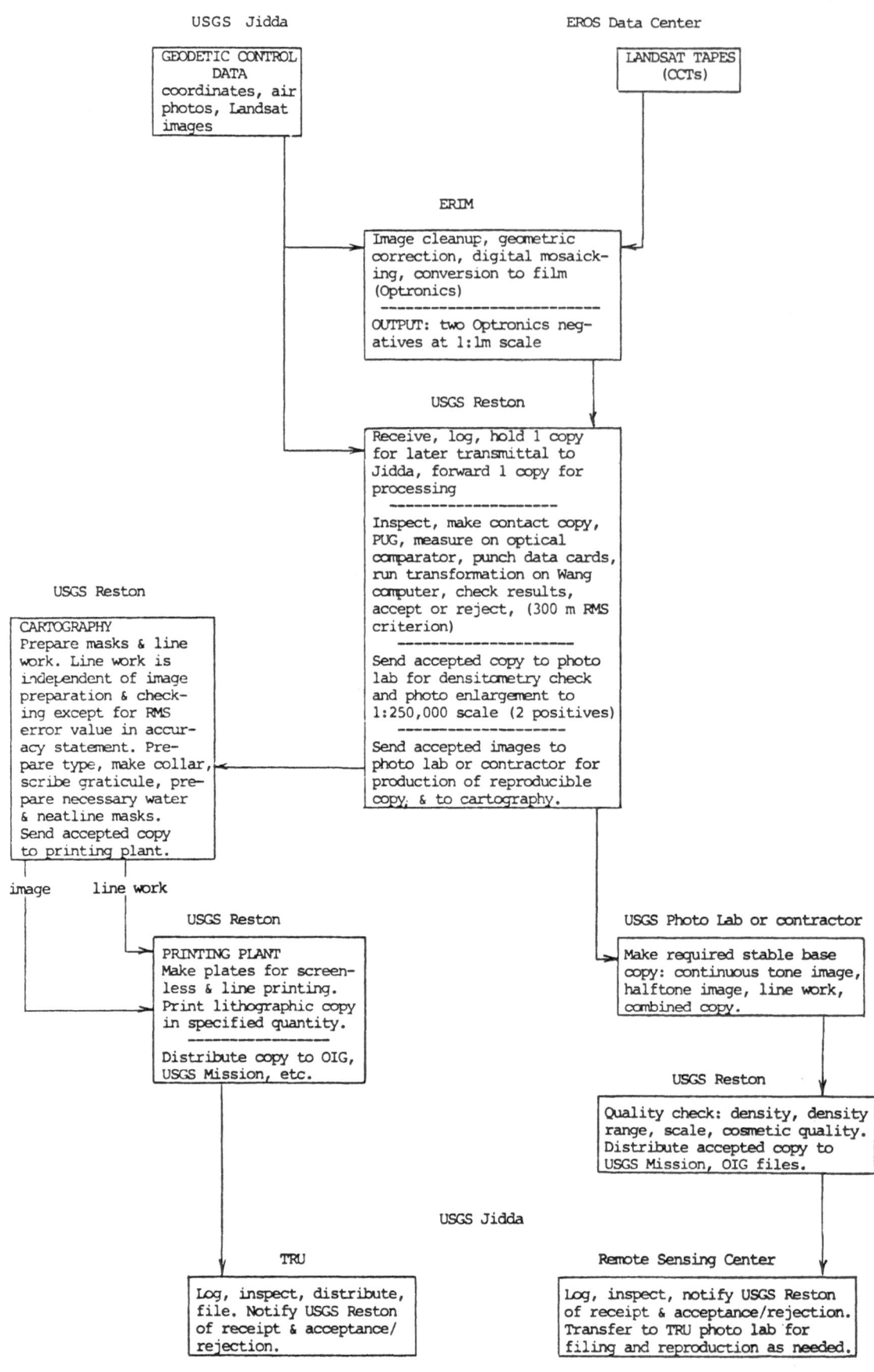

USGS Jidda

GEODETIC CONTROL
DATA
coordinates, air
photos, Landsat
images

EROS Data Center

LANDSAT TAPES
(CCTs)

ERIM

Image cleanup, geometric
correction, digital mosaick-
ing, conversion to film
(Optronics)

OUTPUT: two Optronics negatives at 1:1m scale

USGS Reston

Receive, log, hold 1 copy
for later transmittal to
Jidda, forward 1 copy for
processing

Inspect, make contact copy,
PUG, measure on optical
comparator, punch data cards,
run transformation on Wang
computer, check results,
accept or reject, (300 m RMS
criterion)

Send accepted copy to photo
lab for densitometry check
and photo enlargement to
1:250,000 scale (2 positives)

Send accepted images to
photo lab or contractor for
production of reproducible
copy, & to cartography.

USGS Reston

CARTOGRAPHY
Prepare masks & line
work. Line work is
independent of image
preparation & check-
ing except for RMS
error value in accur-
acy statement. Pre-
pare type, make collar,
scribe graticule, pre-
pare necessary water
& neatline masks.
Send accepted copy
to printing plant.

image   line work

USGS Reston

PRINTING PLANT
Make plates for screen-
less & line printing.
Print lithographic copy
in specified quantity.

Distribute copy to OIG,
USGS Mission, etc.

USGS Photo Lab or contractor

Make required stable base
copy: continuous tone image,
halftone image, line work,
combined copy.

USGS Reston

Quality check: density, density
range, scale, cosmetic quality.
Distribute accepted copy to
USGS Mission, OIG files.

USGS Jidda

TRU

Log, inspect, distribute,
file. Notify USGS Reston
of receipt & acceptance/
rejection.

Remote Sensing Center

Log, inspect, notify USGS Reston
of receipt & acceptance/rejection.
Transfer to TRU photo lab for
filing and reproduction as needed.

Fig. 3 - Stages of production for 1:250,000 scale image maps.

270

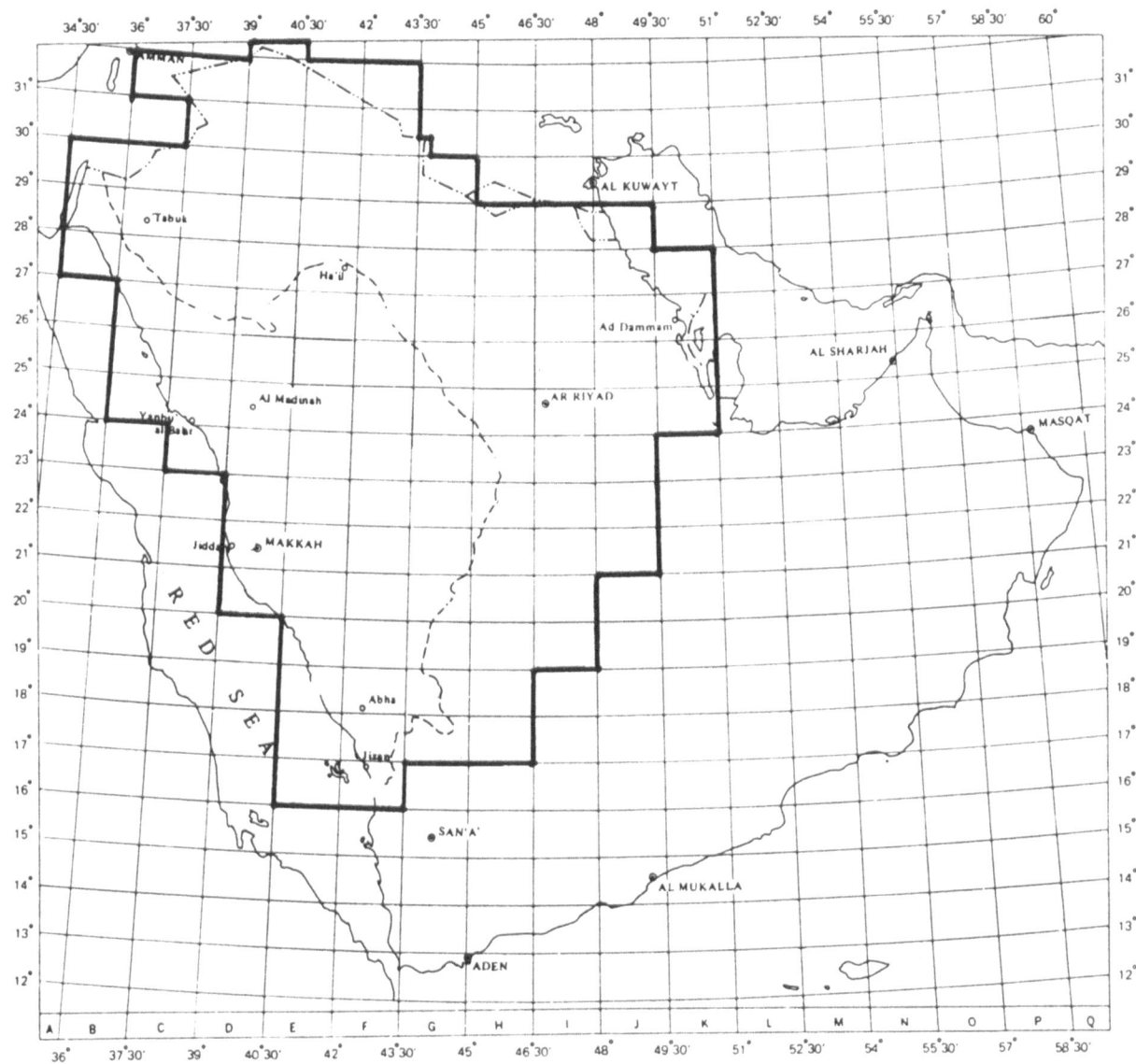

Fig. 4 - Index map showing location of geodetically controlled Landsat base maps at 1:250,000 scale.

distance to study sites. All this information was clearly evident on the images. With few exceptions, helicopters were assigned to each camp. Since the Landsat images enabled the geologists to make accurate distance measurements, they could also estimate the amount of flying time and, therefore, the fuel necessary to complete a work day.

Navigation

The inventory of geometrically corrected image maps was extensively used for air navigation by. helicopter pilots and field geologists. The 1;500,000 scale color image maps were most popular because they provided a reasonable reproduction of actual terrain conditions, enabled users to accurately plot their position by geographical coordinates, and could be folded into a convenient size for use in cramped helicopter cockpits.

In 1981, Landsat bathymetric charts were produced for the entire Red Sea coast of Saudi Arabia. These were used by Seltrust, Ltd., to navigate a research vessel conducting a seismic reflection survey. There are numerous, treacherous atolls off the Saudi coast. The only navigational aids available in 1981 were old, highly generalized British Admiralty Charts published, in some cases, from surveys conducted in the nineteenth century.

This research vessel had a long cable trailing behind it which contained emitters used for the reflection survey. The ship had to maintain a constant speed in order to prevent the cable from sinking and snagging on coral structures. It was crucially important to have an accurate depiction of navigational hazards and the location of channels for safe, uninterrupted traverses. The Landsat bathymetric charts were produced at ERIM and calibrated with depths

271

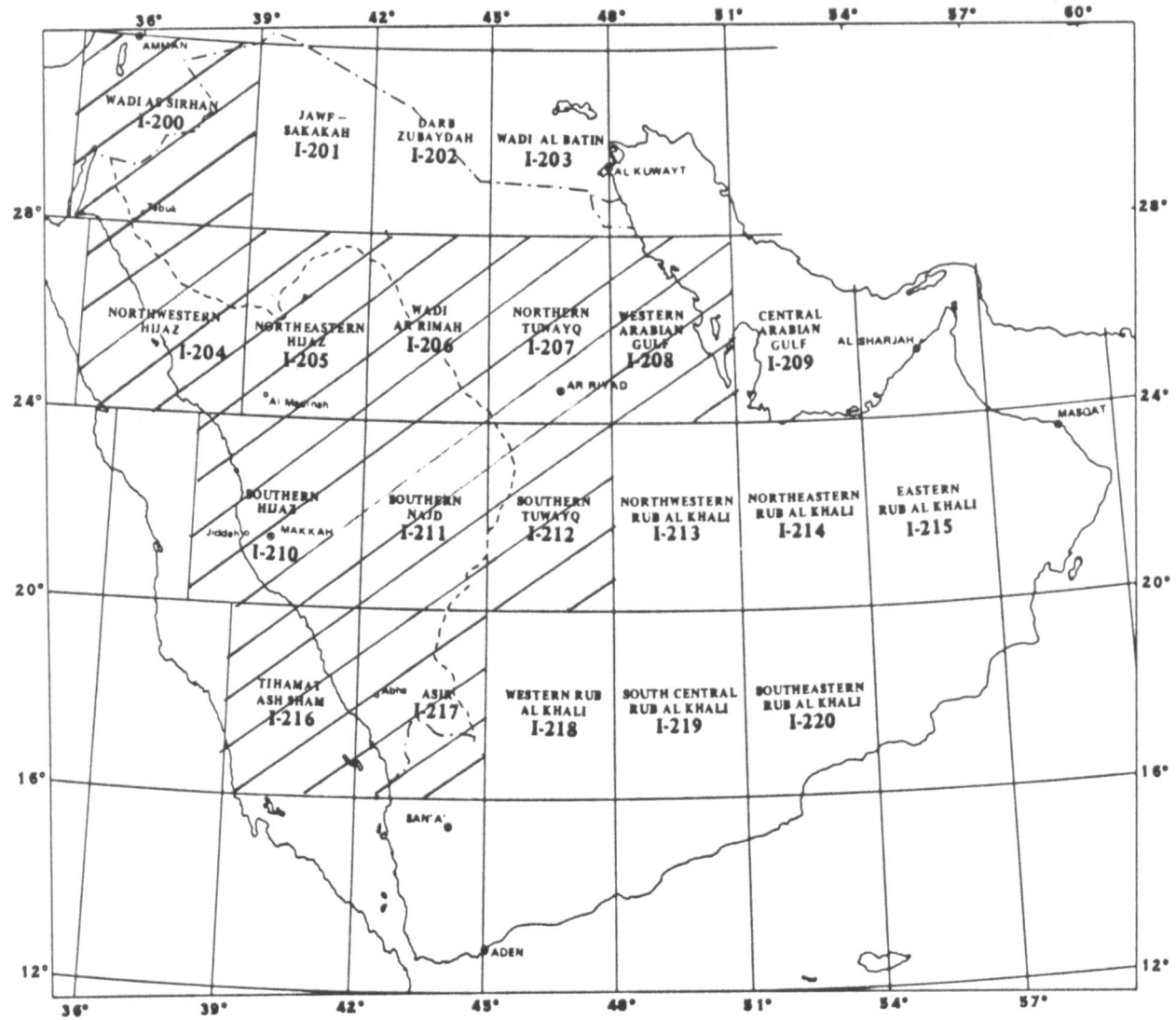

Fig. 5 - Index map showing 1:500,000 scale geodetically controlled quadrangles produced by ERIM for the DMMR.

determined to be reliable on the Admiralty Charts. The program was successfully completed on schedule without damage to the cable or vessel.

REMOTE SENSING APPLICATIONS FACILITY

A remote sensing centre was established in Jiddah, Saudi Arabia, in 1977 and continues to operate today. The purpose of this facility is to provide a data archive and user services, such as consultation, analysis and training. The center was initially designed to provide manual remote sensing services. In response to substantial user demand, a digital image processing system was added in 1985.

Data archive

A total of 113 Landsat images provide complete coverage of Saudi Arabia. A few years after the centre was established, certain types of MSS images were found to be particularly useful to earth scientists. The color, high pass filtered bands 7,5,4 product, and a black and white, high

pass filtered band 7 product were in constant demand.

Files of these images at scales of 1:1,000,000, 1:500,000 and 1:250,000 were maintained in multiple copies. Larger images were stored in flat map cases; smaller images were stored in vertical file units. The entire collection was indexed using the Landsat Worldwide Reference System. Affiliated users simply stated the type, scale and path/row number when placing an order, and the images were immediately provided. An average of 800 image products were distributed annually.

In addition to images, a file of geographic maps published by the Defense Mapping Agency was maintained for use on the premises. These maps included the 1:250,000 scale Joint Operations Graphic (JOG), 1:500,000 scale Tactical Pilotage Charts (TPC) and 1:1,000,000 scale Operational Navigation Charts (ONC). Users frequently referred to these maps to obtain supplementary information, such as place names and approximate elevations.

Files of positive transparencies of the black and white and color images were maintained at the DMMR photographic laboratory. Replacement prints for the archive were reproduced from these positives. Positive rather than negative reproducibles were used because reproduction is simpler and the positives can be used as accurate guides for achieving correct color balance or gray scale values.

Consultation, analysis and training services

Saudi counterpart staff at the centre were trained in remote sensing by the author and then later sent to the USGS Earth Resources Observation Systems (EROS) Data Center (EDC) for more intensive training. In addition, over 25 Saudi earth scientists were sent to the EDC for similar training. Remote sensing center staff became skilled in interviewing users to determine user applications problems. Project managers and field scientists were provided with as much information and explanation as they needed to solve their specific problems. The highly technical language of remote sensing was translated into everyday terminology. Complicated theoretical explanations and discourses on remote sensing were kept to a minimum.

SUMMARY

Remote sensing data were introduced to Saudi Arabia as a practical and inexpensive way to satisfy some informational needs of the earth science community and to expediently achieve work objectives. Remote sensing research was the exception rather than the rule. The highly technical terminology of remote sensing was simplified to attract the maximum number of users. In Saudi Arabia in 1977, there were only two data users from one government organization. By 1985, there were 96 users from 15 government and private organizations. Many of these users were applying Landsat Thematic Mapper (TM) data which became available in 1983 after the launch of the first Tracking and Data Relay Satellite (TDRS). Now in 1990 there are six remote sensing centers with digital processing capability located around the country, and an advanced ground station located in Riyadh receiving Landsat and other remote sensing data.

REFERENCES

Ferguson, K.P. Jr., The Saudi Arabian Deputy Ministry for Mineral Resources Remote Sensing Center: Services and Applications of Landsat Data, U.S. Geological Survey Saudi Arabian Mission Open-File Report, OF-01-2 Jiddah, 1982.

Ferguson, K.P. Jr., "Application of Landsat Data to Earth Science Investigations in the Kingdom of Saudi Arabia", in Exploration and Peaceful Uses of Outer Space, National Paper of the Kingdom of Saudi Arabia, Saudi Arabian National Center for Science and Technology, Riyadh, 1982.

Greenwood, W.R. and Ferguson, K.P. Jr., Instructions for Compilation of 1:250,000 Scale Geologic and Geographic Maps, U.S. Geological Survey Saudi Arabian Mission Administrative Document, Jiddah, 1980.

Ferguson, K.P. Jr., Jackson, R.O. Hadley, D.G., Ramirez, L.F., Mandaville, J.P., and Bowers, S.D., 1984, Geographic Map of the Arabian Peninsula: Saudi Arabian Deputy Ministry for Mineral Resources Arabian Peninsula Map AP-5-B2, Scale 1:2,000,000.

Recent and future developments in producing exploration base maps from satellite imagery

N. P. Press B.Sc., A.R.S.M., M.I.M.M.
Nigel Press Associates, Edenbridge, Kent, England
J-C. Rivereau L.Es/Sc., A.E.I.F.P.
Technical and Development Department, SPOT Image S.A., Toulouse, France

SYNOPSIS

One of the most important prerequisities for an exploration programme is to have accurate and up-to-date topographic maps, yet less than 40% of the world's land areas are mapped at adequate scales, let alone kept up-to date.

Satellite imagery has been routinely available for most areas since the early 1970's, with significant technical improvements occurring in the early and mid-1980's. However, the use of satellite image derived mapping, although gradually becoming more widely adopted is still by no means commonplace, despite the obvious practical advantages and benefits.

This paper is concerned purely with planimetric and topographic (rather than geological, but including offshore bathymetry) mapping and outlines the main applications, data sources, production methods, specifications, budget costs and near future trends in order to acquaint explorationists with the possibilities open to them and encourage wider use of satellite derived mapping in the future.

INTRODUCTION; WHY BASE MAPS ARE IMPORTANT

Explorationists, whether searching for industrial minerals, ores and gemstones or coal, oil and gas require accurate maps, both topographic and geological, on- and off-shore. Onshore, and to a limited extent offshore such maps can now to a high degree be produced from satellite remote sensed imagery.

This paper will consider solely the question of topographic mapping. Current U.N. statistics show that only around 40% of the world's landmass is mapped at scales of 1:100,000 or better. Details on the accuracy and, particularly, currency of such cartographic data are less clear. There are many parts of the world where urban and infrastructure development is proceeding apace or dramatic changes are occurring in agricultural and forestry patterns yet the available topographic maps, although correct in elevation detail may still represent the land cover situation pertaining several decades ago.

Explorationists wishing to conduct geochemical sampling or cost access for seismic crews would find information about such changes of crucial importance. Great time and budget savings could result if an area expected to be forested proved to have been cleared. Enormous access costs or difficulty for seismic work would be incurred if an area has recently urban or agricultural development or contains a military airfield (commonly not shown on published maps). Information such as this can be readily obtained from satellite imagery.

Offshore, in clear water areas satellite imagery can be used to map relative bathymetry to depths of 20-30m. For many areas, such bathymetric maps are not available, and those maps that are available often derive largely from 19th century lead-line measurements. The resulting bathymetric patterns are interpolated from sparsely spread points, whereas satellite image data has the benefit of even distribution in the horizontal plane. Areas of shallow water and reefs can be detected and such information is of great value for both safety and logistical planning in conducting offshore seismic surveys or locating possible drilling platforms.

Returning on-shore, but remaining in the realm of elevation data, some types of satellite imagery are now available in stereoscopic form and so can be used for elevation determination and the preparation of terrain contour maps. This is, of course, highly valuable in areas where no contour maps exist and finds applications ranging from construction of geological cross-sections to measuring the heights of sand dunes to cost and plan logistics and seismic acquisition in remote areas. Of growing importance is the role of stereo satellite imagery as a rapid and cost effective means of generating a D.E.M. (Digital Elevation Model) which is one of the fundamental imputs required for a G.I.S. (Geographical Information System). Although D.E.M.'s can be produced by other means such as digitising from large scale maps or through aerial photogrammetry, these methods are quite time consuming and costly.

G.I.S. techniques are becoming increasingly valuable for co-registering data-sets such as topography, terrain cover, bedrock geology, sub-surface contour maps, surface geochemistry etc. to apply computer-based statistical methods and graphical presentation tools in searching for subtle, yet significant data inter-relationships which may have relevance to exploration. In such a process, satellite imagery may play an important role as the basic raster underlay for the data set, as well as being the source for several of the different "layers" involved.

DATA AVAILABLE

Operational Sources

Medium to high resolution digital imagery from automatic unmanned satellites has been routinely available for most areas of the world, with the exception of parts of the tropics and the arctics (due to high cloud cover and paucity of data transmission arrangements) since the early 1970's. The principal operational systems are described briefly in Table 1.

In addition to these digital sources, the photographic material recently made available from the Russian Sojuz missions should also be mentioned. This provides high resolution (5-10m.) cover of many areas on large format film, much in stereo.

Limited data are available from Russian, Japanese and Indian experimental satellites but these are not considered viable at present for operational work.

There have also been experimental missions with radar systems (Seasat, SIR-A, SIR-B and Almaz) and high resolution photographic equipment (metric cameras etc.) but none of these provides sufficient consistent cover to be considered operational. However, these missions have provided useful information to enable evaluation of the utility of future planned missions, particularly with radar, in comparison with existing sources.

Choice of suitable data: Comparing between operational sources

The photographic material from the Sojuz mission stands apart from the Landsat and SPOT data because it is not so easy to manipulate digitally. As outlined below, digital processing is an important aspect of mapping production techniques. Although the Soyuz imagery can be scanned to create a digital record, this can result in some loss of definition and the dynamic range recorded on the film types used can already be quite restricted in some terrains. On the other hand the stereo capability and the relatively lower cost are advantages to be taken into consideration.

In comparing the other data sources (for the purposes of topographic mapping and producing basemaps), the spectral response and dynamic range of the various satellite systems are fairly similar (Landsat TM however, has a distinct advantage in its spectral response for certain types of geological investigation). SPOT is the only real source of digital stereo data; although a limited stereo effect can be obtained on the overlap between adjacent Landsat passes this is not normally of viable practical use.

Thereafter the choice of data is a trade-off between the area of cover required, suitable scale of presentation and level of ground detail necessary. Since SPOT images cover an area 60x60km. or 60x120km. in twin mode as opposed to the 185x185km. of Landsat, for large areas more individual SPOT scenes can be required as opposed to Landsat with a commensurate increase in data acquisition and processing costs. Against these factors has to be balanced the way the available images fall relative to the area of interest and of course the availability of cloud free archived data for suitable seasonal conditions depending on the climatic zone.

NPA have developed a computer mapping system "LANDSPOT" to help in this process, and computer listings of archived scenes with geographical co-ordinates of corner points and estimates of cloud cover are available for Landsat, SPOT and Soviet data.

TABLE 1: PRINCIPAL OPERATIONAL SOURCES OF DIGITAL IMAGERY

SOURCE	AVAILABLE SINCE	RESOLUTION	OPTIMUM SCALE	ATTRIBUTES
Landsat MSS	early 70's	80m	1:250,000	
Landsat TM	early 80's	30m	1:100,000	mid i/r bands
SPOT	mid 80's	10-20m	1:25/50,000	stereo

For areas where no suitable data are yet available from archive, each of the satellite operators offers an acquisition service on demand. Before embarking on this it is necessary to consider why data may not already be available. Since the Sojuz data were originally intended for Soviet interests and their commercialisation programme has only recently made the imagery available to non-Soviets, coverage is mostly concentrated on areas of recent Soviet strategic concern. Acquisition of data on other areas is a comparatively new development and requests for new data must be scheduled on future missions (data covering Soviet territories are not available to non-Soviet organisations).

Landsat and SPOT, on the other hand have concentrated for some years on building a comprehensive worldwide archive although there are still many areas where no SPOT or Landsat (particularly TM) data exist. This may be due to several factors:

a. lack of anticipated demand.
b. persistently cloudy weather conditions.
c. distance from adequate ground reception facilities.

In the case of areas with persistent cloud, SPOT does have a theoretical advantage over Landsat, because of its more frequent overpass capability due to the pointable sensor. In practice this has clearly worked, since the SPOT archive contains cloud free imagery of many areas which have never been captured cloud free with Landsat.

Within direct range of ground receiving stations there is a greater chance of finding archived cloud free Landsat data for difficult areas but outside these ranges, Landsat must rely on the TDRSS relay satellites to return data to earth, whereas SPOT uses on-board tape recorders which replay the data to earth later. From operational experience it has been found that SPOT can generally acquire new data for difficult areas more rapidly but nevertheless there are still a few parts of the world such as equatorial areas in Africa, central South America and the S.E. Asia archipalego where the only solution will be the forthcoming generation of radar imaging satellites.

Finally, in this complicated equation, there is the technical question of geometric and geographic accuracy. Inherently the Sojuz and SPOT data have higher internal geometric fidelity than Landsat, Sojuz because it is a photographic exposure, in some cases onto a resau marked film plane, SPOT because it is a more recently developed platform and sensor system than Landsat and so has fewer sources of mechanical disturbance.

Externally, and perhaps of greater importance, the geographic registration of SPOT data (ie. the exact location of the image relative to the geographical grid as determined from satellite ephemeris data) appears from operational experience to be superior to Landsat and Sojuz data. In areas lacking geographic registration from either existing mapping or ground control this might be an over-riding factor in data selection.

PRODUCTION METHODS

Image Processing

Landsat and SPOT data are already in digital form (although photographic products can be purchased separately). High resolution Sojuz data is supplied in photographic form and so must be scanned if digital image processing is required for mosaicing, geographic rectification and enhancement. Since the Sojuz films are often on larger format than conventional aerial photography or satellite imagery this must be taken into account in scanner hardware. As mentioned above there will be some loss of fine detail in scanning from film to digital format, but this could be partly made up with image processing techniques.

Image processing is commonly used to join images from different satellite passes to make one contiguous image of an area (it is surprising how many areas of interest fall at the junction of four nominal frames!). SPOT images can be supplied repositioned along the satellite track at no extra cost. Landsat TM sub-scenes can be supplied positioned anywhere in the nominal frame. These services help to reduce the mosaicing effort, and the most common problem is in side-ways joining images. This can be most acute where images have been acquired on widely different dates, but image processing techniques can be used to compensate to a certain extent for the different radiance ranges due to seasonal effects.

Image processing is also used to warp images to fit different map projections and correct to ground control points to remove distortion caused by perturbations in the satellite orbit. To a certain extent such corrections are already applied to the imagery, but, particularly with older Landsat data, rectification to ground control points may be required to achieve acceptable accuracy at 1:100,000 or larger scales.

This type of geometric correction is achieved by identifying points on the image whose precise ground co-ordinates are known either from accurate published mapping or terrestrial or satellite field survey work.

277

The most notable development in this area is the GPS (Global Positioning Satellite) system which enables real time triangulation between members of a constellation of satellites to compute a highly accurate geographic position for the ground receiver. Very recently, low cost (around $4,000) hand held receivers which are extremely easy to operate have been introduced and can produce direct readings of sufficient accuracy (approximately 20m.) to be used in geographic rectification and registration of satellite imagery. At present the GPS constellation is incomplete, so readings can only be made in certain time-windows of a few hours, although several may occur in a 24 hour cycle. When the constellation is complete in 1991, operation should be continuous.

Where possible, conventional survey or GPS ground control points are recommended for correcting and rectifying satellite imagery because of the difficulties in unambiguously recognising common points between the suitable map and the image, and lack of knowledge of the absolute positional accuracy of recognisable points on the map. It is very rare that triangulation point monuments used in national surveys can be identified on images.

Correcting a satellite image to dubious ground control points is likely to cause internal distortion in the end product far worse than not correcting the image at all.

Finally image processing techniques can be used to "sharpen" the appearance of the satellite image by edge enhancement or high pass filtering and to balance out the contrast and tonal or colour ranges to produce the best visual effect.

Printing Methods

In general, custom produced exploration base maps will be in limited editions, so printing will be restricted to photographic methods. If longer runs are required lithographic methods can be applied in the normal way.

Photographic printing falls into two main categories:

1. where the entire product; image, graticule, annotation and cartographic detail are written to a film negative and printed in one exposure.

2. a registration system where the image is projected into a series of masks or overlays to achieve the required composite map.

Using method 1 has the advantage that no particularly special photographic equipment is required and the work can be carried out by any properly equipped large-format printing laboratory. However, with this method it is very difficult to produce a map with a clear white border and sharp cartographic detail, furthermore specialised software is required to merge the raster image with the cartographic detail which is basically in vector format. Method 1 is the only method suitable for large format colour electrosatic plotters which are increasingly being used to generate maps that, although not as high quality as photographic products are very acceptable for base mapping purposes.

A final drawback of method 1 is that if changes need to be made to the final product, new information needs to be added or more accurate ground control becomes available then further image processing and a new film writing procedure have to be initiated before re-printing.

Method 2 requires more sophisticated photographic equipment in the form of process cameras and registration systems as well as technical skills but has the advantages of:

a. producing maps with more conventional presentation and finer cartographic detail.

b. the ability to directly incorporate existing cartography without re-digitising.

c. the possibility of updating the detail of the maps through modification of the cartographic overlays and improving the geographic positioning of the image without having to reprocess or re-film-write the image data.

d. it is possible to produce cartographic style products from Sojuz photo-data without scanning the film into digital format (although without geometric rectification).

SPECIFICATIONS: ACCURACY AND CONTROL

Map Projections

The ground tracks of polar orbiting satellites lie at an oblique angle to lines of longitude due to rotation of the earth, and this causes the orientation of satellite images to be skewed relative to true north.

The basic geographical projection of satellite imagery is Space Oblique Mercator. Locally this is very close to Universal Transverse Mercator and other related or similar projections, that is within the extent of one or two satellite scenes. Only over larger areas would contiguous mosaiced satellite images need to be geometrically rectified to achieve

U.T.M. projection. For other map projections geometric rectification of the imagery through computer resampling techniques will be necessary. This adds to the cost of map production.

Ground Control and Geodetic Extrapolation

Mention has already been made of the need for ground control information and the means of incorporating it. Examples of suitable ground control points are intersections of roads and rivers and rock bluffs which can be unequivocably identified. In certain terrains other man-made features such as oil well-heads, pipe lines and other engineering installations and even traces of old seismic lines can be recognised, and since such features are often already positioned with reliable accuracy they can be used in the geographic registration and rectification process. With SPOT imagery it has been found possible to extrapolate control data over long distances up or down track such that the control points may be some distance from the target area. This clearly has advantages in rugged and remote areas where access costs may be higher than satellite data acquisition, or due to local political circumstances fieldwork may be temporarily inadvisable.

Positional Accuracy

Given accurate ground control information satellite images can theoretically be resampled and corrected to sub-pixel accuracy. Considering a map scale of 1:100,000 this means that in the ideal case positional accuracy of a point on the image relative to the geographical grid on the map would be around 3mm. with TM data and 1-2mm. with SPOT data.

In practise it is found that higher numbers of ground control points are required to approach such accuracies with Landsat TM data than with SPOT, probably due to the older design and characteristics of the Landsat system.

With SPOT imagery, the ephemeris data alone give a geographical accuracy better than 500m. One reliable control point can improve the positioning of the geographic grid to better than 100m. and three points will will reach better than 50m. At 1:100,000 scale this could still produce an offset on the map of 5mm. in the worst case, but within other confines such as paper shrinkage etc. this may be acceptable for initial exploration basemap purposes.

If large areas are to be covered at smaller scales, for example using Landsat MSS or TM data, in the absence of other ground control it is worthwhile considering using a network of SPOT scenes to introduce a primary level of geodetic control since the satellite ephemeris data for Landsat are less reliable in quality, especially for areas remote from receiving stations.

In operational practice, our experience has been that it is not always possible to obtain new ground control and some sort of compromise must be reached in producing satellite image maps in terms of geographic accuracy. The normal situation is to achieve the best fit to existing published maps, in which case if there are significant discrepancies between the map and the basic image it is generally a mistake to force fit the image to the map since although obvious features may be made to coincide internal distortions often end up being greater. There are very many parts of the world where the only available mapping is 1:1 million scale air navigation charts and it is clear that these cannot be regarded as a source of ground information for producing maps at 1:250,000 scale or larger.

Elevation Data and DEM Generation

Elevation data can be extracted from stereo satellite imagery either using digital image processing techniques to measure the offset between identical ground points on the two images or with analogue, photogrammetric techniques. In theory stereo-Sojuz data could be digitally scanned for the former method but no such operational projects are recorded.

Photogrammetric techniques require the use of digital stereo-comparators or analogue photogrammetric equipment upgraded with digital encoders and special software. Given suitable equipment the procedure can be quite rapid and cost-effective, especially if only terrain profiles are required.

The approach using digital stereo-correlation, on the other hand, is demanding on computer processing resources and may therefore appear quite costly. However, it does provide a new 'image' where every pixel represents an elevation value and may be a prerequisite for G.I.S. compilation where satellite imagery is the base.

Although maps or profiles of relative elevation can be prepared, to obtain absolute elevations and contour maps requires good ground control information, with a minimum of 20 points evenly spread around the image, accurate in the x,y and z dimensions.

Given good ground control data elevation accuracies in the order of 10m. have been claimed with SPOT data.

An additional benefit of D.E.M. generation is the resulting ability to produce perspective views of terrain as well as 'fly-past' sequences, both of which are helpful for logistical purposes to acquaint management and field personnel alike with terrain conditions and hazards, as well as having value in interpreting other data such as geochemical dispersion patterns.

COSTS OF MAP PREPARATION

A very general outline of likely costs for different types of maps will be given in U.S.$ at 1990 rates. The main data costs are well known, in simplistic terms these approximate to the following figures per sq.km.

Landsat MSS	$0.03
Landsat TM	$0.12
Sojuz (photographic)	$0.23
SPOT XS	$0.63
SPOT Pan	$0.75

Although this shows the trade-off between area covered and ground resolution, it does not give a true picture of relative mapping costs with each data source since processing and production costs are not considered, neither is the size of the target area. For example if an area falls on one SPOT scene (which can be shifted up and down track at will), then data acquisition costs for SPOT XS will actually be cheaper than buying a quadrant or moveable sub-scene of TM data. Because of the smaller area covered by each scene, SPOT data can be less costly to process than TM data. A break-point is likely to arise when an area requires two to three or more SPOT scenes to complete coverage, but then only if the area falls entirely on one Landsat TM scene.

In operational practice this complex equation is often overridden by other technical considerations, not the least the fact that the cost of data acquisition is generally no more and often far less than 50% of the total base map production budget.

As a very approximate guide it can be said that processing a basic image from digital data to enhanced photographic print without cartographic correction or formatting costs in the order of 50% of the data acquisition cost. Cartographic correction and formatting may cost a similar sum. This does not take into account other factors such as the requirement for digital mosaicing or derivation of ground control.

With elevation mapping from stereo data, costs are more sensitive to the actual area required, although data acquisition will require a minimum of two scenes. Under these circumstances the use of photogrammetric techniques on Soyuz data should show a strong cost advantage, given reasonable ground control. Due to complication of different types of terrain only approximate figures can be given for photogrammetric work: in the order of $6,000 to $16,000 per stereo model.

For fully digital stereo processing, the costs are likely to fall in the range from $12,000 to $24,000 per scene.

Although such costs may seem high, given a definite need for an accurate contour map or D.E.M., in the absence of other available data, they must be balanced against the costs of mobilizing a survey aircraft, acquiring stereo air photos and compiling maps photogrammetrically. Such costs may not be enormous in areas which are already well developed, but they will certainly be quite high for remote areas of the world, which are precisely those where mapping is absent or inadequate.

FUTURE DEVELOPMENTS

New satellites are planned for launch in the coming decade, and the trend will be towards improved resolution and more stereo-coverage. For example Landsat 6 is scheduled for late 1991 with a 15m. resolution panchromatic band as well as the current middle infrared bands and SPOT 3 is in preparation to replace SPOT 1 or 2 as required. Later in the decade more advanced cameras and instruments will be mounted on the Polar Platform.

In the short term, there is assurance of a continued data flow from systems such as Landsat and SPOT, although these do not solve the problem of data acquisition over cloudy sites. For this reason considerable interest will focus on future radar imaging satellites which have the ability to produce terrain images through cloud. Experimental data has already been examined from Seasat, the Shuttle Missions (SIR-A and SIR-B) and the Soviet Almaz programme. Airbourne radar systems are currently in use where cloud cover precludes acquisition of satellite imagery. Based on experience from recent projects, acquisition of airborne radar data is at least five times as costly as the most expensive satellite imagery, nevertheless this cost has clearly been justified by oil companies exploring in areas such as Gabon, Congo, Sumatra and Papua New Guinea. In comparing the quality of radar imagery with that from conventional or optical sensors, it should be realised that whilst radar images give a more dramatic impression of terrain relief to the eye, they are less sensitive to ground cover conditions, and the radar reflectance or return is more difficult to interpret. Over project areas studied recently in West Africa, geological detail was revealed on the radar data that was not so obvious on SPOT imagery, but the SPOT images showed important fine details of forest roads, tracks and clearings which could not be discerned on the radar data.

The Soviets are expecting to launch a new Almaz radar with 15m. ground resolution in late 1990/early 1991 and will be programming acquisition on demand and selling data commercially. Also in 1991 the European Space Agency's ERS1 should be launched although its imaging radar is not fully optimised for land observation and world-wide data reception is not assured. The Canadian Radarsat should be in commercial operation by the middle of the decade and the Japanese have similar plans. These systems are of primary interest for repetitive monitoring in areas of difficult weather conditions, but they should at least ensure that those parts of the earth not yet seen cloud free will finally be revealed.

Due to the competition from at least three major international sources it is not expected that data costs will increase markedly above normal inflation in the coming years. Radar image data are expected to be in the same order of magnitude cost as other satellite image data.

CONCLUSIONS

Satellite image base maps may appear quite costly in comparison with purchase of published printed maps (when these are available).

However, satellite image maps can reveal important, up to date information not found or omitted from published maps, and, in some parts of the world, may be the only source of mapping anyway. New radar imaging systems about to be launched should ensure that even perpetually cloudy regions can now be mapped.

All too often, we have found that satellite image maps are prepared as an after-thought rather than as a precursor to planning and budgeting for an exploration programme, or else are considered a luxury. Bearing in mind that onshore siesmic acquisition can cost from $7,000 to $20,000 or more per line km. with typical programmes costing hundreds of thousands of dollars or that a 2000 sample geochemical programme must budget in excess of $50,000, working without the best base maps from the start may be seen as a false economy.

Landsat and SPOT data—important tools for environmental assessment and hydrogeological applications

Peter Volk Dr. rer.nat. (Geol.)
Rupert Haydn Dr. rer.nat. (Geol.)
Stefan Saradeth Dr. rer.nat. (Geophys.)
GAF—Gesellschaft für Angewandte Fernerkundung mbH, München, Germany

SYNOPSIS

Remote sensing in the mining business has developed during the last decade into an operational technique. Environmental assessment of mining activities is an excellent example of how the monitoring capabilities of operational satellite systems can contribute to economical solutions. Hydrological and hydrogeological mapping is often a crucial factor in the early stages of mining projects. In such cases satellites also provide the basic tool of historical data.

The paper discusses the constraints and advantages of Landsat and SPOT data during several projects in these applications. A pilot project in the NW German Ruhr area for detection and monitoring of subsidence effects due to large-scale coal mining is presented. Mapping of vegetation, landuse, structure and drainage systems for a feasibility study on the development of a Zn-mine is also demonstrated. The role of satellite data within a groundwater exploration programme in an arid environment is explained.

BACKGROUND

Mining activities always have impact on the environment. It is the size and character of the operations planned which determine the kind and degree of the impact.

Crowning importance is attached to minimizing these impacts. In many cases no attempt is made at their assessmen or they are underestimated in feasibility studies, which are normally executed in a stage of a mine development project when spatial data on the recent status of vegetation, soil, landuse and hydrology are urgently needed. Today high-resolution satellite data can provide this vital information, especially if their limitations are appreciated. Implementation of an environmental monitoring data base for continuing surveillance is another important application.

MAPPING BY MEANS OF LANDSAT AND SPOT DATA

Landsat Thematic Mapper and SPOT data are the most widely used for land applications. More than 2.5 million TM scenes are archived at NOAA and EOSAT and a similar number of SPOT scenes at SPOT Image. Many more scenes are doubtless available at ground receiving stations around the world. Experience at GAF suggests that these earth observation programmes have exceeded early expectations of their operationality. Recently nearly all land areas are covered by at least one scene.

For environmental application, the availibility of the most recent imagery is in many cases very important. This need is filled by Eosat, SPOT Image and some ground receiving stations which undertake to fill user requests for the acquisition of specific images. There are however limitations to this service for high latitudes and in tropical areas were persistant cloud cover makes acquisition within a reasonable time frame unlikely.

Despite of a high ground resolution, images acquired by the Soviets are used less widely than SPOT and Landsat TM. They are not available on a worldwide basis and acquisition requests cannot be filled in a short time frame.

Numerous scientific investigations and projects have contributed to a thorough understanding of the mapping potential of satellite data. As a result, standard processing concepts have been developed for enhancing or extracting the information in the images, for example:

* supervised classification of crops and forested land using SPOT multispectral bands and TM bands 3,4,5,
* generation of multitemporal vegetation indices for mapping of landuse intensity and degradation of natural vegetation.
* ratioing of visible and infrared bands for mapping of soil units (only in areas with low vegetation density).
* enhancements of selected individual Landsat bands and SPOT panchromatic data for the mapping of geological structure.

Satisfactory results can only be achieved in these and other examples where the data in the imagery is complemented by other information.

REQUIREMENTS FOR ENVIRONMENTAL AND HYDROGEOLOGICAL MAPPING WITH SATELLITE DATA

Successful and economic use of satellite data for environmental and hydrogeological mapping and monitoring is conditioned by a number of factors.

Firstly, the selection of the satellite system to be used depends on the matching of the information content of the imagery with the application. Compared to other commonly used remotely sensed data SPOT panchromatic data are excellently suited for infrastructural and structural mapping because of its better spacial resolution. SPOT XS provides information on the vegetation cover, whereas Landsat-TM with its infrared bands has the potential to map the spectral details and the variation of vegetation vitality. Maximum mapping scale for Landsat TM and SPOT XS data is

1:50.000; SPOT panchromatic data can be analyzed at scales up to 1:25.000.

Secondly, appropriate processing and data extraction methods must be selected. A fundamental decision is whether computer-based methods are to be used throughout, or whether the imagery is to be enhanced, followed by classical photo interpretation. If the former, then the most widely used method is classification, based on a statistical analysis of the spectral brightness of individual pixels. If the latter, then various methods of enhancement may be used (ratios, principal components analysis), followed by interpretation based on the criteria of tone, hue, texture, shape, size and context.

For vegetation classes characterized by seasonal variations of spectral signals, multitemporal scenes can greatly increase the reliability of mapping.

In all cases, additional field information and experience in comparable projects is vital for a high mapping accuracy.

GROWING IMPORTANCE OF DATA BASE OPERATIONS

A thematic data base including information derived from satellite data, existing thematic maps and available auxiliary data is a prerequisite for any medium and long term monitoring purpose. Today PC-based Geographic Information Systems are capable of performing the necessary input, analysis and output procedures.

Such instruments allow a flexible environment for the analysis of many problems associated with minig development, such as the impact of planned dumps, roads and processing plants. It is also possible to simulate developments and to draw conclusions about the future environmental pressure on a given area.

PROJECT EXAMPLES

Mapping of subsidence effects due to coal mining

Extensive mining of the coal seams in the Ruhr region has caused subsidence in large areas for more than 100 years. Subsidence of a few meters in flat terrain under humid climatic conditions can cause disturbances in the hydrological and hydrogeological system. Groundwater and surface water flow directions may be changed; soils can be degraded and even new ponds develop. Because the Ruhr region is one of the most populated areas in Europe, housing and industrial areas as well as agriculture may be affected by such changes.

In order to minimize possible environmental impacts, it is important for the mining companies to monitor such effects at an early stage. Besides aerial photogrammetic and other measurements, the analysis of satellite data can be regarded as an important complementary approach to identifying and mapping soil moisture effects.

Satellite data were selected as an information source because the investigation areas are spread over a 50 km x 200 km region and therefore continuous monitoring can be established economically. Since the soil or vegetation is affected by a high groundwater table and is sensitive to specific spectral absorptions in the infrared bands, Landsat TM data of the spring months were chosen for the digital processing.

Available aerial photographs were digitized and combined with colour-coded infrared band ratios. Principal

components analysis also yielded good results. However, the ratio of TM bands 4/5 most exactly outlines the areas affected by different soil humidity and open water. The mapping was performed at 1:10.000 scale for selected targets. however this must be regarded as an exceptional example made possible by the opportunity to correlate high-resolution aerial photographs with recent, thematically processed satellite data.

Mapping of landuse, natural vegetation and hydrogeologically important structures for a feasibility study

The feasibility study, performed for a planned Zn-mine in northern Tunisia, required recent data on the landuse and the distribution of natural vegetation. There also exists a lack of structural data and information on the drainage system in the vicinity of the mine. Winter SPOT panchromatic data were processed using High-pass filtering and contrast enhancement designed for displaying the image on black and white photographic paper. Landsat TM data from April (vegetation period) and August (dry season) were selected to map the dynamics of seasonal landuse and the permanent natural vegetation. Vegetation indexing and infrared false colour images enabled the outlining of the more prominent vegetation units and estimation of landuse intensity. Additionally, valuable lithological information was obtained from the summer data, processed by means of the enhancement of the colour saturation component.

Hybrid evaluation of SPOT pan and Landsat-TM spring data resulted in a tectoni map, which was a large improvement on existing data. Additionally, in the alluvial plains separating morphologically prominent anticlines, fossil and recent drainage patterns were mapped in the SPOT data.

This information is particularly of interest for the siting of the processing plant and wells. The processed satellite data was interpreted in the light of field data and using spectral criteria based on experience with comparable projects. A vectorized data base was constructed incorporating the available maps and processed image data. At least 3 new thematic maps were plotted at 1:25.000 scale (geological base map, landuse/vegetation, drainage pattern).

Groundwater exploration in an arid karst environment

The remote karst plateau on the eastern slope of the central highlands of the Yemen Arab Republic lacks proven groundwater resources which could serve as a basis for a rural water supply.

Since no reliable base data on geology and hydrogeology were available, satellite and reconnaissance aeromagnetic data were digitally processed, analyzed and field checked. SPOT panchromatic data were enhanced for geological structures, and from this the detailed tectonic pattern was obtained. Strongly contrast enhanced TM infrared bands clearly show contact metamorphic effects due to basaltic intrusions. These features are important for the delineation of potential fractured aquifers in a marl and limestone sequence. Processed total intensity aeromagnetic data provided depth to the basement/sediment interface and led to the conclusion that a sandstone aquifer, formerly thought to be accessible for water supply, was in fact to deep.

After having performed evaluation and compilation of this information (remote sensing data and existing maps), it was transfered to a GIS data base. Subsequent analysis and overlay of different data sets (contact metamorphism, structures and catchment boundaries) resulted in the selection of 28 target areas (approx. 200 m x 200 m) with high potential for groundwater. A detailed inventory of existing wells proved that at least some of the above mentioned geological criteria are fulfilled by most of the successful wells in the area.

In a second project phase, detailed ground geophysics (electromagnetics, magnetics and refraction seismics) verified the presence of deep-seated and hydrogeologically promising structures and yielded indicators for siting boreholes.

SOME ECONOMIC CONSIDERATIONS

Mapping supported by satellite data has become more costly during the past 10 years. But digital satellite data are not expensive considering the large areas covered by one scene. However, the situation and shape of a project area may require the acquisition of image data covering a much larger area. Subscenes movable within the swath can sometimes minimize the consequently increased costs.

In many cases the use of data from two different sensors (e.g. SPOT panchromatic and Landsat TM) is necessary to increase the reliability of maps. This further increases the cost of mapping based on satellite data. Digital image processing dedicated to improved delineation of lithology, structural aquifers, and vegetation types is an indispensable tool fully to exploit the potential information contents of digital satellite data. Therefore computer intensive techniques need to be applied, which can cost as much as the raw digital data. Despite an improvement in quality compared to old EROS MSS images, the new image products of the data centres generally do not meet the specific mapping requirements of environmental and hydrogeological assessment.

A PC-based GIS system is a valuable and complementary tool to further proces information derived from satellite data. All existing maps, sample data and statistical mining data can be stored and analysed. Recently turn-key solutions in hard- and software are available on the market.